CONSTRUCTION PLANNING AND TECHNOLOGY

CONSTRUCTION PLANNING AND TECHNOLOGY

SECOND EDITION

Rajiv Gupta

Professor of Civil Engineering;
Dean, Engineering Services Division,
Birla Institute of Technology & Science,
Pilani (Rajasthan)

CBS Publishers & Distributors Pvt. Ltd.

New Delhi • Bengaluru • Chennai • Kochi • Kolkata • Mumbai
Hyderabad • Nagpur • Patna • Pune • Vijayawada

ISBN: 978-81-239-1611-8

First Edition: 1994
Second Edition: 2008
Reprint: 2009, 2010, 2012, 2013, 2014, 2016

Published by:
Satish Kumar Jain for CBS Publishers & Distributors Pvt. Ltd.,
4819/XI Prahlad Street, 24 Ansari Road, Daryaganj, New Delhi - 110002
delhi@cbspd.com, cbspubs@airtelmail.in • www.cbspd.com
Ph.: 23289259, 23266861, 23266867 • Fax: 011-23243014

Corporate Office: 204 FIE, Industrial Area, Patparganj, Delhi - 110 092
Ph: 49344934 • Fax: 011-49344935
E-mail: publishing@cbspd.com • publicity@cbspd.com

Branches:
• *Bengaluru:* 2975, 17th Cross, K.R. Road, Bansankari 2nd Stage,
  Bengaluru - 70 • Ph: +91-80-26771678/79 • Fax: +91-80-26771680
  E-mail: cbsbng@gmail.com, bangalore@cbspd.com
• *Chennai:* No. 7, Subbaraya Street, Shenoy Nagar, Chennai - 600030
  Ph: +91-44-26681266, 26680620 • Fax: +91-44-42032115
  E-mail: chennai@cbspd.com
• *Kochi:* Ashana House, 39/1904, A.M. Thomas Road, Valanjambalam,
  Ernakulum, Kochi • Ph: +91-484-4059061-65
  Fax: +91-484-4059065 • E-mail: cochin@cbspd.com
• *Kolkata:* 6-B, Ground Floor, Rameshwar Shaw Road, Kolkata - 700014
  Ph: +91-33-22891126/7/8 • E-mail: kolkata@cbspd.com
• *Mumbai:* 83-C, Dr. E. Moses Road, Worli, Mumbai - 400018
  Ph: +91-9833017933, 022-24902340/41 • E-mail: mumbai@cbspd.com

Representatives:

• Hyderabad: 0-9885175004	• Nagpur: 0-9021734563
• Patna: 0-9334159340	• Pune: 0-9623451994
• Vijayawada: 0-9000660880	

Printed at:
J.S. Offset Printers, Delhi

PREFACE

Planning, an important and essential phase in any project, is more or less lacking in our country. Due to lack of planning, money, manpower and time are lost. The concept of planning should be emphasised in the curriculum at all levels of education.

This is an attempt to tell the young friends the necessity of the planning, planning methods and the topics which are required to know for planning. The author hopes that these planning methods would help the readers in achieving a successful planning.

The book is a supplementary to the course in which all the building materials, components of a building (doors, windows, finish, staircase etc.) are taught. The author feels that these components can be understood if one goes through them. This is the reason for not including these topics in this book. These topics can be found in any book dealing with construction and building materials.

In the text, "Costing and Estimation" is covered in chapter one which is the first phase of planning to execute a project. After estimation, CPM or PERT networks can be drawn with proper duration of the activities involved. Once network is prepared, resource levelling can be done. These three phases are interlinked. In chapter 2, different planning techniques are discussed, which can be used effectively for planning for different projects.

In chapter 3, "Civil Engineering Systems", different Civil Engineering projects are dealt and analysed by systems approach to get optimum results.

Chapter 4 deals with the latest construction materials and technology along with the different constraints. The emphasis of using these materials is to save the energy.

I wish to express my thanks to our Vice-Chancellor, Prof. S. Venkateswaran, for giving the opportunity, environment and facilities to work on the project.

I am grateful to Prof. H.S. Moondra, for his moral support and encouragement.

I thank Prof. P.S.V.S.K. Raju, Chief I.P.C., for his timely help and support; Dr. S. Ghoshal, Group Leader, Civil Engg., for his constructive criticism. The most of material of chapter 4 is provided by Dr. J. Venkataramana and his team members. I sincerely thank him and his team members for the support and help.

Special thanks are due to Mr. C.V. Prasad, who helped me a lot in writing this book. At times he took pain to share my work, so that I can complete my work in time. Thanks a lot.

I thank Mr. Sanjeev Jain, Mr. R.V. Ramanan, Mr. Sarvanan D. and Mr. Deeraj Gupta.

I thank CBS Publishers & Distributors for taking pain in publishing the work.

All the users of this book are most welcome for any comments and suggestions and they are thanked in anticipation.

Gupta Rajiv

CONTENTS

Chapter I

Section A

ESTIMATION AND COSTING

A. 1 General

For preparing the estimate, the quantities of various items of works are calculated by using simple mensuration method. The worked out quantities are used to find out its cost. For the correct estimation, the correct dimensions are to be taken from the drawing. The length, width, thickness or height are taken from the drawing if estimate is to be prepared. There are no hard and fast rules for determining the dimensions from the drawings, but to avoid errors and for systematic approach, we follow certain principles. In the beginning one may find it bit difficult in finding out the dimensions from the drawings but by a little practice they can master this subject. In this chapter different examples are taken which will help to understand the basic dimensions of the structure.

A. 2 What is an estimate

Before taking up any work for its execution the executer should have a thorough knowledge about the probable cost that may be required to complete the work, otherwise it may happen that the work has to be stopped before the completion due to shortage of funds, or unavailability of resources.

Thus an estimate for any construction work may be defined as the process of calculating the quantities and costs of the various items required in connection with the work. To prepare an estimate, drawings consisting of plan, elevation and the sections through important points along with a detailed specifications giving specific description of all workmanship, properties and proportion of materials are required.

A. 3 Types of estimates

An estimate prepared from the detailed drawings, specifications and based on the present market prices of material is never the actual cost of the work. The cost of materials and labours may vary during the period of its actual execution or due to variations and modifications of actual dimensions shown in the drawing or due to some other contingencies. The different types of estimates are as follows :

1. **A Detailed estimate :** The detailed estimate is the most accurate and reliable estimate and it includes the cost and quantities of everything required for completion of the project. It also has
   (a) Report
   (b) Specifications
   (c) Detailed drawings
   (d) Design data and calculations
   (e) Basis of rates adopted in estimate.

2. **A rough or preliminary or approximate estimate :** As the name indicates, this is an estimate made to find out an approximate cost in a short time to consider the financial aspect of the scheme for sanctioning the money. Such an estimate is formed after knowing the rates of similar works and for that, one of the following methods is used :
   (a) **Unit rate estimate**
   (b) **Plinth area estimate**
   (c) **Cubic rate estimate**
   (a) **Unit rate estimate:** In this method the cost of a unit quantity is considered first and the total estimate is prepared by multiplying the cost by the number of units in the structure.
   (b) **Plinth area estimate :** In this method the plinth area should be calculated by taking the external dimensions of the building at the plinth. Total floor area of all the rooms, corridors, _ verandah (opening areas are not included) , kitchen, WC and bath are found out. Sometimes a percentage for walls and waste is also added to the total areas to get the approximate total plinth area.This plinth area is multiplied by the plinth area rate for similar type design and specification of building at the locality.
   (c) **Cube rate estimate:** This is more accurate method than the method of estimating cost by plinth area.The cost of **structure**

depends not only on its plinth area but also on its height. In this method, the volume of the building is found out and then the total cost is found out by multiplying the volume by the local cubic rate for similar type of building. Length and breadth should be measured external to external excluding plinth offset, corbelling, string course, etc. The height should be measured from the top of the flat roof (or half way of the sloped roof) to half the depth of the foundation below the plinth.

3. **A Quantity estimate or quantity survey:** It is a complete estimate of quantities of materials that may be required to complete the work concerned.

4. **Revised Estimate:** When a sanctioned estimate is likely to be exceeded by more than 5% due to any cause whatsoever (except important structural alterations) another estimate is prepared which is called a revised estimate.

5. **A suppplementry estimate:** While work is being executed some additional work may be thought necessary for the project and the expenditure for such supplemantry work is estimated which is known as supplemantry estimate.

6. **A complete estimate:** This is an estimated cost of all items which are related to the work. A picture of a complete estimate is shown in Fig. A.1.

COMPLETE ESTIMATE

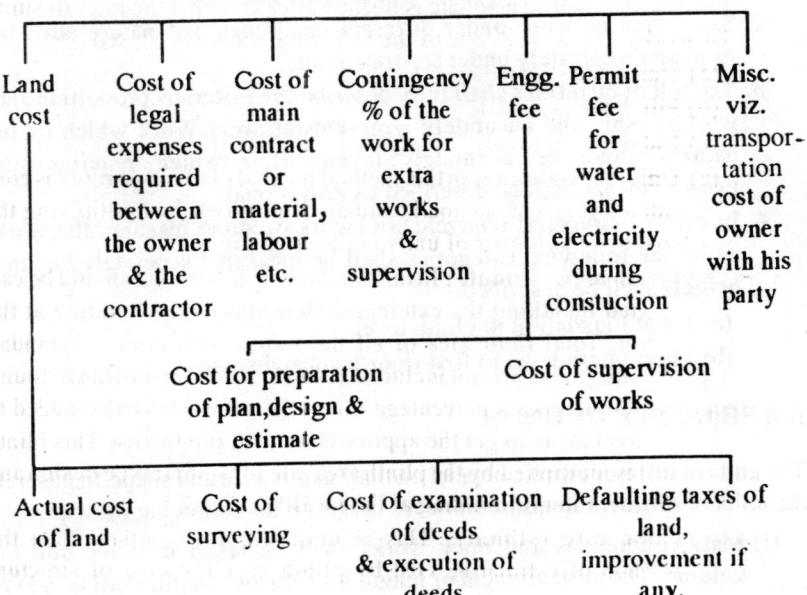

Fig. A.1. Items to be considered for complete estimate.

7. **Annual Maintenance or Repair Estimate:** After completion of a work it is necessary to estimate the same for its proper function and an estimate is prepared for the items which require renewal, replacement, repairs, etc. in the form of a detailed estimate which is known as annual maintenance or a repair estimate.

A.4 RULES & METHODS OF MEASUREMENT OF WORKS AND TAKING OUT QUANTITIES

Measurement of works occupies a very important place in the planning and execution of any work or project.

The methods followed for the measurement are not uniform and the practices as prevalent differ considerably in states. Even in the same state the different departments follow different methods. For convenience, the uniform method of measurement is to be followed which is applicable to the preparation of the estimates and bill of quantities and to the site measurement of completed works.

GENERAL RULES

1. Measurement shall be taken itemwise for the finished item of work and the description of each item shall include materials, transport, labour, fabrication, hoisting, tools and plants, overheads and other incidental charges for finishing the work to the required shape,size,design and specifications.The nomenclature of each item shall be fully described so that the work involved in item is self-explainatory.

2. Same type of work under different conditions and nature shall be measured seperately under seperate items.

3. The bill of quantities shall fully describe the materials,proportion and workmanship and accurately represent the work.Work which by its nature cannot be accurately taken off,or which requires site measurements,shall be described as Provisional.

4. In case of structural concrete,brickwork or stone masonry,the work under the following categories shall be measured seperately and the heights shall be described:-

   (a) From foundation to plinth level

   (b) From plinth level to first floor level and so on.

A.5 PRINCIPLE OF UNITS

The units of different works depend on their nature,size and shape.In general the units of different items of work are based on the following principles:-

   (i) Mass,voluminous and thick works shall be taken in cubic unit or volume. The measurement of length,breadth and height shall be taken to compute the volume.

   (ii) Shallow,thin and surface work shall be taken in square units or in area.The measurement of length and breadth or height shall be taken to compute the area(sq m) .

     (iii) Long and thin work shall be taken in linear or running unit and linear measurement shall be taken in running metre.

     (iv) Piece work,job work,etc shall be enumerated ie.,taken in number.

A.6 ABSTRACT OF ESTIMATED COST

A percentage (usually 5 to 10 %) cost of the estimate is considered as contingency and 25 % cost is added for work charged to establishment.

Sl No	Description of Item	Quantity	Unit	Rate	Amount
					Total =
					Contingency 5%=
					Work charge 25%=
					Grand total =

Bill of Quantities

It is a complete list of all items of work involved in connection of the estimate for a work,giving all the items as Abstract or estimated cost.It is prepared from drawings and specifications and arranged in a tabular form,without filling up the columns of rate and amount.It is particularly required for inviting item rate tender where the contractors fill up the rate and amount columns.

Item No	Description of item	Quantity	Unit Rate Amount
1.	Earthwork	10 cum	per cum

Example 1: A R.C.C lintel is provided over a door. The door size is 1.10 m by 2.50 m. The lintel is 20 cm × 15 cm in cross-section (see Fig. A.2). Work out in detail the quantity of reinforcement required for the lintel from the following data :

Main Bars	: 4 Nos of 12 mm diameter of which two are straight and two are bent up.
Anchor bars	: 2 Nos of 10 mm diameter.
Stirrups	: 6 mm diameter at 20 cm c to c.
Bearing for the lintel on either side	: 15 cm

Solution

1. As cover is not mentioned,we take standard cover of 25 mm.

2. Angle for bending bars is not mentioned so we take standard bend of 45 degrees.

3. Bar bending schedule

Bar Mark	Dia mm	Shape of bending bars	Length (mm)	Nos	Total length (mm)	Qty (kg)
1	12		1566	2	3132	2.779
2	12		1576.08	2	3152.16	2.78
3	10		1566	2	3132	1.94
4	6		576	6	3456	0.76

Total quantity = 8.259 kg

Length of different types of bars:

Type 1 = 1100 (clear span) + 300(bearings) - 50 (cover) + $2 \times 9 \times 12$
= 1566 mm

Type 2 = 1100 (clear span) +300(bearing) - 50 (cover)
+ $2 \times 0.42 \times 12$ (extra length for cranks)
+ $2 \times 9 \times 12$ (two bends) = 1576.08 mm

Type 3 = as of Type 1

Type 4 = $2(200 - 2 \times 25 - 2 \times 3) + 2 (150 - 2 \times 25 - 2 \times 3)$
+ extra length of 100 mm (assuming 10 cm)
= 288 + 188 + 100 (length is calculated according to mid dimensions)
= 576 mm

Density of steel = 7.85 g/cm^3

Weight = volume × density = area × length × density
= ($\pi/4 \times d \times d \times l \times$ density) where d is the dia of bar

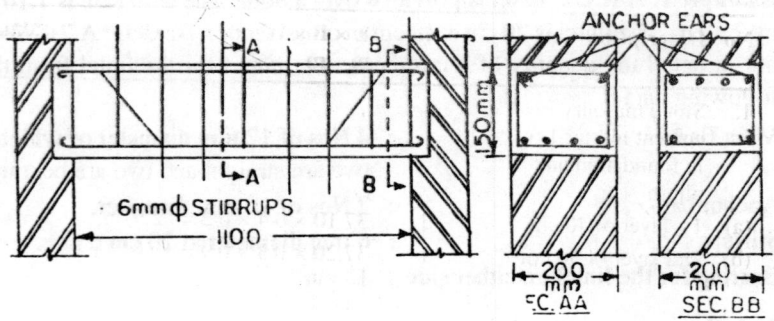

Fig. A.2.

Example 2: A building with a semi-circular portion as reading room and a rectangular portion as library room is shown in Fig. A.3.

Estimate cost of the following items of work at present rates:

(a) Stone masonry in cement mortar 1:6 for foundation and plinth.

(b) 10 cm thick cement concrete flooring with 2 cm thick cement mortar finishing complete. Cement concrete being of 1:2:4 proportions and cement mortar being of proportion 1:4.

Solution

There are three different methods to calculate the quantities. They are:

1. Center line method
2. Long and short wall method
3. Crossing method.

1. Reading room - From Fig.A.3(Here, example is solved by first method only

The inner radius of circular portion	= (6.7 - 4 - 0.2) m
	= 2.5 m
Mean radius = (2.5 + thickness/2)	= (2.5 + 0.10) = 2.6 m
(a) Center line for curved portion	= π × 2.6 = 8.17 m
(b) Center line for long walls	= 2 (6.7 - 2.5 + 0.1) m
(long walls)	= 8.6 m
(c) Center line for short wall	= (5 + 0.2) = 5.2 m

2. Library room:

(a) Center line for long walls	= 2 (5.5 + 0.20) m = 11.4 m
(b) Center line for short side	= (4 + 0.20) m = 4.20 m
3. Total length of center line	= (8.17+8.6 + 5.20 + 11.4 + 4.2)
	= 37.5 m

DETAILS OF MEASUREMENT AND QUANTITIES

No	Description	No	L × B × H m m m	Qty m³	Remarks
1.	Stone masonry in cement mortar 1:6 in foundation and plinth				37.10 = 37.50
(a)	1st layer of 40 cm	1	37.10 × 0.4 × 0.2 =	2.97	2 × (4/2);
(b)	2nd layer of 30 cm	1	37.20 × 0.3 × 0.9 =	10.04	37.20 = 37.50
					2 × (3/2)
				13.01	

(Contd.)

No	Description	No	L x B x H m m m	Qty m³	Remarks
2.	Brickwork in cement				
	mortar (1:6) in superstructure	1	37.37x0.2x3.3=24.66		
	Deductions for				
	doors D	2	1.2x0.2x2.1 =1.01 -ve		
	Windows W1	7	1.0x0.2x0.5 =2.10 -ve		
	W2	2	1.2x0.2x1.5 =0.72 -ve		
	W3	1	2.0x0.2x1.5 =0.60 -ve		
	Lintel	1	37.37x0.2x0.15=1.12-ve		
			= 19.11		
3.	10 cm thick cement concrete flooring				
(a)	For library room	1	3.9 x 3.9 = 15.21		3.9=
(b)	For reading room				4.0-2x0.5
i)	Circular portion	1/2	x 4.9x4.9/2 =9.43		
ii)	Rectangular portion	1	4.15 x 4.19 = 20.43		4.15=6.7-
					2.5-0.05
			44.98 m²		
4.	2 cm thick cement mortar				
	(1:4) finishing etc		44.98 m²		same as
					item 3

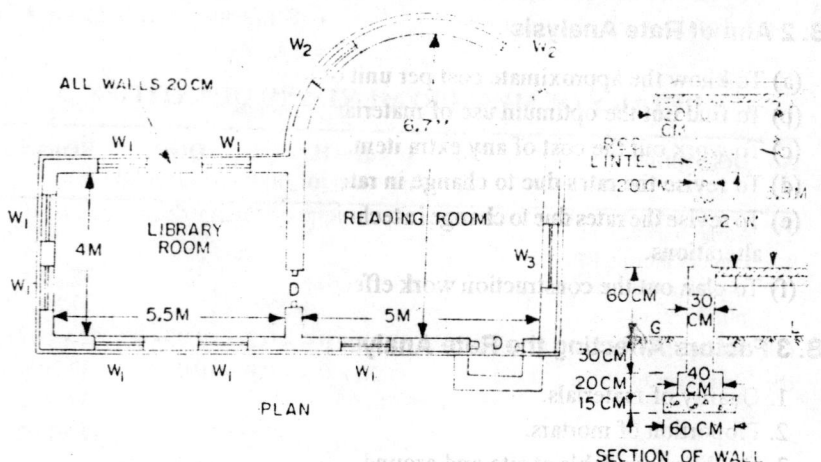

Fig. A.3

Section B

ANALYSIS OF RATES

B. 1 Introduction

While filling the tender if a contractor wants to quote the rate of a particular item, first he has to workout the rate. The procedure for working out the rate of an item is known as rate analysis. For example, he wants the rate for RCC slab, from his experience or by other means (drawing etc.) he works out the quantities of materials, that is how many bags of cement, aggregates and steel are required for the slab. Along with materials he also calculates how many labours are required to complete the work in stipulated time and then adds his percentage of profit. At the same time he keeps other factors also, which can influence his work, in mind and arrives at a figure for each item and quotes it in tender which is the rate for that particular item at that place in that particular condition and time.

B. 2 Aim of Rate Analysis

(a) To know the approximate cost per unit of item.
(b) To find out the optimum use of materials, procedures and labours.
(c) To work out the cost of any extra item.
(d) To revise the rates due to change in rates of materials and labours.
(e) To revise the rates due to change in technique or change due to technical alterations.
(f) To plan out the construction work effectively.

B. 3 Factors Affecting the Rate Analysis

1. Quality of materials.
2. Proportion of mortars.
3. Facilities available at site and around.
4. Location of the site work.
5. Transportation charges etc.

6. Overhead charges.
7. Miscellaneous expenditure.
8. Profit desired.
9. Experience of workers and supervising staff.
10. Management.

B. 4 Fixing up Rate/Unit of Item

Following factors should be taken into account to fix up the rate per unit of item.

(a) **Quantity of materials and cost:** It includes cost, transportation charges, sales tax, insurance charges, etc.

(b) **Labour cost: The number and wages of the different categories** of labours (skilled, unskilled) should be known per unit of item.

(c) **Cost of equipments:** Operating cost, rent charges etc. for each tool to be used. The total cost should be divided for the whole duration of the work and should be added to the particular item.

(d) **Overhead charges:** It includes office rent, depreciation of equipment, postage, lighting, travelling, telephone bills, small tools etc. It is normally from two and half percent to five percent.

(e) **Profit:** Generally a profit of 10% is considered reasonable, but it varies from 8% for big projects to 15% for small projects.

Table B.1: Labour Requirement

Description	Qt. of work per day (8 hours a day)
1. Earthwork in excavation in foundation trenches in ordinary soil, lead up to 50m and lift upto 1.5 m.	2.75 m^3 per Mazdoor
2. Excavation rock by blasting, lead upto 50 m and lift upto 1.5 m.	0.55 m^3 per Mazdoor
3. Sand filling in plinth, consolidating and dressing	4.00 m^3 per Mazdoor
4. Single layer brick flat soling including ramming and dressing the bed etc.	9.00 m^2 per Mazdoor
5. Lime concrete in foundation	10.00 m^3 per Mason
6. Cement concrete	5.00 m^3 per Mason
7. Cement concrete (1:2:4) for R.C.C. work	3.25 m^3 per Mason
8. Brick work in foundation and plinth	1.25 m^3 per Mason
9. Brick work in superstructure ground floor	1.25 m^3 per Mason

(Contd.)

Description	Qt. of work per day (8 hours a day)
10. 2.5 cm thick cement concrete D.P.C.	12.50 m² per Mason
11. 20 mm thick D.P.C., with cement mortar	20.00 m² per Mason
12. Random rubble masonry in foundation and plinth	1.00 m³ per Mason
13. Random rubble masonry in superstructure	0.90 m³ per Mason
14. Ashlar masonry in superstructure	0.40 m³ per Mason
15. Brick on edge floor with cement mortar	1.00 m³ per Mason
16. 7.5 cm thick cement concrete floor (1:4:8)	10.00 m³ per Mason
17. Terraced flooring 7.5 cm thick	20.00 m³ per Mason
18. Terrazzo floor 6 cm thick mosaic work over 2 cm thick cement concrete (1:2:4)	5.00 m³ per Mason
19. Terrazzo skirting or dado 6 cm thick Terrazzo layer over 12 mm thick cement plaster	3.85 m³ per Mason
20. Precast Terrazzo tiles 20 cm thick in skirting and rises of steps on 12 mm cement plaster	3.20 m³ per Mason
21. 10 cm average thick lime terracing on R.C. roof	9.10 m³ per Mason
22. Flat terrace roofing average 10 cm thick over two layers of tiles (with 2.5 cm mortar)	5.00 m³ per Mason
23. Mangalore Tiles roofing including wooden battens, tiles set in cement mortar	20.00 m³ per Mason
24. Corrugated galvanised iron sheet roofing	10.00 m³ per Carpenter
25. 6 mm thick cement plaster to R.C. ceiling	10.00 m³ per Mason
26. Single coat white - washing over old white-washed surface	133.00 m³ per Painter
27. White-washing two coats on a coat of primer	66.70 m³ per Painter
28. Distempering two coats to new cement plaster	20.00 m³ per Painter
29. Water-proofing cement paint to new cement plaster	20.00 m³ per Painter
30. Primer coat with ready mixed primer on wood or steel-work	40.00 m³ per Painter
31. Painting two coats (excluding printer coat) with ready mixed paint for wood-work	18.00 m³ per Painter
32. Breaking of overburnt brick to ballast 40 mm down	0.75 m³ per Mazdoor

Table B.2: Team of labours required

(I) Name of gang	(II) Strength of gang	(III) Any one item of work which may be completed by the gang. Volume of work is as in column (IV)	(IV) Volume of work	(V) Adjustment for column (II)
(A)	1/4 Head Mason	1. Lime concrete in foundation	10 m³	
	1 Mason	2. Lime punning over plastering	10 m³	decrease 18 mazdoors and
	20 Mazdoors (Beldars)	3. Cement concrete in foundation	10 m³	one head mason
				Increase 1 Mason
		1. Brickwork in ground floor	10 m³	For plinth and foundation reduce 2 Masons and increase 2 Mazdoor
(B)	1/2 Head	2. Brick work in plain arches	10 m³	(Beldars) & for first floor reduce 1 Mason and increase 5 Mazdoors
	10 Masons			
	15 Maz-doors (Beldars)	3. Brick on edge floor	100 m³	Add 4 more Mazdoors. For ground floor, add 1 mason and 2 mazdoors
		4. Reinforced concrete brick	10 m³	overplinth
		5. Random rubble masonry in foundation and plinth	10 m³	
		6. 6 mm thick cement plastering to ceiling	100 m²	Reduce 3 mazdoors

(Contd.)

(I)	(II)	(III)	(IV)	(V)
Name of gang	Strength of gang	Any one item of work which may be completed by the gang. Volume of work is as in column (IV)	Volume of work	Adjustment for column (II)
		7. 12 mm thick cement plastering on walls	10 m²	7. Reduce 5 mazdoors
		8. Rule pointing on brick wall	100 m²	8. Reduce 5 mazdoors
		9. 7.5 thick cement concrete floor	100 m²	9. Increase 5 mazdoors
(C)	1/2 Head mason	1. Ashlar masonry structure at ground floor	10 m³	1. Add 10 masons and 10 mazdoors
	15 masons	2. Course rubble stone masonry in	10 m³ 10 m³	2. Add 4 mazdoors
	20 mazdoor	superstructure at ground floor		
		3. Half brick work at ground floor	100 m²	3. Reduce 1 mason, increase 1 mazdoor

Table B.3: Quantity of materials required for different items of work

Name of work			Quantity of material
		Bricks 20 cm × 10 cm × 10 cm	5000 nos
1. Brick work	per 10 cu m	Bricks 25.4 cm x 12.7 cm × 7.6 cm	4100 nos
		Bricks 22.9 cm × 11.4 cm × 7.6 cm	5000 nos
		Dry cement mortar	3.5 cu m
		Dry lime mortar	4.0 cu m
2. Half-brick work	per 100 sq m	Brick 20 cm × 10 cm × 10 cm	5000 nos
		Bricks 25.4 cm × 12.7 cm × 7.6 cm	4100 nos
		Bricks 22.9 cm × 11.4 cm × 7.6 cm	5000 nos
		Dry mortar	3.15 cu m

(Contd.)

Name of work			Quantity of material
3. Random rubble stone masonry	per 10 cu m	Stone	12.5 m³
		Lime mortar	4.4 m³
		Cement mortar	4.2 m³
4. Ashlar masonry	per 10 cu m	Stone	10.0 m³
		Dry cement mortar	3.0 m³
		Dry lime mortar	3.2 m³
5. Course rubble stone	per 10 cu m	Stone	12.5 m³
		Dry cement mortar	4.2 m³
		Dry lime mortar	4.0 m³
6. Single brick flat soiling	per 10 sq m	Bricks 20 cm × 10 cm × 10 cm	425 nos
		Bricks 25.4 cm × 12.7 cm × 7.6 cm	320 nos
		Bricks 22.9 cm × 11.4 cm × 7.6 cm	425 nos
7. Brick on edge floor with cement mortar	per 10 sq m	Brick 20 cm × 10 cm × 10 cm	500 nos
		Brick 25.4 cm × 12.7 cm × 7.6 cm	410 nos
		Bricks 22.9 cm × 11.4 cm × 7.6 cm	500 nos
8. 20 mm thick DPC with cement mortar	per 100 sq m	Cement	27 bags
		Sand	1.8 m³
		Water-proofing compound	27 kg
9. Reinf. brick work	per 10 cu m	Bricks	4500 nos
		Dry cement mortar	4.8 m³

10. Precast Terrazzo tiles 20 mm thick on 12 mm thick cement plaster per 100 sq m

(a) For Terrazo work		Tiles	110 sq m
		Cement	13 bags
		Pigment	46 kg
(b) For 12 mm thick cement plaster			14 bags
		Cement	14 bags
		Sand	1.31 m³

11. Lime terracing in roof with brick ballast (proportion 2 : 2 : 7) per 100 m²

(a) For 7.5 cm thick Lime	2.1 m³		
		Surki	2.1 m³

(Contd.)

Name of work		Quantity of material
	Brick ballast	7.5 m³
(b) For 10 cm thk Lime	2.9 cu m	
	Surki	2.9 m³
	Brick ballast	10.2 m³
12. Raniganj tile roofing in lime mortar per 10 sq m		
	Raniganj tiles	124 nos
	Dry lime mortar	0.15 m³
	Cement mortar	0.14 m³
13. CGI roof sheeting per 10 sq m	G I sheet	12.8 m²
14. AC corrugated roof sheeting per 10 sq m	A C sheet	11.50 m²
15. 12 mm tkh cement plaster per 100 sq m	Dry mortar	1.92 m²
16. Neat cement punning (about 1.5 mm) per 100 sq m	Cement	5.5 bags
17. Flush pointing in cement per 100 sq m	Stone lime (unslaked)	10 kg
18. Single coat white washing to old work per 100 sq m	Stone lime (unslaked)	30 kg
20. Distempering two coats to new cement plaster per 100 sq m	Tropic dry distemper	
	1st coat	12 kg
	2nd coat	7.5 kg
21. Lime punning over plastered surfaces per 10 sq m	Slaked stone lime	18 kg
	Shell lime	8 kg
22. Snow-cem washing on plastered surfaces per 100 sq m	1st coat	30 kg
	2nd coat	20 kg
23. Primer coat on steel work with ready mixed primer per 100 sq m	Primer	5.5 litres
24. Primer coat on woodwork with ready mixed primer per 100 sq m	Primer	7.5 litres
25. Painting two coats on new work with ready mixed paint per 100 sq m	Ready mixed paint	12.5 litres

(Contd.)

Name of work		Quantity of material
26. Water proofing cement painting two coats to new plaster to exterior walls		
per 10 sq m	Mixed cement paint	2 litres
27. Spray painting with wall paints on new work including primer coat		
per 100 sq m	Primer coat	8.1 litres
	Wall paint	11 litres
28. Painting with synthetic enamel paint on new work (excluding primer coat)		
per 100 sq m	Enamel paint	11.6 litres
29. Varnishing with Copal varnish on new work including under coat		
per 100 sq m	Under coat varnishing	7.0 litres
	Copal varnish	11.6 litres
30. Wax polishing on new woodwork with ready made polish		
per 100 sq m	Ready-made wax polish	5.0 kg
31. Floating coat of cement		
per sq m	Cement	2.2 kg

Table B.4: Quantity of materials for 10 cu m of concrete (variation 5% allowed) based on assumption that dry sand with necessary allowance for bulking is used.

Volumetric proportion	Cement cum	Cement bags	Dry sand cu m	Stone chips 20 mm to 6 mm cu m	Water per bags of cement (lit)
1 : 1 : 2	3.0	112.4	3.9	7.8	7.5 litres
1 : 1.5 : 3	2.8	80.7	4.2	8.4	8.75 litres
1 : 2 : 4	2.2	63.4	4.4	8.8	9.5 litres
1 : 2 : 5 : 5	1.8	51.9	4.6	9.2	10 litres
1 : 3 : 6	1.57	45.2	4.7	9.4	11 litres
1 : 4 : 8	1.20	34.6	4.8	9.6	13.5 litres
1 : 5 : 10	0.98	28.2	4.9	9.8	14 litres
1 : 1 : 2		Not to be recommended			
1 : 1.5 : 3		Not to be recommended			
1 : 2 : 4	2.25	64.8	4.5	9.0	9.5 litres

(Contd.)

Volumetric proportion	Cement cum	bags	Dry sand cu m	Stone chips 20 mm to 6 mm	Water per bags of cement (lit)
1 : 2.5 : 5	1.86	53.6	4.7	9.3	10 litres
1 : 3 : 6	1.60	35.4	4.9	9.8	11 litres
1 : 4 : 8	1.23	35.4	4.9	9.8	13.5 litres
1 : 5 : 10	1.00	28.8	5.5	10.0	14 litres

Table B.5: Rate of different materials

Description of item	Unit of Rate	Rate Rs.
(A) Materials as per approved specifications delivered at work site including stacking		
1. Bricks first class (kiln burnt) 20 cm × 10 cm × 10 cm or 9" × 4.5" × 3"	1000	950
2. Bricks first class (kiln burnt) 10" × 5" × 3"	1000	1000
3. Bricks 2nd class (kiln burnt) 12 cm × 10 cm × 10 cm or traditional 9" × 4.5" × 3"	1000	780
4. Bricks 2nd class (kiln burnt) 10" × 5" × 3"	1000	980
5. Overburnt brick 20 cm × 10 cm × 10 cm or traditional brick	1000	675
6. Overburnt bricks 10" × 5" × 3"	1000	595
7. Overburnt brick bats	m^3	150
8. Sand (coarse)	m^3	115
9. Sand (medium)	m^3	130
10. Sand (local)	m^3	110
11. Surki	m^3	166
12. Cement	per bag	130
13. Lime (slaked stone or white)	m^3	398
14. Lime (unslaked stone or white) free lime 60% and above	qunital	150

(Contd.)

Description of item	Unit of Rate	Rate Rs.
15. Stone ballast 40 mm down	m³	304
16. Stone chips 20 mm down	m³	304
17. Stone chips 12 mm down	m³	280
18. Hard stone ballast 40 mm down	m³	140
19. Overburnt brick chips 25 mm down	m³	166
20. Brick ballast 40 mm down	m³	280
21. Marble chips (grit Dehradun)	quintal	154
22. MS bar upto 16 mm diameter	quintal	960
23. MS bar from 16 upto 32 mm dia	quintal	670
24. Black iron wire	kg	10
25. H.B. wire netting	m²	25
26. Water-proffing compound	kg	11 to 17
27. Gum	kg	18
28. Distemper primer	kg	15 to 55
29. Tropic distempering	kg	14
30. Raniganj Pattern tiles	% Nos	300
(B) Labour working period (8 hours) in day time		
1. Head Mason	each/day	45
2. Mason (ordinary)	each/day	40
3. Mazdoor (Beldar)	each/day	22
4. Mazdoor (Female)	each/day	22
5. Carpenter	each/day	45
6. Black-smith and Fitter	each/day	35
7. Painter	each/day	40
8. Plumbing mistry	each/day	40

Note: For a a short period work increase the rate @ Rs. 3 for each kind of labourer.

* These rates (Table B.5) are old rates and vary from place to place and with time.

ANALYSIS OF RATES FOR DIFFERENT ITEMS

Item No 1: Analysis of rate of Excavation in foundation in ordinary soil including lift upto 1.5 m and lead upto 30 m including filling, watering and ramming of excavated earth and removal and disposal of surplus earth as directed by the engineer-in-charge upto a distance of 30 m.

S.No.	Particulars	Qnty. or Nos.	Rate	Amount (Rs.)
1.	Mistri	0.5 No	35/day	17.50
2.	Beldars	21 Nos	22/day	462.00
3.	Coolie	25 Nos	22/day	550.00
4.	Blacksmith for tools sharpening etc.	0.5 No	30/day	15.00
			Total	1044.50
5.	Tools, Plants, Baskets etc.		Lump sum	5.50
			Total	1050.00
	Add 10% for Contractors profit			105.00
		Grand Total		1155.00
				for 100 m^3

Rate per cu m = 1155/100 = Rs. 11.55

Item No 2: Analysis of Rate of sand filling in plinth including supply of sand from a distance not exceeding 8 km including watering, dressing, etc., including cost of all materials, labour, tools and plants, etc. for completion of work. (for 100 cu m).

S.No.	Particulars	Qnty. or Nos.	Rate (Rs.)	Amount (Rs.)
	Labour :			
1.	Beldar	11 Nos	22/day	242.00
2.	Coolie	18 Nos	22/day	396.00
3.	Bhisti	2.5 Nos	24/day	58.00
4.	Tools and plants	Lumpsum	-	4.00
	Material :			
1.	Local sand	100 cu m	50/cu m	5000.00
		Totals of materials & labour		5700.00
		Add 10% for contractors profit		570.00
		Grand total		6270.00
			per	100 m^3

Rate per cu m = 6270/100 = Rs. 62.70

Item No. 3: Analysis of rate of lime concrete with 4 cm gauge brick ballast and kankar lime in foundation and under floors, including supply of all materials, labour and Tools and plants (T & P) etc. required for proper completion of work. (For 10 cu m).

S.No.	Particulars	Qnty. or Nos.	Rate Rs.	Amount Rs.
	Materials :			
1.	Brick-ballast	10 cu m	110/cum	1100.00
2.	Kankar lime	3.3 cum	225/cum	742.50
			Total	1842.50
	Labour :			
1.	Mistry	0.5 Nos	40/day	20.50
2.	Mason	1 No	40/day	40.00
3.	Beldar	11 Nos	22/day	242.00
4.	Coolie	11 Nos	22/day	242.00
5.	Bhisti	2 Nos	25/day	50.00
6.	Blacksmith	0.5 Nos	30/day	15.00
7.	T & P, Sundries, etc.	-	LS	20.00
			Total	629.00
		Total of materials & labour =		2471.50
		Add 1.5% for water charges =		37.00
		Add 10% for contractors profit =		247.00
		Grand Total =		2755.50
				per 10 m³

Rate per cu m = 2755.50/10 = Rs. 27.55

Item No. 4: Analysis of Rate of Cement Concrete with 4 cm gauge with brick ballast, fine local sand of 1.25 F.M. and cement in proportion of 12 : 6 : 1 in foundations and under floors, including supply of all materials, labour and T & P etc., complete item of work. (For 10 cu m)

S.No.	Particulars	Qnty. or Nos.	Rate Rs.	Amount Rs.
	Materials :			
1.	Brick ballast	10 cu m	110/cum	1100.00
2.	Fine local sand of 1.25 F.M.	5 cu m	11/cum	550.00

(Contd.)

S.No.	Particulars	Qnty. or Nos.	Rate Rs.	Amount Rs.
3.	Cement (0.84 cu m)	24 bags	130/bag	3120.00
	Add 2.5% wastage on cement and sand		Total	4770.00
	Labour :			
1.	Mistri	0.5 Nos	40/day	20.00
2.	Mason	1 No	40/day	40.00
3.	Beldar	11 Nos	22/day	242.00
4.	Coolie	11 Nos	22/day	242.00
5.	Bhisti	2 Nos	25/day	50.00
6.	Blacksmith	0.5 Nos	35/day	17.50
7.	T & P and Sundries and hire of mixer lubricants	-	Lump sum	100.00
			Total	711.50
		Total of materials and labour =		5481.50
		Add 1.5% for water charges =		82.20
		Add 10% for contractors profit =		548.10
		Grand total =		6111.80
				for 10 m³

Rate per cu m = 6111.80/10 = Rs. 611.20

Item No. 5: Analysis of rate of Cast Cement Concrete door sills, chaukats, boundary posts, shelves and similar works with cement, coarse sand and 2 cm stone ballast in proportion of 1 : 2 : 4, excluding supply of reinforcement and its binding but including its fixing and binding, including cost of binding wire, moulds, etc., complete.   (for - 10 cu m)

S.No.	Particulars	Qnty. or Nos.	Rate Rs.	Amount Rs.
	Materials :			
1.	Stone ballast	8.8 m³	300/cum	2640.00
2.	Coarse sand	4.4 m³	115/cum	506.00
3.	Cement (2.2 cu m)	66 bags	130/bag	8580.00
4.	Binding wire	1.5 kg	40/kg	60.00
			Total	11786.00

(Contd.)

S.No.	Particulars	Qnty. or Nos.	Rate Rs.	Amount Rs.
	Labour :			
1.	Mistri	0.5 Nos	40/day	20.00
2.	Mason	3 Nos	40/day	120.00
3.	Beldar	15 Nos	22/day	330.00
4.	Coolie	15 Nos	22/day	330.00
5.	Bhisti	7 Nos	25/day	175.00
6.	Sundries, T & P, hire of mixer etc.	-	Lump sum	150.00
				1125.00
	Framework (both erection and dismantling)			
7.	Timber planks for moulds	-	LS	250.00
8.	Carpenter	2 Nos	45/day	90.00
9.	Nails, T & P etc.	-	LS	25.00
				170.00
		Total of materials and labour =		13276.00
		Add 1.5% for water charges =		199.10
		Add 10% for contractor's profit		1327.00
		Grand total	14802.14	

Rate per cu m = 14802.14/10 = Rs. 1480.21

Item No. 6: Analysis of Rate of First class brickwork in kankar lime in foundation and plinth supply of all materials, labour, T & P etc., complete work. (For 10 cu m)

S.No.	Particulars	Qnty. or Nos.	Rate Rs.	Amount Rs.
	Materials :			
1.	Bricks (500/cu m)	5000 Nos	100/100	5000.00
2.	Kankar Lime	3 cu m	400/cum	1200.00
			Total	6200.00
	Labour :			
1.	Mistri	0.6 Nos	40/day	24.00
2.	Mason	7 Nos	40/day	280.00

(Contd.)

S.No.	Particulars	Qnty. or Nos.	Rate Rs.	Amount Rs.
3.	Beldar	7 Nos	22/day	154.00
4.	Coolie	7 Nos	22/day	154.00
5.	Bhisti	2 Nos	25/day	50.00
6.	T & P etc	-	LS	50.00
			Total	712.00
		Total of materials and labour		6912.00
		Add 1.5% for water charges		103.70
		Add 10% for contractor's profit		691.20
		Grand total		7706.90

Rate per cu m = 7706.90/10 = Rs. 770.70

Item No. 7: Analysis of Rate of 1st class brick work in 1 : 6 cement sand mortar including supply of all materials, labour, T & P, etc. complete item of work. (For 10 cu m)

S.No.	Particulars	Qnty. or Nos.	Rate Rs.	Amount Rs.
1.	Bricks	5000 Nos	100/100	5000.00
2.	Cement (0.45 cu m)	13.5 bags	130/bag	1755.00
3.	Local sand	2.7 cu m	50/cum	135.00
			Total	6890.00
	Labour :			
	Same as item no 6			712.00
		Total of materials and labour =		7602.00
		Add 1.5% for water charges =		114.00
		Add 10% for contractor's profit		760.2.0
		Grand Total		8476.20

Rate per cu m = 8476.20/10 = 847.60

Item No. 8: Analysis of Rate of Third Class brick-work in superstructure in mud mortar, including supply of all materials, labour, T & P etc., for complete work. (For 10 cum)

S.No.	Particulars	Qnty. or Nos.	Rate Rs.	Amount Rs.
	Materials :			
1.	Bricks (3rd class)	5000 Nos	70/100	3500.00
2.	Loamy soil earth	4 cu m	70/cum	280.00
			Total	3780.00
	Labour :			
	Same as item no 7			712.00
		Total of materials and labour =		4492.00
		Add 1.5% for water charges =		69.38
		Add 10% for contractor's profit =		449.20
		Grand Total		5008.58

Rate per cu m = 5008.58/10 = Rs.500.85

Item No. 9: Analysis of rate of reinforced brick-work in slabs, lintels etc. in 1 : 3 cement coarse sand mortar including supply of all materials, labour, T & P etc. required for complete work. (For 10 cu m)

S.No.	Particulars	Qnty. or Nos.	Rate Rs.	Amount Rs.
	Materials :			
1.	Bricks	4500 Nos	100/100	4500.00
2.	Cement (1.2 cu m)	36 bags	130/bag	4680.00
3.	Coarse Sand	3.6 cu m	115/cum	414.00
4.	Mild steel bars @ 0.8% = 0.08 cu m @ 78.5 q/cum = 6.25 qtl	6.25 quintal	960/qtl	6000.00
			Total	15594.00
	Labour :			
1.	Mistri	0.6 Nos	40/day	24.00
2.	Mason	10 Nos	40/day	400.00
3.	Beldar	10 Nos	22/day	220.00
4.	Coolie	10 Nos	22/day	220.00
5.	Bhisti	4 Nos	25/day	100.00
6.	Sundries, T & P etc	-	LS	50.00
			Total	1014.00
	Centering and shuttering both erection and dismantling			
1.	Timber planks and ballies		LS	200.00

(Contd.)

S.No.	Particulars	Qnty. or Nos.	Rate Rs.	Amount Rs.
2.	Carpenter	8 Nos	45/day	360.00
3.	Beldar	8 Nos	22/day	112.00
4.	Nails, T & P etc	-	LS	50.00
			Total	432.00
	Bending and cranking of steel bars			
1.	Blacksmith	7 Nos	35/day	245.00
2.	Beldar	7 Nos	22/day	98.00
3.	T & P etc.	-	LS	50.00
			Total	449.00
		Total of materials and labour =		17843.00
		Add 1.5% water charges =		376.64
		Add 10% for contractors profit		1784.60
		Grand total		19894.90

Rate per m^3 = Rs. 1989.5

Item No. 10: Analysis of rate of 2 cm thick DPC with cement and coarse sand in 1 : 2 proportion, including water-proofing material, including supply of all materials, labour, T & P etc., for complete work.   (For 100 sq m)

S.No.	Particulars	Qnty. or Nos.	Rate Rs.	Amount Rs.
	Materials :			
1.	Cement (0.9 cu m)	27 bags	130/bag	3510.00
2.	Coarse sand	18.0 m^3	115/cum	2070.00
3.	Water proofing compound	27 kg	27/kg	729.00
			Total	6309.00
	Labour :			
1.	Mistri	1.7 Nos	40/day	68.00
2.	Mason	9 Nos	40/day	360.00
3.	Beldar	9 Nos	22/day	198.00
4.	Bhisti	4 Nos	25/day	100.00
5.	Sundries, T & P etc.	-	LS	40.00
			Total	766.50
	Total of materials and labour			7075.00
	Add 1.5% for water charges			106.10
	Add 10% for contractor's profit			707.50
			Grand total	7888.60

Rate per sq m = Rs. 78.88

Item No. 11: Analysis of rate of supplying and fixing of 5 cm thick stone fine dressed in floor with 1 : 6 cement sand mortar including supply of all materials. (for 10 sq m)

S.No.	Particulars	Qnty. or Nos.	Rate Rs.	Amount Rs.
	Materials :			
1.	5 cm thick stone	10 sq m	150/sqm	1500.00
2.	Cement (0.01 cu m)	0.3 bag	130/bag	39.00
3.	Local sand	0.06 cu m	110/cu m	6.60
			Total	11545.00
	Stone Dressing Work			
1.	Stone Mason	11 Nos	45/day	440.00
2.	Blacksmith	3 Nos	35/day	105.00
3.	Beldar	2.2 Nos	22/day	48.40
4.	Coolie	2.2 Nos	22/day	48.40
			Total	641.80
	Fixing of stone work			
1.	Stone Mason	2.2 Nos	45/day	88.00
2.	Beldar	3.5 Nos	22/day	77.00
3.	Bhisti	1.1 Nos	25/day	27.50
4.	Blacksmith	0.33 Nos	35/day	11.50
			Total	204.05
	Total of materials and labour		2391.45	
	Add 1.5% for water charges		35.87	
	Add 10% for contractor's profit		239.14	
	Grand total			266.46

Rate per sq m = Rs. 266.64

Item No. 12: Analysis of rate of 2.5 cm thick 1 : 2 : 4 plain cement concrete floor with coarse sand and 2 cm stone ballast, laid in panels finished with 3 mm flooring coat of neat cement or cement and marble dust in 5 : 1 including 8 cm thick base concrete in 1 : 4 : 8 with cement, local sand and 4 cm stone ballast, for completion of work.

(for 100 m²)

S.No.	Particulars	Qnty. or Nos.	Rate Rs.	Amount Rs.
	Material :			
1.	Stone ballast 2 cm gauge	2.4 cu m	150/m³	360.00
2.	Coarse sand	1.2 cum	115/m³	138.00
3.	Brick ballast 4 cm gauge	7.5 cum	110/m³	600.00
4.	Cement	50.4 bags	130/bag	6552.00
5.	Local sand	3.5	110/m³	385.00
		Grand total		7900.00
	Labour :			
1.	Mistry	11 Nos	40/day	440.00
2.	Mason	11 Nos	40/day	440.00
3.	Beldar	15 Nos	22/day	330.00
4.	Coolie	15 Nos	22/day	330.00
5.	Bhisti	3 Nos	25/day	75.00
6.	T & P sundries etc.	-	LS	50.00
7.	Side forms	-	LS	70.00
		Total		1735.00

Total of materials and labour = 9635.00
Add 1.5% for water charges = 144.50
Add 10% for contractor's profit = 963.50
Grand total 10743.00

Rate per cu m = Rs. 107.43

Item No. 13: Analysis of Rate of Mosaic or terrazo tile flooring over 20 mm lime mortar, including supply of all materials, labour, T & P etc., including and polishing the surface, complete work. (for 100 sq m).

S.No.	Particulars	Qnty. or Nos.	Rate Rs.	Amount Rs.
	Materials :			
1.	Tiles (20 cm x 20 cm)	2500 Nos	500/100	12500.00
2.	Cement	6 bags	130/bag	780.00
3.	White Lime	1 cu m	150/cum	150.00
4.	Surkhi	2 cu m	90/cum	180.00
		Total		13610.00

(Contd.)

S.No.	Particulars	Qnty. or Nos.	Rate Rs.	Amount Rs.
	Labour :			
1.	Mistry	1 No	40/day	35.00
2.	Tile laying mason	14 Nos	40/day	490.00
3.	Beldar	14 Nos	22/day	196.00
4.	Bhisti	2 Nos	25/day	50.00
5.	Tile cutter and polisher	40 Nos	50/day	1000.00
6.	Polishing stone	one set	400/set	200.00
7.	Polishing material	-	LS	100.00
8.	T & P, Sundries, etc.	-	LS	50.00
9.	Hire of floor cutting m/c	20 days	50/day	500.00
			Total	2621.00
	Total of materials & labour =		18268.00	
	Add 1.5% for water charges =		274.00	
	Add 10% for contractor's profit =		1826.80	
		Grand total		20368.82
	Rate per sq m = Rs. 203.68			

Item No. 14: Analysis of Rate of 12 mm thick 1 : 4 cement and local sand mortar plaster on brick wall, including supply of all materials, labour, T & P etc. for completion of work including raking and watering.(for 100 sq m)

S.No.	Particulars	Qnty. or Nos.	Rate Rs.	Amount Rs.
	Materials :			
1.	Cement (0.375 cu m)	11.25 bags	130/bag	1462.50
2.	Local sand	1.5 cum	110/cum	165.00
			Total	162.50
	Labour :			
1.	Mistri	0.33 Nos	40/day	13.20
2.	Mason	10 Nos	40/day	400.00
3.	Beldar	11 Nos	22/day	242.00
4.	Bhisti	3.8 Nos	25/day	95.00
5.	Scaffolding	-	LS	50.00
6.	T & P, Sundries	-	LS	30.00
			Total	830.20

(Contd.)

S.No.	Particulars	Qnty. or Nos.	Rate Rs.	Amount Rs.
	Total of materials and labour			245.70
	Add 1.5% of water charges			36.86
	Add 10% for contractor's profit			245.70
				2740.26
	Rate per sq m = Rs. 27.40			

Item No. 15: Analysis of Rate of supplying and fixing 6 mm thk corrugated A.C. sheets roofing in position including supply of all materials, labour, T & P etc., for completion of work. (for 10 sq m).

S.No.	Particulars	Qnty. or Nos.	Rate Rs.	Amount Rs.
	Materials :			
1.	A.C. corrugated sheet	11 m²	100/m²	1100.00
2.	G.I.J Bolts (10 x 0.6 cm)	24 Nos	80/100	19.20
3.	Lampet Washer	24 Nos	10/100	2.40
4.	Bitumen washer	24 Nos	10/100	2.40
			Total	1124.00
	Labour :			
1.	Mistry	0.1 No	40/day	3.50
2.	Carpenter	1 No	35/day	30.00
3.	Beldar	1 No	22/day	14.00
4.	T & P, Sundries, etc.	-	LS	5.00
			Total	52.50
	Total of Materials & Labour			1210.00
	Add 10% for contractor's profit			121.00
			Grand total	1331.00
	Rate per sq m = Rs. 1331.			

Item No. 16: Analysis of rate of supply and fixing of 2.5 cm Deodar wood planks in ceiling, including supply of 2.5 cm x 4.0 cm, beading over plank joints, but without ceiling frame work, including cost of materials, labour, T & P, for completion of work. (For 10 sq m).

S.No.	Particulars	Qnty. or Nos.	Rate Rs.	Amount Rs.
	Materials :			
1.	Deodar wood			
	(a) Planks including overlap and wastage - 0.34 cu m	0.40 m^3	10000/m^3	4000.00
	(b) Beading - 0.06 cu m			
2.	Nails	3 kg	25/kg	75.00
3.	Screws 35 mm	300 Nos	35/100	105.00
			Total	4180.00
	Labour :			
1.	Mistri	2 Nos	40/day	80.00
2.	Carpenter	2 Nos	35/day	70.00
3.	Beldar	4 Nos	22/day	88.00
4.	Scaffolding	-	LS	50.00
5.	T & P, Sundries	-	LS	55.00
			Total	313.00
	Total of materials and labour			4493.00
	Add 10% for contractor's profit			449.30
			Grand Total	4942.30
	Rate per sq m = Rs. 494.20			

Item No. 17: Analysis of rate of white washing one coat, including supply of all materials, labour, T & P etc., for completion of work.(for 100 sq m)

S.No.	Particulars	Qnty. or Nos.	Rate Rs.	Amount Rs.
	Materials :			
1.	White lime	11 kg	2/kg	22.00
2.	Gum or glue powder	-	LS	5.00
3.	Indigo (blue pigment)	-	LS	5.00
			Total	32.00
	Labour :			
1.	White Washer	0.73 Nos	25/day	13.25
2.	Coolie	0.73 No	22/day	16.00

(Contd.)

S.No.	Particulars	Qnty. or Nos.	Rate Rs.	Amount Rs.
3.	T & P, ladder, Sundries, etc.	-	LS	15.00
			Total	44.31
	Total of materials and labour		76.31	
	Add 10% for contractor's profit		7.31	
			Grand total	83.94

Rate per sq m = Rs. 0.84

Item No. 18: Analysis of Rate of colour washing one coat, using approved colouring material of bulf, grey or brown shade, including supply of all material, labour, T & P etc., for completion of work. (for 100 sq m)

S.No.	Particulars	Qnty. or Nos.	Rate Rs.	Amount Rs.
	Materials :			
1.	White Lime	11 kg	2/kg	22.00
2.	Gum or glue powder	-	LS	5.00
3.	Colouring pigment	2.5 kg	30/kg	75.00
			Total	102.00
	Labour :			
1.	White washer	0.75 No	25/day	18.75
2.	Coolie	0.75 No	18/day	13.50
3.	T & P, ladder	-	LS	15.00
			Total	47.25
	Total of materials and labour		149.25	
	Add 10% for contractor's profit		14.92	
			Grand total	164.17

Rate per sq m = Rs. 1.64

Item No. 19: Analysis of rate of one priming coat and one coat of dry distemper on new work, including supply of all materials, labour and T & P, etc., for completion of work. (for 100 sq m)

S.No.	Particulars	Qnty. or Nos.	Rate Rs.	Amount Rs.
	Materials :			
1.	Dry distemper	8.1 kg	20/kg	162.00
2.	White Lime	11 kg	2/kg	22.00
3.	Gum or glue powder	-	LS	10.00
			Total	194.00
	Labour :			
1.	White powder	0.73 No	25/day	6.25
2.	Distemper or painter	3.5 Nos	40/day	140.00
3.	Coolie	4.23 Nos	18/day	76.14
4.	T & P, Sundries, etc.	-	LS	25.00
			Total	140.98
	Total of materials and labour			441.40
	Add 10% for contractor's profit			44.10
			Grand total	485.50

Rate per sq m = Rs. 4.85

Item No. 20: Analysis of rate of painting or varnishing on iron work in small areas or on wood work with one coat of ready mixed priming paint and one coat of superior quality ready mixed paint or synthetic varnish including supply of all materials, labour, T and P, etc., for completion of work. (for 100 sq m)

S.No.	Particulars	Qnty. or Nos.	Rate Rs.	Amount Rs.
	Materials :			
1.	Primer paint	7.5 litres	30/lit	225.00
2.	Paint or varnish	6.7 litres	50/lit	201.00
			Total	351.00
	Labour :			
1.	Painter	5.4 Nos	40/day	135.00
2.	Coolie	5.4 Nos	18/day	54.00
3.	Brushes, T & P etc.		LS	10.00
			Total	199.00
	Total of materials and labour			923.20
	Add 10% for contractor's profit			92.30
			Grand total	1015.50

Rate per sq m -= Rs. 10.15

Item No. 21: Analysis of rate of wood work in C.P. Teak wood wrought framed and fixed including simple mouldings, door, chaukats, including supply of all materials, labour, T & P etc., for completion of work. (for one chaukat).

S.No.	Particulars	Qnty. or Nos.	Rate Rs.	Amount Rs.
	Assuming size of door opening as 100 x 200 cm and section of chaukats as 10 x 7 cm and also the chaukat without sill.			
	Materials :			
1.	Wood for posts	$2 (2 \times 0.1 \times 0.07) =$	0.028 m^3	
2.	Wood for head	$1 (1 \times 0.1 \times 0.07) =$	0.007 m^3	
			0.035 m^3	
3.	Wastage		0.002 m^3	
			0.037 m^3	
			$8500/\text{m}^3$	314.50
	Labour :			
1.	Mistri	0.06	40/day	2.40
2.	Carpenter	0.5	45/days	22.50
3.	Beldar or helper	0.25	22/day	5.50
4.	T & P, Sundries		LS	25.00
			Total	54.40
	Total of materials and labour			370.00
	Add 10% for contractor's profit			37.00
			Grand total	407.00
	Rate per cum = 407.00/0.037 = Rs. 11000.00			

Item No. 23: Analysis of Rate of panelled and glazed door shutter of shisham wood without frame, including fixing in chaukat with all brass fittings, labour, materials, T & P, etc., for completion of work.

(for 118 x 209 cm shutter size)

S.No.	Particulars	Qnty. or Nos.	Rate Rs.	Amount Rs.
	Materials :			
1.	Timber for styles	$4 (2.85 \times 0.11 \times 0.04) = 0.036 m^3$		$0.036 m^3$
2.	Timber for bottom rail	$2 (0.60 \times 0.23 \times 0.04) = 0.011\ m^3$		$0.011\ m^3$
3.	Timber for back rail	$2 (0.6 \times 0.15 \times 0.04) = 0.007\ m^3$		$0.007\ m^3$
4.	Timber for shash bars	$4 (0.50 \times 0.05 \times 0.04) = 0.004\ m^3$		$0.004\ m^3$
5.	Timber for top rail	$2 (0.60 \times 0.14 \times 0.04) = 0.007\ m^3$		$0.007\ m^3$
6.	Timber for panels	$2 (0.40 \times 0.55 \times 0.02) = 0.009\ m^3$		$0.006\ m^3$
	Add 10% for wastage and for overlapping of joints			$0.008\ m^3$
				$0.093\ m^3$
		@ 8500/cum	=	790.00
7.	4 mm thk glass panels	$4 (0.40 \times 0.33)$	=	$0.528\ m^2$
		@ 50/m²	=	26.40
			Total	816.90
	Fittings (Brass) :			
1.	Tower bolt 30 cm	2 Nos	6/each	12.00
2.	Tower bolt 15 cm	2 Nos	4/each	8.00
3.	Hinge 10 cm	6 Nos	3/each	18.00
4.	Brass Handle	2 Nos	4/each	8.00
5.	Wooden cleat	2 Nos	1/each	2.00
6.	Hinges 2.5 cm for cleat	2 Nos	0.5/each	1.00
7.	Screws 30 mm	20 Nos	20/100	4.00
8.	Screws 20 mm	100 Nos	15/100	15.00
9.	Sliding bolt	1 No	15/each	15.00
				83.00
	Labour :			
1.	Mistri	0.7 Nos	40/day	28.00
2.	Carpenter	3.5 Nos	45/day	157.50
3.	Helper	3.5 Nos	22/day	77.00
		Total		183.50
		Total of materials, fittings & labour		1162.40
		Add 10% for contractor's profit		116.20
		Grand total		1278.60

$$\text{Rate per sq m} = \frac{1278.60}{1.18 \times 2.09} = \text{Rs. } 518.50$$

Item No. 23: Analysis of rate of providing and fixing 200 litre capacity storage tank of 1.25 mm thk G.I plane sheet to 40 x 40 x 6 mm angle iron, complete and flat iron 40 x 6 mm, fixed with 40 mm diaa G.I. pipe with overflow pipe fittings and mosquito proof cowl, ball valve and mosquito proof G.I. covered with flat iron rim and locking arrangement, including supply of all materials, labour, T & P etc. for completion of work.

S.No.	Particulars	Qnty. or Nos.	Rate Rs.	Amount Rs.
	Materials :			
1.	1.25 mm thk GI sheet (10.3 cm)	22.6 kg	21/kg	474.60
2.	40 x 40 x 6 cm angle iron (7.4 rm)	25.3 kg	9.50/kg	240.35
3.	40 x 6 mm flat iron (1.8 rm)	3.1 kg	9.50/kg	29.45
4.	Rivets etc.		LS	40.00
5.	Mosquito proof GI cover with locking arrangement	1 No	50/each	50.00
6.	40 mm dia screw pipe with fitting	1 No	15/each	15.00
7.	25 mm dia GI overflow pipe with fitting	1 No	8/each	8.00
8.	25 mm dia brass mosquito proof coupling	1 No	8/each	8.00
9.	Specials for pipe connection	-	LS	40.00
10.	White lead, putty, etc.	-	LS	25.00
			Total	930.40
	Labour :			
1.	Fitter (Plumber)	3 Nos	40/day	120.00
2.	Coolie	6 Nos	22/day	132.00
3.	T & P, Sundries, etc.	-	LS	25.00
			Total	277.00
		Total of materials and labour =		1207.40
		Add 1% for water charges =		12.07
		Add 10% for contractor's profit =		120.70
			Grand total	1340.17

Item No. 24: Analysis of rate of providing and fixing white vitrous Indian type W.C. pan size 580 mm, with C I high level flushing cistern with fittings, C.I. brackets, 32 mm dia steel telescopic flush pipe and C.I. (a gauge) with fittings, including all cuttings, threadings etc., including cistern painting, for completion of work.

S.No.	Particulars	Qnty. or Nos.	Rate Rs.	Amount Rs.
	Materials :			
1.	White glazed 580 mm W.C.	1 No	200/each	200.00
2.	CI high level flushing cistern with fittings GI chain and Pull	1 No	225/each	225.00
3.	C.I. bracket	1 pair	10/pair	10.00
4.	32 mm GI telescopic flush pipe with brass union	1 No	45/each	45.00
5.	32 mm dia M.S. holder clamp	1 No	2/each	2.00
6.	10 mm CI P or S trap	1 No	60/each	60.00
7.	20 mm GI overflow pipe	2 m	20/metre	40.00
8.	Whitelead, etc.	-	LS	25.00
9.	Cement, sand and grit	-	LS	40.00
10.	Painting of cistern and other fittings	-	LS	75.00
			Total	722.00
	Labour :			
1.	Plumber	1.25 Nos	40/day	50.00
2.	Mason	0.50 Nos	40/day	20.00
3.	Beldar	1.00 No	22/day	22.00
4.	T & P, Sundries etc.	-	LS	25.00
			Total	71.50
	Total of materials and labour			839.00
	Add 1% for water charges			8.39
	Add 10% for contractor's profit			83.90
			Grand total	931.30

Section C

SPECIFICATIONS

C. 1 Introduction

In structural drawings, shape and size of the different components are clearly mentioned but it is not possible to give details about the materials and the workmanship of the work or the methods, tools and plants to complete the work. For example, in the drawing it is clear that plastering to be done on the walls but it is not clear from the drawing what should be the thickness of the plastering, how many no. of coatings of white wash and colour wash etc. should be done.

Specifications are the instructions, meant to supplement the drawings and provide technical requirement of the work. The drawings with the specifications will give the complete requirement of the the structure.

C. 2 Aim of Specifications

1. It checks the strength of materials for a work involved in a project.
2. The cost of an unit quantity of work is governed by its specification.
3. It is required for court cases.
4. Specifications gives detailed information regarding quantity and quality of the material which enables a contractor to estimate the cost correctly.
5. Specifications give detailed information regarding the equipments, tools and plants to complete the project which enables us to procure them before hand in time.
6. Cost of work depends mainly on the specifications. If specifications are changed, the cost of work is also changed. Specifications give correct idea about the estimated cost so that funds can be arranged accordingly.
7. It also serves as a guide.

C. 3 Principles of Specification writing:

(a) Description of materials
(i) The quantity and quality of materials required should be given.
(ii) The proportion of mixing or treatment of materials before use should be clearly mentioned.

(b) Workmanship
(i) The method of mixing to the proportion
(ii) Method of laying
(iii) Method of preparation of surface
(iv) Compaction, finishing and curing etc. should be clearly mentioned in the specifications.

(c) Equipments : The equipments, tools and plants should be described clearly to carryout the work. The method of operation also should be stated in the specifications.

(d) Expressions: The following points should be kept in mind while preparing specifications:
(i) The sentence should be brief, simple and concise without any specifications.
(ii) The writing style and tense should be kept same throughout.
(iii) There should not be any grammatical mistake.
(iv) The omission or misplacement of comma, fullstop etc. should not change the meaning of the sentence.
(v) There should not be any ambiguity in the sentence.

(e) Practical limitations of materials, workmanship etc. should be kept in mind while writing the specifications.

(f) The specifications should be fair to both the parties and it should not favour any party.

(g) In case of any foreseen difficulty or hazards, it should be specified.

(h) Specifications are legal documents and they should be prepared keeping constitution in mind.

C. 4 Types of specifications

The specifications are broadly classified as:
(a) General or brief specifications
(b) Detailed specifications

General specifications: In general specifications, nature of class of works and materials that should be used are mentioned in brief. Only a brief description of each and every item is given which serve as a good guide to prepare the estimate and to execute the work. These specifications do not form the part of the contract document.

Detailed specifications: It gives the detailed description of each item and forms a part of the contract document. It can be subdivided as follows:

   (i) General provisions
   (ii) Technical provisions
   (iii) Standard specifications

   (i) **General provisions:** These are the general conditions of the contract.

   (ii) **Technical provisions:** It specifies the quality of the final product and the inspections and tests which should be conducted during construction.

   The technical provisions are of the following types:

   (a) Specifications for materials and workmanship
   (b) Specifications for performance
   (c) Specifications for proprietory commodoties

   (iii) **Standard specifications:** Most of the works are usually standardised which are numbered serially so while writing the contract bond only serial numbers of the standard specifications are written. This saves lots of time, labour and stationary expenditure.

C. 5 Cement Concrete

The cement concrete shall be prepared by mixing graded stone or brick-aggregate of nominal size as specified with fine aggregate and cement in specified proportions with required quantity of water. The grading and quantity of aggregate shall be such as to give maximum compressive strength

Proportioning: It shall be done by volume. Boxes of suitable size generally 35x25x40 cm. shall be used for the measurement of sand and aggregate. The unit of measurement for cement shall be a bag of 50 kg and this shall be taken as 0.0347cum.

Mixing: Brick aggregate shall be well soaked with water for a minimum period of 2 hours and stone aggregate shall be washed with water to remove dirt, dust or any other foreign materials. The mixing shall be done in mechanical mixer. Mixing by hand shall be employed only in special case with the prior permission.

Machine mixing: Measured quantity of dry coarse aggregate shall be first placed in the hopper. This shall be followed with measured quantity of fine aggregate and then cement. In case damp sand is used, add half the quantity of coarse aggregate followed by cement and sand. Finally add balance quantity of the coarse aggregate. The total quantity of water for mixing shall be introduced before 25% of the mixing time has elapsed and shall be regulated to achieve the specific water cement ratio. When the mixer is closed down for the day or at any time exceeding 20 minutes, the drum should be washed with water.

Mixing time: The material shall be mixed for a period of not less than 2 minutes and until a uniform colour and consistency are obtained.

Hand mixing: The mixing shall be done on a smooth, water tight, clean platform of suitable size in the following manner:

(i) Measured quantity of sand shall be spread evenly.

(ii) The cement shall be dumped on the sand and distributed evenly

(iii) The sand and cement shall be mixed with spade, turning the mixture over and over again, until it is of even colour throughout and free from streaks.

(iv) The sand cement mixture shall be spread out and measured quantity of coarse aggregate shall be spread on its top. Alternatively the measured quantity of coarse aggregate shall be spread out and the sand, cement mixture shall be spread on its top.

(v) This shall be mixed at least three times by shovelling and turning over by twist from centre to side, then back to the centre and again to the sides.

(vi) Three quarters of the total quantity of water required shall be added while the material is turned in towards the centre with spades. The remaining water shall be added by a water can fitted with rose head, slowly turning the whole mixture over and over again until a uniform colour and consistency is obtained throughout the pile.

(vii) The mixing platform shall be washed at the end of the day's work.

Consistency: The quantity of water to be used for each mix of 50 kg cement, to give the required consistency shall not be more than 34 litres for 1:3:6 mix, 30 litres for 1:2:4 mix, 27 litres for 1:1.5:3 mix and 25 litres for 1:1:2 mix.

Laying: The entire concrete used in the work shall be laid gently (not thrown) in layers not exceeding 15 cm, shall be thoroughly vibrated by means of mechanical vibrators till a dense concrete is obtained. Hand compaction shall be done with the help of tamping rods so that concrete is thoroughly compacted and completely worked into toe corners of the formwork. Compaction shall be completed before the initial setting starts i.e. within 30 minutes of addition of water in the dry mixture.

During cold weather concreting shall not be done when the temperature falls below 4.5 C. The laid concrete shall be protected against frost by suitable covering. During hot weather, precuations shall be taken to see that temperature of wet concrete does not exceed 38°C.

Where it is found to be necessary to deposit any concrete under water, the method, material, equipment and mix shall first to be approved. The concrete shall be deposited under water by one of the approved methods such as Treamic method, Drop bottom bucket, bags etc.

Curing: After 1-2 hours of laying the concrete, it shall be protected with moist gunny bags,sand or any other material approved, against drying. 24 hours after laying the concrete, its surface shall be cured by flooding with water of minimun 25 mm depth, or by covering with wet absorbent materials. The curing is done for atleast 14 days. In special cases the curing period can be increased.

In case of foundation concrete the masonary work may be started after 48 hours of its laying, but the curing of cement concrete shall be continued along with the masonary work for a minimum period of 14 days.

Form work: If centering and shuttering are required to be done for this work, it shall be done in accordance with the specifications for form work and shall be paid for separately unless otherwise specified.

Rate: The rate is inclusive of the cost of labour and materials involved in all the operations described above except for "Form Work" in case of cost-in-situ work.

C. 6 Steel Reinforcement

Steel Reinforcement shall be conforming to the I.S. Codes given in Table C-1.

Table C-1

Description	Confirming to I.S.I.Code no
1. Mild steel & medium tensile	432 (Pt.II) - 1966
2. Hard drawn steel wire	432 (Pt.II) - 1966
3. Cold twisted bars	1786-1966
4. Hot rolled deformed bars	1139-1966
5. Hard drawn steel wire fabric	1566-1967
6. Structural steel sections	226-1962

Bending and Overlapping:- Bars shall be bent cold, correctly and accurately to the size and shape as shown in the detailed drawings. The overlapping bars shall not touch each other and these shall be kept apart by 25mm or 1.25 times the maximum size of the coarse aggregate whichever is greater, with concrete between them. But where this cannot be done, the overlapping bars shall be bound together at intervals not exceeding twice the diameter of such bars with two strands of annealed steel wire of 0.90mm to 1.6mm thickness twisted tight. In case of mild steel the ends of rods shall be bent into semi-circular hooks, having clear diameter equal to four times the diameter of the bar, with a length beyond the bend equal to four times the diameter of the bar. In case of deformed bars the hooks are not required.

Placing in position: The bars crossing one another shall be tied together at every intersection with two strands of annealed steel wire 0.90 to 1.6mm diameter and twisted tight to make the skeleton of the steel work rigid so that the reinforcement does not get displaced doing the deposition of concrete.

The reinforcement bars shall be kept in position by the following methods:

(a) In case of columns and walls, the vertical bars shall be kept in position by means of timber templates with slots accurately cut in them, or with block cement mortar (1:2) suitably tied to the reinforcement.

(b) In case of R.C.C beams and slab construction, precast cover blocks in cement mortar 1:2, about 4x4 cm. section and of thickness equal to the specified cover shall be placed between the bars and shutterings so as to secure and maintain the requisite cover of concrete over reinforcement.

(c) In case of cantilever and double reinforced beams or slabs the vertical distance between the horizontal bars shall be maintained by introducing chain spacers or support bars of steel at 1m or at shorter spacing to avoid sagging.

(d) In case of other R.C.C. structure such as arches, domes etc., cover blocks, spacers, and templates shall be used as directed.

Measurement: Reinforcement including authorised spacer bars shall be measured in lengths of different diameters, as actually used in the work nearest to a centimetre and their weight calculated on the basis of standard tables. Wastage and unauthorised overlaps shall not be paid for.

Rate: The rate for reinforcement shall include the cost of labour and materials required for all operations described above except welding in lieu of overlaps which shall be paid separately.

C. 7 Terrazo(Marble Chips) Flooring Laid in Site

The panels shall be of uniform size, not exceeding 2 sqm in area and 2m length for inside situations. Incase of exposed situations, the length of any side of the panel shall not be more than 1.25m. Cement slurry at 2kg/sqm shall be applied before laying of under layer over the cement concrete/R.C.C. surface.

Strips: 4 mm to 6 mm thick glass strips or 2 mm thick aluminium strips shall be fixed with the top at the proper level and floor slope.

Top-layer: The cement and marble powder shall be mixed in the proportion of 3:1 (cement:marble powder) by weight. Table gives the proportions of aggregate to binder mix (concrete: marble powder) by volume.

Table C-2

Grade of aggregate(marble chips)	Proportion of aggregate to binder mix
1. Grade 0 and	1 1.75 parts
2. Grade 2 and	3 1.50 parts
3. Grade 4 and 5	1.25 parts
4. Mixed grade (mixed size aggregat(e)	1.50 parts

The marble chips shall be white or pink Makrana, Jaisalmer yellow, Chittor black, yellow Patan, Cherola(Madras), grey Gadu(Surat), Dehradun white, Baroda green, Alwar black etc. as specified. The maximum thickness of the top layer for various sizes of marble aggregates(marble chips) shall be given in table.

Table C-3

Sl.no.	Code no.	Size of marble chips in mm	Minimum thickness of top layer in mm
1.	00	1-2	6
2.	0	2-4	9
3.	1	4-7	9
4.	2	7-10	12

If aggregate of size larger then 10 mm are used the minimum thickness of topping shall not be less than 1.5 times the minimum size of the chips where large size chips such as 20 mm or 25 mm are used. They shall be used only with a flat shape and bedded on the flat face so as to keep the minimum thickness of wearing layer.

The terrazo topping shall be laid when the under layer is still plastic, but has hardened sufficiently to prevent cement from rising to the surface, this is normally achieved between 18-24 hours after laying the under layer.

Polishing, Curing and Finishing: Polishing shall be done by machine. After about 36 hours of laying the top layer, the surface shall be watered and ground evenly with machine fitted with special rapid cutting grit blocks (Carborundum stone of coarse grade no. 60) till the marble chips are evenly exposed and the floor is smooth. After first grinding the surface shall be thoroughly washed to remove all grinding mud and covered with a grout of cement or/and colouring matter in same mix and proportion as the topping in order to fill any pin holes that appear. The surface shall be allowed to cure for 5 to 7 days and then ground with machine fitted with grid blocks no. 120. The surface is cleaned and repaired as before and allowed to cure again for 3 to 5 days. Finally the third grinding shall be done with machine fitted with fine grades grid block no. 320 to get even and smooth surface without pin holes.

Where use of machine for polishing is not feasible or possible, rubbing and polishing shall be done by hand, in the same manner as specified for machine polishing except that carborundum stone of coarse grade no. 60 shall be used for the first rubbing. Stone of medium grade no. 80 for second rubbing and stone of the grade no. 120 for final rubbing and polishing is used.

After the final polish either by machine or hand, oxalic acid shall be dusted over the surface 33gm/sqm sprinkled with water and rubbed hard with a pad of woolen rags. The following day, the floor shall be wiped with a moist rag and dried with a soft cloth and finished clean.

Measurement: Length and breadth shall be measured correct to a cm before laying skirting & wall plaster. The area as laid shall be calculated in sqm correct to two places of decimal. No deductions shall be made or extra paid for any opening in floor of area upto 0.1 sqm.

Marble chips(terrazo) flooring laid as floor borders, margins and similar bends upto 30 cm width and on staircase treads, shall be measured under the item of terrazo flooring, but extra shall be paid for such work.

Rate: The rate shall include the cost of all materials and labour involved in all the operations described above including cleaning of surface of R.C.C. slab or sub-grade and application of cement slurry but shall not include the cost of sub-grade concrete and strips of glass or aluminium used for making panels.

C. 8 Wood Work for Door and Window Framers

Materials: Timber shall be of teak, sal, deodar etc. as mentioned, well seasoned, dry, free from sap, knots, cracks or any other defects or diseases. It shall be sawn in the direction of the grains.Sawing shall be truly straight and square. The scantling shall be plane, smooth and accurate to the full dimensions, rebates, roundings and mouldings as shown in the drawing made, before assembling. Patching or plugging of any kind shall not be permitted except as provided.

Joints: These shall be of mortise and tenon type, simple, neat and strong. Mortise and tenon joints shall fit in fully and accurately without wedging or filling. The joints shall be glued, framed, put together and pinned with hardwood or bamboo pins not less than 10 mm dia. after frames are put together pressed in position by means of a press.

Surface treatment: Wood work shall not be painted, oiled or otherwise treated before it has been approved by the Engineer-in-charge. All portions of timber abutting against masonary or concrete or embedded in ground shall be painted with approved wood primer or with boiling coaltar.

Gluing of joints: The contract surface of tenon and mortise joints shall be treated before putting together with bulk type synthetic resin adhesive of a make approved by the Engineer-in-charge.

Fixing in position: The frame shall be placed in position truly vertical before the masonry reaches half the highest of the opening with iron clamps or as directed by the Engineer-in-Charge. In case of door frames without sills, the vertical members shall be embedded in the flooring to a depth of 40 mm or as directed by the Engineer-in-Charge. The door frames without sills while being placed in position, shall be suitably strutted and wedged in order to prevent warping during construction. The frames shall also be protected from damage, during construction.

C.9 Wood Work for Door and Window Shutters

Materials: Specified timber shall be used, and it shall be well seasoned, dry, free from sap, knots crack or any other defects or diseases. Patching or plugging of any kind shall not be permitted except as provided.

Joinery work: All pieces shall be accurately cut and planed smooth to the full dimension. All members of the shutters shall be straight without any warp or bow and shall have smooth, well planed faces at right angles to each other. In case of panelled shutters the corners and edges of panels shall be finished as shown in drawings, and these shall be framed into groovers to the full depth of the groove leaving an air space of 1.5 mm and the faces shall be closely fitted to the sides of the groove. In case of glazed shutter, slash bars shall have mitred joints with styles. Styles and rails shall be properly and accurately mortised and tenoned.Rails which are more than 180 mm in width shall have two tenons. Styles and end rails of shutters shall be made out of one piece only. The tenons shall pass through styles for at least 8/4 th of the width of the styles. When assembling a leave, styles shall be left projecting as a horn. The styles and rails shall have 12 mm groove in panelled portion for the panel to fit in.

The depth of rebate in frames for housing the shutters shall in all cases be 1.25 cm and the rebate in shutters for closing in double shutter doors and windows shall not be less than 2 cm. The rebate shall be splayed. The joints shall be placed, and secured by bamboo pins of about 6 mm diameter. The horns of styles shall be sawn off.

The case of battened shutters, planks for battens shall be 20 mm thick unless otherwise specified and of uniform width of 125 to 175 mm. These shall be planed and made smooth, and provided with minimum 12 mm rebated joints. The joint lines shall be chamfered. Unless otherwise specified fixed with the battens on the inside face of shutter with minimum two number 50 mm long wood screws per batten. The ledges shall be 225 mm wide and braces 175 mm wide, unless otherwise specified. The braces shall incline downwords towards the side on which the door is being hung.

Glueing of Joints Panelled or Glazed Shutters

The contact surface of tenon and mortise joints shall be treated before putting together with bulk type synthetic resin adhesive of a make approved by the Engineer-in-chief. Shutters shall be not be painted, oiled or otherwise treated, before these are fixed in position and passed by the Engineer-in-Chief. For glazed shutters, mounting and glazing bars shall be tub-tenoned to the maximum depth which the size of the member would permit or to a depth of 25 mm, whichever is less.

Fittings: Details of fittings to be provided shall be as per the schedule of fittings supplied by the Engineer-in-Charge in each case. The cost of providing and fixing shutters shall include the cost of hinges and necessary screws for fixing the same. All other fittings shall conform to their respective IS specifications. All other fittings shall be enumerated and paid separately. Where fittings are stipulated to be supplied by the department free of cost, screw for fixing the fittings shall be provided by the contractor and nothing extra will be paid for the same.

C.10 Roads

W.B.M with stone aggregate sub-base: Stone aggregate of size 30 mm to 40 mm is used.Quantities of coarse aggregate screening and binding material required to be stacked for 100 mm approximately compacted thickness of W.B.M. sub-base course for 10 sq m shall be as follows:-

	Loose Qnty
1. Grading I 90 mm to 40 mm Coarse aggregate	1.35 cu m
2. Type a 12.5 mm size stone screenings	0.42 cu m
3. Building materials	0.10 cu m

Foundation Preparation: In the case of an existing road,where new material is to be laid,the surface shall be scarified and reshaped to the required grade,camber and shape as necessary.Weak places shall be strengthened, corrugations removed and depressions and pot holes made good with suitable materials,before spreading the aggregate for W.M.B.

Spreading aggregate: The coarse aggregate shall be spread uniformly and evenly upon the prepared base in the required quantities with a twisting motion to avoid segregation.The aggregate shall be spread uniformly to proper profile by using templates placed across the road,six metre apart where specified approved mechanical devices may be used to spread the aggregate uniformly.

The W.B.M sub-base shall be normally constructed in layers of 100 mm compacted thickness.Normally the coarse aggergate shall not be spread in lengths exceeding three days average work ahead of the rolling and blending of the preceding section.

Rolling: Immediately after spreading the coarse aggregate,it shall be compacted to the full width by rolling with either a three-wheel power roller of 8 to 10 tonnes or an equivalent vibratory roller.Initially,light rolling is done,which shall be discontinued when the aggregate is partially compacted with sufficient void space in them to permit application of screenings.The roller shall then progress gradually from the edges to the center,parallel to the centre line of the road and overlapping uniformly each preceeding rear wheel track by one-half width and shall continue untill the area of the coarse has been rolled by the rear wheel.Rolling shall continue untill the road metal is thoroughly keyed with no creeping of metal ahead of the roller.Slight sprinkling of water shall be done while rolling,if required.

When rolling develops irregularities that exceed 12 mm when tested with a three metre straight edge,the irregular surface shall be loosened and then aggregate added to or removed from it as required and the area rolied untill it gives a uniform surface conforming to the approved cross-section and grade.

Application of Screenings: After rolling the coarse aggregate the screening shall be applied gradually over the surface to completely fill the interstices. While spreading the screenings, dry rolling shall be continued. The screenings shall not be dumped in piles on the coarse aggregates,but shall be spread uniformly in successive thin layers by the sprading motion of the hand,shovels or a mechanical spreader.

The screenings shall be applied in three or more applications at a slow rate,so as to ensure filling of all voids.Rolling and brooming shall continue with the screenings spreading.Mechanical brooms or hand brooms or both may be used for this purpose.The spreading,rolling and brooming of screenings shall be performed on sections which can be completed within one day's operation and shall continue untill no more screenings can be forced into the voids of the coarse aggregate.

Sprinkling & Grouting: After spreading the screening and rolling,the surface shall be copiously sprinkled with water,swept and rolled.Hand brooms shall be used to sweep the screenings into the voids and to distribute them evenly.The sprinkling,sweeping and rolling operations shall be continued and well bonded and firmly set for the entire depth and untill a grout has been formed of screenings and water that will fill all voids and form a wave of grout ahead of the wheels of the roller.The quantity of water to be used during the construction shall not be excessive so as to cause damage to the sub-base or sub-grade.

Application of Binding materials: After the application of screenings and rolling,a suitable binding material shall be applied at a uniform and slow rate in two or more successive thin layers.After each application of binding material,the surface shall be copiously sprinkled with water and the resulting slurry swept in with hand brooms or mechanical broom;or both so as to fill the voids properly. The surface shall then be rolled by a 8-10 tonne roller. Water being applied to the wheels in order to wash down the binding material that may get stuck to the wheels.The spreading of binding material,sprinkling of water,sweeping with brooms and rolling shall continue untill the slurry that is formed well, form.

Setting & Drying: After final compaction of the course,the road shall be allowed to cure overnight.Next morning defective spots shall be filled with screenings or binding materials,lightly sprinkled with water (if necessary) and rolled.No traffic shall be allowed till the macadam sets.

Surface Evenness: The surface evenness of completed W.B.M sub-base in the longitudinal and transverse direction shall be as specified.The maximum permissible undulation when measured with a 3 meter straight edge should not be more than 15 mm for longitudinal profile.The maximum permissible undulation when measured with a camber template should not be more than 12 mm for cross profile.

Rectification of Defective Constuction: Where the surface irregularity of the W.B.M. sub-base course exceeds the tolerances given above,or where the course is otherwise defective due to sub-grade soil mixing with the aggregates,the layer to its full thickness shall be scarified over the affected area,reshaped with added material as applicable and recompacted.The area so treated shall not be less than 10 sq m.In no case shall depressions be filled up with screenings and binding material.

Measurements: The length and breadth shall be taken to the nearest cm and thickness to the nearest half cm.The consolidated cubical contents shall be calculated in cu m correct to two places of decimals.The quantity shall be the same as the net quantity of the aggregates in stacks used in the work.

Rate: Rate shall include the cost of all labour and materials involved in all the operations described above except cost of stone aggregates, Kankar, moorum, screenings and red bajri,if specified,for which seperate payment shall be made.In case the W.B.M is to be laid over an existing road,scarifying and consolidation of the aggregate recieved from scarifying shall be paid for seperately.

Section D

EARTHWORK

D.1 Introduction

For the construction of any structure, earthwork in cutting and filling is required. For the construction of small structure or where the big area is not required, the earthwork can be easily carried out but earthwork requires a lot of attention for the construction of roads, canals, railway tracks etc. because of the variation in ground level.

D.2 Principles of Earthwork Calculation

1. Earthwork shall be measured in cubic meters , otherwise mentioned.
2. Earthwork for different kinds of soil should be classified seperately.
3. No separate measurement should be taken for the hinderances met during the excavation.
4. Measurement of earthwork should be taken carefully by applying a proper and appropriate method.
5. Dressing and levelling should be included in the item.
6. The lead should be taken into account according to table.
7. While calculating the earthwork, lift should also be considered.

D.3 Earth Computations

The cross-section of a road, railway and canal in cutting, filling or banking is normally trapezoidal. To find out the quantity of earthwork, average cross-sectional area is multiplied by the length. This method is very approximate method but by reducing the interval between two cross-sectional area a better result can be determined.

(a) Method of calculating cross-sectional area

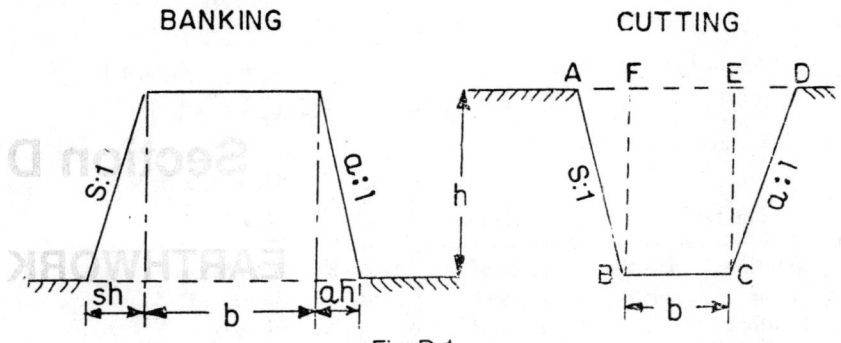

Fig. D.1

Cross-sectional area   =   Area of rectangle BCEF
+ Area of triangle ABF
+ Area of triangle CDE

=   h (b + 0.5 × s × h + 0.5 × a × h)

where   b = width of shorter side of trapezoid
h = height or depth of trapezoidal section

In case of a = s ; area of cross sectional area = h × b + s × h × h
s :1 or a :1 is the ratio of side slopes.

There are three methods to calculate the earthwork

(a) Average cross-sectional area method or by trapezoidal formula.

(b) Mid-sectional method

(c) Prismoidal formula method

(a) **Average end area method :-** In this method the cross- sectional area
of two ends is calculated which is multiplied by the length of the
section. This method does give good results in case length is very large
between two ends. In such cases the whole length is divided into many
small lengths and for each length, cross-sectional area is determined.

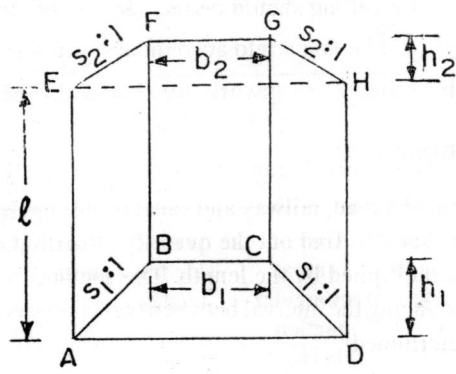

Fig. D.2

Volume of earthwork	=	(1/2) (area of ABCD) + (area of EFGH × 1
Area of ABCD	=	$(b_1 \times h_1 + s_1 \times h_1 \times h_1)$
Area of EFGH	=	$(b_2 \times h_2 + s_2 \times h_2 \times h_2)$
Volume of earthwork	=	$0.5 \times L \times (b_1 \times h_1 + s_1 \times h_1 \times h_1 + b_2 \times h_2 + s_2 \times h_2 \times h_2)$

The table which is used for calculating earthwork is

Station or Chainage	Depth or ht (st)	Area of central portion	Area of side triangles	Total Mean sectional area	Length	qty

(b) **Mid section formula:** In this method the area of mid-section is calculated by determining the mean depth which is multiplied by the length of the section to find out the total earth work.

Mean depth of cutting H_m	=	$(h_1 + h_2)/2$
Area of mid section A_m	=	$b \times H_m + 0.5 \times s \times H_m \times H_m + 0.5 \times a \times H_m \times H_m$
or if a = s; area A_m	=	$b \times H_m + s \times H_m \times H_m$
Quantity of earthwork	=	$A_m \times 1$

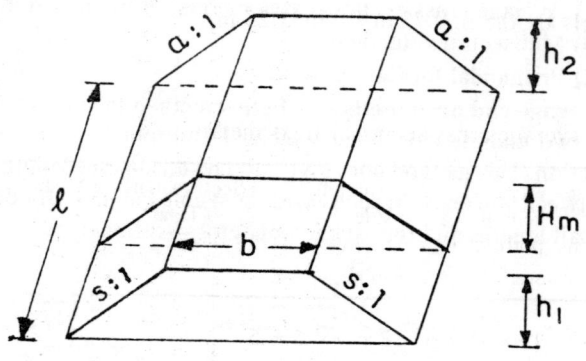

Fig. D.3

The table used for calculating the earthwork is

Station	Depth or ht h	Mean depth H_m	Area of central rectangular portion b H_m	Area of side tri-angles	Total area	Length l	Quantity

where

h = ht of cutting or filling at station

b = width

s = slope

H_m = mean ht or depth of two stations $(h_1 + h_2)/2$

(c) **Prismoidal formula method:** The area of cross section at both the ends and in the middle are calculated,then,the quantity is worked out by the following formula:

Total quantity of earthwork $= (A_1 + A_2 + 4\ A_m) \times 1/6$ (1)

where A_1 = area at station 1

A_2 = area at station 2

A_m = mid section area

L = length between two stations

In general the following formula is used

Total quantity of earth   = (first area + last area + 4 even areas +2 odd areas)   (2)

In the second case the number of sections should be odd.If the number of sections is even,for first or last we apply eqn (1) and rest we apply eqn (2) .

Example 1: Calculate the quantity of earthwork for the cutting 150 m long,12 m wide at crest and whose side slopes is 2.5:1.The central heights at every 30 m intervals are 0.5 m,0.75 m,1.0 m, 1.25 m,1.5 m.

Solution

(a) By average cross-sectional areas method:

Stn	Depth at station	Area of central portion sq m	Area of side triangles sq m	Sectional area Total sq m	Mean sq m	Length m	Qty m³
0	0.5	$12 \times 0.5 = 6$	$2 \times .5 \times .5 = .5$	6.5			
					8.3125	30	249.3
1	0.75	$12 \times .75 = 9$	$2 \times (.75)^2 = 1.125$	10.125			
					12.06	30	361.8
2	1.0	$12 \times 1 = 12$	$2 \times 1 \times 1 = 2$	14			
					16.06	30	481.8
3	1.25	$12 \times 1.25 = 15$	$2 \times (1.25)^2 = 3.125$	18.125			
					20.312	30	609.38
4	1.5	$12 \times 1.5 = 18$	$2 \times 1.5 \times 1.5 = 4.5$	22.5			

Total qty = 1702.26 cu m

(b) By Mid-sectional method:

Stn	Depth	Mean depth	Area of central portion	Area of side triangles	Total area	Distance between station	qty
0	0.5						
		0.625	7.5	0.781	8.281	30	248.4
1	1.0						
		0.875	10.5	1.53	12.03	30	360.9
2	1.0						
		1.125	13.5	2.53	16.03	30	480.9
3	1.25						
		1.375	16.5	3.78	20.28	30	608.4
4	1.5						

Total Qty = 1698.6 cu m

(c) By Prismoidal Formula:

No of stations = 5

Applying formula:

$V = L/3$ (1st area + last area + 4 even areas + 2 odd areas)

$= 30/3$ (6.5 + 22.5 + 4 (10.125 + 18.125) + 2 (14)

$= 10$ (6.5 + 22.5 + 113 + 28)

$= 1700$ cu m. '

D.4 Prismoidal Correction

The best results for earthwork is obtained by Prismoidal formula but total quantity of earthwork is often determined by using end area formula.To obtain the better results by end area method a correction is applied to the method which is known as Prismoidal Correction.According to this correction the difference between the quantity of earthwork is determined by using Prismoidal formula and by Trapezoidal formula.The correction for the prismoidal is always subtractive.

Prismoidal correction for

(a) Level sections = $\dfrac{l \times s}{6} (h_1 - h_2)^2$

where h_1 & h_2 are the depths or heights of earthwork on the centre line,

s is the side slope

l is the distance between the stations

(b) For a two level section

(i) for wholly in cutting or banking

$$PC = 1 \times (W_{11} - W_{12})(W_{21} - W_{22})/6 \times s$$

$$PC = \frac{1 \times s}{6}\left(\frac{r^2}{r^2 - s^2}\right)(h_1 - h_2)^2$$

(ii) For partly in cutting and partly in banking ie.,for a hillside section

for cutting $PC = \dfrac{L \times r\,(h_1 - h_2)}{12\,(r - s)}$

for filling $PC = \dfrac{L \times s\,(h_1 - h_2)}{12\,(r - s)}$

where

PC = Prismoidal correction

1 = Length between two successive stations

$W_{11}, W_{12}, W_{21}, W_{22}$ = the side widths of two adjacent sections

h_1 & h_2 = depths at the centre of a road at two adjacent sections

r = transverse slope

D.5 Curvature Correction

If the end sections are on a straight line,the assumption that end planes are parallel to each other and perpendicular to the centre line is correct but if the end planes lie on the curved surface, the assumption is no more valid and correction called Curvature Correction is applied.

The correction is applied when the cross-sectional area is unsymmetrical about the centre line.It could be positive or negative according to the situation of the section.

(i) For level section - No correction is required because the area is symmetrical about the centre line.

(ii) For 2-level and 3-level sections

$$CC = L/6R\,(W_{11} - W_{21})\,(h + b/2s)$$

where R is radius of curve,b is width of formation

(iii) For a two level section,the corrrection to the area =

=unit length $\dfrac{A\,e}{R}$

h= a constant distance

e= the eccentricity = $\dfrac{W_{11}\,W_{21}\,(W_{11} + W_{21})}{3\,A\,s}$

A = sectional area

s = slope

(iv) For side hill,

2-level section CC = $(W_{11} + b/2 - sh)/3$ for large area

and CC = $(W_{21} + b/2 + sb)/3$ for small area.

Correction is additive if the centroid and the centre of curvature are to the opposite side of the centre line, and it is negative if the centroid and the centre of the curvature are to the same side of centre line.

(a) **Lead:** It is the average horizontal straight practicable distance through which the excavated soil is carried from the sources to the place of spreading or dumping. For the purpose of calculation or even in field, the area is divided into a number of blocks and for each block the lead is measured from the centre of the block to the dumping place.

1. The unit of lead is 50 m for a distance upto 500 m and is considered as a seperate item if,
    (a) 0 m to a distance not exceeding 250 m
    (b) distance exceeding 250 m but less than 500 m
2. The unit of lead is 500 m for a distance between 500 m to 5 km and shall be measured as seperate items if
    (a) 500 to 1000 m
    (b) 1000 to 1500 m
    (c) 1500 to 2000 m etc. upto 5 km.
3. The unit of lead is 1 km where it exceeds 5 km. Half or more than half km shall be taken as 1 km and less than half should be ignored.

(b) **Lift:** Lift means the average vertical distance (height) above which the soil has to be lifted from the source. The unit of lift is 1.5 m and for each unit it should be considered as separate work.

In case where excavated earth shall having to be carried over a bank and dumped on the top of bank, lift shall be measured as the difference between the centre of gravity of the excavated earth and the formation level of the bank in successive stages of 1.5 m starting commencing level.

D.7 Mass Diagram

This is a graph showing the accumulation of cut and fill with distance from a starting point, or origin(reference point) .Cut is usually is considered positive and fill negative. The volume of each Section is plotted in cubic yards or cubic metre. Distance normally is measured along the centre line of the construction, in 100 metres apart, starting with origin as 0.S well factors are applied to the cuts (normally 20% and shrinkage factors to the embankments (normally 10%) to obtain cubic yards excavated and compacted fill respectively.

Some important definitions are:

Haul distance: The distance at any time from working face of an excavation to the tip end of the embankment formed.

Haul: Sum of the products of each load by its haul distance.

Free haul distance: In contract, the price for the schedule of item 'earthwork' includes the cost of haul within a specified distance. This limit of distance to which material is hauled without any extra cost is called 'Free haul distance'.

Overhaul: The excess of haul distance from the free haul distance is called overhaul distance. The sum of the products of volumes by the respective overhaul distance is termed overhaul. The extra payment is made for overhaul.

Economical haul: When the cost of excavation and hauling earth, obtained from cutting a road section, is used in filling is not more than the cost of wasting the excess earth of excavation and the cost of excavation and hauling earth from borrow pit, it is called economical haul.

Balancing line: It is the horizontal line (parallel to x-axis denoting the distance) that intersects mass curve at two points and indicate that the total volume of cutting and filling are same.

If a mass curve is horizontal between stations, the implication is that no material has to be moved in that stretch. Actually there may be cuts and fills but they balance. If work consists of side hill cuts and fills, the mass diagram tends to flatten because the cuts can be moved into the fills and not moved from one station to another. Moving excavation from one side of the centre line to the other is called "Cross Haul."

The slope of the mass curve increases with volume between stations. An ascending mass curve indicates cut a descending diagram fill. The curve reaches a maximum where cut end and fill begins, and a minimum where fill ends and cut begins. If a mass diagram is inserted by a horizontal line, cuts balance fills between the points of intersection. Total haul is the product of the total amount of excavation hauled and average haul distance. Centre of mass of cut and fill can be determined from the mass diagram in following manner:

(a) Draw the maximum ordinate between a balancing line and the curve.

(b) Draw a horizontal line through the midpoint of that ordinate and note the stations at the points of intersection with the curve.

(c) The station on the increasing portion of the diagram is the centre of mass of cut and the station on the decreasing portion, the centre of fill. The centre between the station is the haul distance.

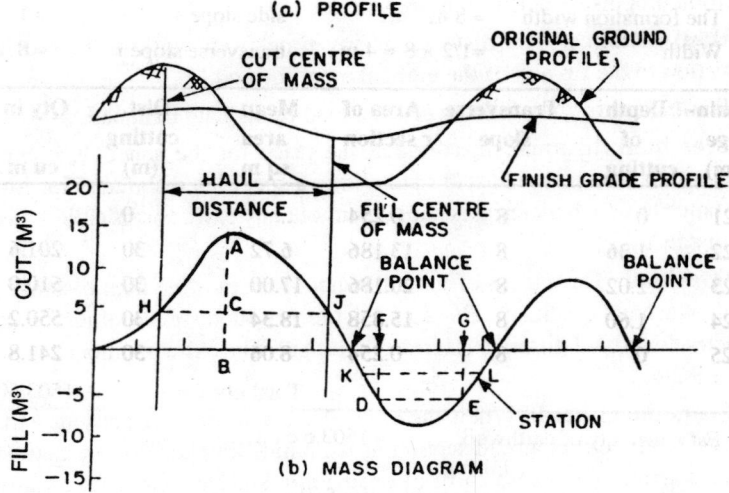

(a) PROFILE

(b) MASS DIAGRAM

Fig. D.4

Example 2: The ground levels at various chainages along line of a proposed road are as under:

Chainage	21	22	23	24	25
Ground level R.L (m)	180.50	183.36	185.52	187.10	186.50

The ground has uniform cross slope of 1 in 8. The chain is 30 m long. The road formation is proposed at uniform gradient passing through the G.L at end chainages with formation width as 8 m and side slope of cutting as 1:1. Estimate the quantity of earthwork for the proposed road section in a tabular form.

Solution

The difference of G.L at the end chainages = 186.50 - 180.50 = 6.00 m

Distance between the end chainages = 120 m.

Hence, Slope of the uniform gradient = 6/120

The chainage, R.L of ground and R.L of the formation are given below:

Chainage	Distance (m)	R.L of ground at centre (m)	R.L of formation at center (m)
21	0	180.50	180.50
22	30	183.36	182.00
23	60	185.52	183.50
24	90	187.10	185.00
25	120	186.50	186.50

The formation width	= 8 m	side slope s	= 1
Width	=1/2 × 8 = 4 m	transverse slope r	= 8

Chain-age (m)	Depth of cutting	Transverse slope	Area of r section	Mean area sq m	Dist. cutting (m)	Qty in cu m
21	0	8	0.254	...	0	
22	1.36	8	13.186	6.72	30	201.6
23	2.02	8	20.186	17.00	30	510.0
24	1.60	8	15.858	18.34	30	550.2
25	0	8	0.254	8.06	30	241.8
				Total qty =		1503.60

Estimated qty of earthwork = 1503.6 cu m in cutting
Filling = 0

Area of the section is calculated by the following formula:

$$\text{Area} = \frac{sb^2 + 2br^2 h + r^2 s h^2}{r^2 - s^2}$$

(± chose according to the problem)

where s = side slope

r = transverse slope

b = width

h = height

Example 3: Plot the mass haul diagram of the road section from the data given below:

Free haul distance = 90mts

Cost of excavation with free haul distance = Rs. 100/m³

Cost of borrowed earth with each 50m lead = Rs. 150/m³

Cost of overhauling for each additional lead of 50m over free haul distance = Rs. 25/m³

Chainage	0	1	2	3	4	5	6	7	8	9	10	11
Cutting(+) & Filling(-) m³	350	450	250	100	–350	–550	–625	–475	–175	250	525	500

Calculate a) Total Free haul, (b) Total overhaul, (c) Economical haul, (d) Division points in cut for least possible total haul, and (e) Total cost of earthwork

Solution

1) Construction of Mass diagram
   (i) Work out the accumulated volume +ve for cutting and –ve for filling.
   (ii) Draw a horizontal line (ab) as abscissa and mark the chainage.
   (iii) Choose a scale (1cm = 250m³) and mark the accumulative volume against chainage and join them.
   (iv) Draw a horizontal line (balancing line) such that $a_1c_1 = c_1b_1$
   (v) Mark the apex 'm' and 'n' respectively.

2) Free haul volume: According to free haul distance (90m in this case) draw horizontal parallel line (as per scale) to abscissa such that it cuts the loops (pq and rs). From apex ('m' and 'n') draw vertical lines mm_2 and nn_2. These vertical lines cut pq and rs at m_1 and n_1 respectively. mm_1 and nn_1 are free haul.

3) Total Overhaul volume: Construct verticals pp_1, qq_1, rr_1, and ss_1. These are overhaul.

4) Calculate the overhaul distances: Bisect verticals pp_1, qq_1, rr_1, and ss_1 at points k_1, k_2, k_3, and k_4 and draw horizontal lines which cut the mass curve at O_1, O_2, O_3, and O_4. O_1k_1, O_2k_2, O_3k_3, and O_4k_4 are average overhaul distances for overhauls pp_1, qq_1, rr_1, and ss_1 respectively.

Total free haul volume $= mm_1 + nn_1$ (scale 1cm = 250m³)

$$= 1.4*250 + 2.2*250$$

$$= 900$$

Chainage	Cutting	Filling	Accumulative volume
0	0		0
30	350		350
60	450		800
90	250		1050
120	100		1150
150		350	800
180		550	250
210		625	-375
240		475	-850
270		175	-1025
300	250		-775
330	525		-250
360	500		250

Total overhaul volume:

i) $m_1 m_2$ for overhaul distance

$O_1 k_1 + O_2 k_2$ (43.5m) = 2.8 * 250 = 700 m^3

ii) $n_1 n_2$ for overhaul distance $O_3 k_3 + O_4 k_4$ (33 m) = 2.1 * 250 = 525m^3

Volume of borrowed earth from borrowed pit = bb_2 = 250m^3

Total cost of the project = Cost of free haul + cost of overhaul + cost of excavation from borrowed pit

= 900 *100 + (700+250)*150 + 250*25

= 90000 + 142500 + 6250

= 238750

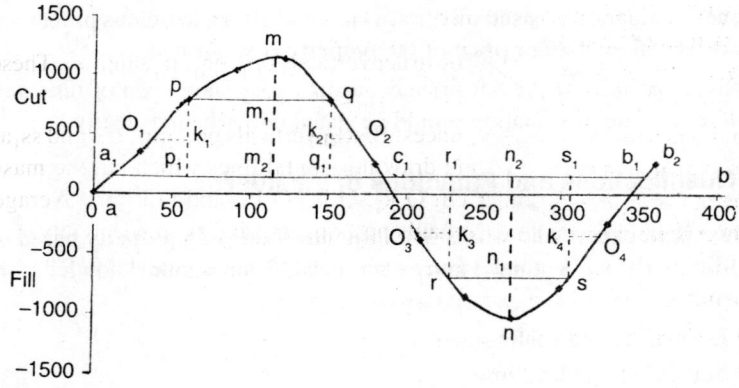

Section E
VALUATION

Valuation is a scientific approach of assessing the present fair value of a property. Valuation of any property or land is an estimate of the value in terms of money. Valuation is based on certain facts and after a judicious processing of facts the value of a fair price of the property is suggested.

Rises and falls of the fair price occur in a very short span of time and therefore the date of valuation should be stated properly and clearly.

E. 1 Qualifications and Functions of a Valuer

A valuer is an expert who works out the actual value of a property based on scientific analysis. A good valuer should have the sound knowledge of following.

(a) Estimating and costing
(b) Surveying and levelling
(c) Planning and designing
(d) Experience in construction works
(e) Building bye-laws
(f) Law of easements
(g) Law of contracts
(h) Arbitration
(i) Fire insurance
(j) Writing reports
(k) Central and local governments taxation
(l) Money market and rate of interest
(m) Zonal importance of land and buildings
(n) Land acquisition and town planning act

E. 2 The Purposes of Valuation

(a) Purchase for investment

(b) Tax fixation

(c) Sale

(d) Rent fixation

(e) Insurance Premium

(f) Compulsory acquisition

(g) Speculation

(h) Betterment charges

(i) Wealth tax and estate duty

(j) Gift tax

(k) Probate

(l) Partition

(m) Assessment of income

(n) Mortgage value or security of loans

E. 3 Definitions

E. 3.1 Freehold and lease hold properties:

(a) **Free hold properties:** The properties which are owned by owner only for an indefinite period and the owner has all the rights to use the property at his own will. He does not have to pay any rent and he can sell, divide, donate or grant it on lease.

(b) **Lease-hold Propertries:** The owner of a freehold property can give permission to another person to use his property for a certain duration under some terms and conditions. The person who grants lease is known as "lessor" and the person who takes the lease is called the "lease holder" or "lessee".

When the lease is given for 99 years it is called "long term lease" and when it is granted for 999 years it is called "lease for endless duration". There are always some terms and conditions for lease which are laid by owner and accepted by lessee.

The principal types of leases are as follows:

(a) Building lease

(b) Occpation lease

(c) Sub lease

(d) Life lease

(e) Perpetual lease

E.3.2 Sinking Fund: When a building, vehicle, machine etc. is used for a number of years, they become useless after a certain period of time and they should be replaced by new building, vehicle and machine. Hence it is necessary to make some arrangement so that the owner can replace the property once it is useless. For this purpose sinking fund is collected periodically (a year, 5 years, 10 years etc.) and deposited in a place (normally bank) to get highest compound interest.

Determination of sinking fund: The calculation of sinking fund depends upon:

(a) the life of the article and

(b) rate of interest

When the life of an article is over the owner can get back some amount by resetting it. For buildings he can sell the material and the value is known as scrap value. This value is considered as 10% of the building cost. Therefore, the calculation of sinking fund is made on 90% cost of the building.

Let s = Total amount of the sinking fund

c = Rate of interest

n = Number of years

I = Annual installment required

If you start accumulating some money (I) for number of years (n) then (I) will earn interest for (n-1) years, the second installment will earn for (n-2) years and so on.

E.3.3 Amortization: This is accumulation of sinking fund at compound interest for payment of debt.

E.3.4 Scrap Value and Salvage Value: Scrap value is the value of dismantled materials of an article once its life time is over. For buildings the old materials like bricks, steel wooden articles (Planks) etc. after demolition of the building is the scrap value. Scrap value is considered as 10% of the cost of construction.

For machines once it cannot be repaired or it parts can not be replaced some amount can be earned by selling its parts and the value of the part is scrap value for machine.

The scrap value is also known as "junk value" or demolition value. On some cases the scrap value can be zero or negative when the dismantling cost is more than or equal to scrap value of a building or machine on any other article.

Salvage value is the value of the article (building, machine etc.) once its life is over. This value is estimated without dismantling the article and it is high when a building or machine becomes useful after repairing, replacement or remodeling.

E.3.5 Market Value: It is the value at which an article can be sold in the market at a particular time. Values of any article in market fluctuates very fast because it depends on many factors. Some of the factors which affect the value of an article are as follows:

  (a) Forces of demand and supply
  (b) Cost of construction
  (c) Different acts like rent restriction act etc.
  (d) Improvement due to public schemes
  (e) Rise in population
  (f) Purpose of purchase
  (g) Abnormal conditions
  (h) Imposition of control of prices of building materials

E.3.6 Book Value: It is defined as the value of the article shown in the account book in that particular year. Book value of any article any year is the original cost less the total depreciation in the cost till that year. It is applicable on building and movable properties but not on the land.

  Difference betwen market and book values

Market Value	Book Value
1. The value is fixed by purchaser and owner.	1. The value is fixed by the rate of depreciation.
2. The value may increase during the subsequent years.	2. The value keeps on reducing.
3. The value may be constant for period.	3. Value falls gradually.
4. Applicable to any type of property.	4. Not applicable in case of lands and metal articles.
5. It is considered for valuation.	5. It is mainly used in accounts book of a company.
6. Market value depends on number	6. Such factors don't affect the book value.

E.3.7 Obsolescence: The value of property or structure becomes less by its belowing out of date in style, in structure, in design in adequecy to growing (present) needs, etc, and it is termed as obsolescence. An old dated building with massive walls, arrangement of rooms not suited to present days and for similar reasons, becomes obsolete even if it is maintair·d in a very good condition and its value obsolescence may be due to

  (a) Poor, odd or eccentric original design
  (b) Change in utility demand

(c) Change in type of construction

(d) Change in kind of construction

and this type of obsolescence is termed as internal obsolescence. Second type of obsolescence which is termed as external obsolescence is due to

(a) Zoning laws

(b) Poor original location

(c) Change in the character of the district

(d) Specific detrimental influences

E.3.8 Depreciation: It is the gradual exhaustion of the usefulness of the property. **It may be defined as the decrease or the loss in the value of the property due to structural deterioration, use, life, wear & tear, decay and obsolescence.**

Types of Depreciation

1. **Physical depreciation:**
   (a) Wear and tear from operation of time and the elements
   (b) Wear and tear due to action

2. **Functional depreciation:**
   (a) Inadequacy or suppression
   (b) Obsolescence

3. **Contingent depreciation:**
   (a) Accidents
   (b) Diseases
   (c) Diminution of supply

Methods of Calculating depreciation:

1. **Straight Line Method:** In this method a fixed amount of the original cost is deducted every year and at the end of the utility period only the scrap value is left.

$$\text{Annual Depreciation} = (D) = \frac{\text{Original cost - scrap value}}{\text{Life in year}} = \frac{C - V}{N}$$

The book value after a number of years,

= Original cost - no of years × Annual depreciation

2. **Constant percentage method or declining balance method:** In this method the decrease in value is increased by a constant percentage at the beginning of every year.

Annual depreciation = (D) = $(1 - (C/V)^{1/N}$

The value of the property at the end of the first year = $V - D \times V = V_1$

The value of the property at the end of the 2 nd year = $V_1 - D \times V_1$

and so on

The value of the propery at the end of the M years

$$= V (C/V)^{M/N}$$

3. **Sinking fund method:** In this method, the depreciation of a property is assumed to be equal to the annual sinking fund plus the interest on the fund for that year, which is supposed to be invested on interest bering investment. If F is the annual sinking fund and b,c,d etc represent interest on the sinking fund for subsequent years, then

At the end of	Depreciation for the year	Total depreciation	Book value
1 st year	F	F	V - F
2nd year	F + b	2F + b	V -(2F+(b)
3rd year	F + c	3F +b+c	V-(3F+b+(c)

4. **Quantity Survey Method:** In this method the property is studied in detail and loss in value due to life, wear & tear, decay, obsolescence etc is worked out. Each and every step is based on some logical ground without any fixed percentage of the cost of the property. This method is generally used by an experienced valuer.

(a) **Gross Income:** It is the total outcome and includes all receipts from various sources. The outgoings operational and collections are not deducted from the total receipts.

(b) **Net Income or Net return:** It is the total amount left after deducting all outgoings operational and collection expenses from the gross income or total receipt.

(c) **Outgoings:** Outgoings are the expenses to be made to keep the property in possession and maintaining the property. The different types of outgoings are as follows:

(i) **Taxes:** To keep the property in possession, various types of taxes such as wealth tax, property tax, municipal tax etc. are paid. These taxes vary from place to place and they are fixed on the basis of the annual rental value of the property.

(ii) **Loss of rent:** If a portion of the whole property is not occupied, the owner will not get any money as rent for that portion but a suitable amount is deducted from the gross income under outgoings and it is known as loss of rent.

(iii) **Sinking fund:** Generally fixed amount from the gross income is kept seperately so that the money can be accumulated to get the cost of construction when the life of property is over. This fixed amount also comes under outgoings.

(iv) **Repairs:** For the maintenance of the building its proper repairing is to be done as and when it is required. The expenditure or repair work depends upon its age, nature of construction, use of building etc. The amount is calculated in one of the following ways -

(a) 10% to 15% amount of the gross income

(b) 1 to 1.5 months rent

(c) 1% to 1.5% of its buildings total cost annually.

(v) **Management and collection charges:** To keep the proper record and maintenance, the owner has to keep different persons like chaukidar, liftman, personal secretary, clerks etc. and all these charges are considered to be outgoings and usually 5 to 10% of the gross rent income is taken as Management and collection charges.

(vi) **Insurance:** If the owner insures his property against fire, accidents etc, then the amount paid to the insurance company as premium of insurance is treated as outgoing charges.

(vii) **Miscellaneous Expenditures:** Electric charges for running lifts, pumps, lighting of common place, cleaning charges of septic tanks, maintenance of approach road, road inside the premises etc. come under miscellaneous expenditures.

E.3.10. Annuity: Now-a-days to make the people of weaker section a house owner, Government is running different schemes. Under these schemes, the house is given to such people and an amount is fixed depending upon the instalment whether it is monthly, six monthly, or yearly. An annuity is a series of such payments.

If the amount of annuity is paid for definite number of years or period, it is called annuity certain.

If the amount of annuity is paid at the beginning of each period or year and payment is made for definite number of periods, it is known as Annuity due.

If the payment starts at some future date after a number of years, it is known as Deferred Annuity.

If the payments of annuity continue for indefinite period, it is known as Perpetual Annuity.

If an annuity of Rs. i per annum is received one year from now, the total amount of annuity (A_m) over m years is

$$A_m = 1 + (1 + i) + (1 + i)^2 + \text{------} + (1 + i)^{m-1} \tag{1}$$

The first payment of Rs. 1 accumulates interest for (n-1) years and become $(1 + i)^{n-1}$ where i is the interest on Re. 1/- per year. The last payment of annuity remains the same i.e. Re. 1/-. Now multiply the expression, by (1 + i).

$$A_m (1 + i) = (1 + i) + (1 + i)^2 + \text{------} + (1 + i^n) \tag{2}$$

Expression (2) - (1) gives

$$A_m (1 + i) - Am = (1 + i)^n - 1$$

$$n A_m = (1 + i)^n - 1) / i$$

for an annuity of Re. 1/- the general expression is

$$A_m = P \times (1 + i)^n - 1/i$$

E.3.11 Capital Cost: It is the total cost of construction including land, boundary walls etc. or in other words it is the actual (original) amount spent to possess the property.

E.3.12 Capitalized Value: It is the amount of money whose annual interest at the highest prevailing rate of interest will be equal to the net income from the property. To understand it, let's assume that a net annual rent of property is Rs. 500/-. If the highest rate of interest is 8% then to get Rs. 100/- interest, capital should be $500 \times 100/8 = 6250/-$.

E.3.13 Year's Purchase: Year's purchase can be treated as multiplier which when multiplied by the net income of property gives its capitalised value.

Year's purchase = capitalised value/Net income

When the income of a property is continuous for a reasonable long period, it may be treated as perpetual or continuous and then year's purchase= (100/rate of interest) capital investment.

E.3.14 Mortgage: When a person wants a loan he can raise the loan on interest against the security of his property and it is called mortgage loan. The party or the financial institution which gives the loan is known as the mortgage, the party which is taking loan is called the mortgager and the conditions of loans are entered in a document which is called mortgage deed. Loan is given upto 60% present value of the property. The party taking loan has to pay back the money in instalments along with the interest. In case the mortgager fails to return the loan along the interest in specified duration of time the mortgage can sell the property and can take back its money.

E. 4 Valuation Tables

To calculate the simple and compound interest on sinking fund, depreciation etc., can be calculated through the respective formulas but for large properties this methods become very lengthy and tedious. Therefore, the different valuation tables are prepared which are based on these formulae. These tables can be used directly which saves a lot of time and energy. There are number of tables but mainly following tables are used.

(a) To find present values of Re.1/- receivable at the end of given time.

(b) Year's purchase table.

(c) To find the amount of Re.1/- at the end of a given number of years.

(d) To find depreciation based on sinking fund method.

(e) To calculate the amount of Re.1/- per annum in a given number of years.

(f) To know the sinking fund.

E. 5 Valuation of Land

While calculating the total cost of building, the cost of land is also included which depends on the following factors.

1. Size - Medium sized plots are the most costly because of high demand.

2. Nature of soil and tanks.

3. Situation - Buildings built at the city centre are more in demands than those which are away from the city centre.

4. Orientation - South facing plots will get more money than the plots facing north.

5. Visits - Plots close to a road will fetch more money.

6. Shape - lands of odd or irregular shape will be less costlier because wastage of land will be more.

7. Frontage and depth - Frontage and depth should not be too small for good piece of land.

8. Width of the roadway - It plays an important role and according to the nature of building, cost is decided.

9. Return frontage - A plot which is situated at the junction of two roads is said to have a return frontage. The front of plot should be situated at the wider or more important road.

E. 6 Methods of Valuation

The valuation of building depends on the type of materials used in construction, shape, size, situation etc. It also depends upon the roof covering, plinth level, type of foundation etc. The value of building mainly depends upon the rent, it can get. Due to many constraints one method of valuation is not sufficient and therefore there are no. of valuation methods namely,

(i) Rental Method

(ii) Profit based method

(iii) Cost based method

(iv) Development based method
 (v) Depreciation method and value
(vi) Plinth area method
(vii) Capital value comparison method.

  (i) **Rental Method:** In this method first the total income from property is
      calculated then outgoings are deducted and thus net income from
      property is foundout. On the basis of this net income the value of
      property is estimated. This method is very useful for a new building. In
      the value of property, depreciation is also calculated depending upon
      the life of building.

 (ii) **Profit based method:** In this method the annual net income is calcu-
      lated after deducting all the expenditures. The net profit per year is
      multiplied by year's purchase to get the value of the building. This
      method is applied to the buildings where capitalized value mainly
      depends upon the profit like hotel, cinema theatres, buildings etc.

(iii) **Cost based method:** In this method the actual cost of building is taken
      during construction and then taking depreciation and other points into
      consideration, the present value of building is calculated.

 (iv) **Development based method:** This method is used for the buildings
      which are in developing or undeveloped colonies. The valuation of the
      building is calculated on the basis of future income, which it will get
      once the colony is fully developed. The capitalized value is calculated
      by multiplying the net income to year's purchase. The total amount for
      renovation and the original cost together should be compared with the
      capitalized value.

  (v) **Depreciation Method:** In this method the cost of each part is calculated
      on the basis of present rates and the depreciated cost of part N is
      calculated by different formulae. Normal $D = P(1 - r/100)^N$ is used
      where D = depreciated value; P = present cost; r = fixed % of deprecia-
      tion; N = present age of the buildings in years. The present value of land,
      water supply, sanitary etc. should be added to get the total cost of
      building.

 (vi) **Plinth area Method:** To get the cost of the building, either old records
      are referred or detailed meaurement are taken and then cost is calcu-
      lated. But if records are not available and detailed measuremnts
      calculations are time consuming then, plinth area method is applied.
      Though it is not very accurate method but if different components of
      building are taken into consideration, it gives fairly good results. By
      applying the depreciation cost formula the actual present cost of
      building can be calculated. As name indicates the value of building is
      calculated on the basis of plinth area rates of building.

(vii) Capital Value comparison method: In this method the capitalized value of the property is worked out by direct comparison with the capitalized value of the similar property in the same locality whose sales records are available. This method is used for the valuation of the property, when the rental value of the property is not available but sale records of similar buildings are available.

E. 7 Fixation of rent

The rent of building is fixed on the basis of certain percentage of annual interest on the capital cost and all possible annual expenditures as outgoings. The capital cost includes following:

(a) Cost of Construction

(b) Cost of Sanitary and Water Supply Works.

(c) Cost of Electric Fittings

(d) Cost of Subsequent additions and alterations

The net return is calculated by dividing the capital cost by years purchase. All outgoings are added to the net return to get gross rent which can be divided by 12 to get the monthly rent. The rent calculated by this method is called the standard rent while the actual rent of the property may be higher or lower than this rent depending upon the situation, shape, construction demand etc. of the property.

E. 8 Rules of Capital Cost Fixation for Govt. Buildings

For the purpose of the assessment of rent, the capital cost of a residence owned by the Govt. shall include the cost of sanitary, water-supply and electric installations and fittings, and shall be either -

(a) the cost of acquring or constructing the residence and any capital expenditure incurred after acquisition or construction, or when this is not known,

(b) the present value of the residence.

Provided that -

(i) the present value of residences shall be determined in the manner which the Governor may prescribe by rules or orders.

(ii) the expenditure which is to be regarded, for the purpose of subclause (a) above, as expenditure upon the prepartion of site, shall be determined in accordance with such rules or orders as the Governor may issue;

(iii) The Government may, after recording the reasons, authorize a revaluation of all residences of a specified class or classes within a specified area to be conducted under the rules referred to in provision (i) above, and may revise the capital cost of any or all such residences on the basis of such revaluation.

(iv) The capital cost howsoever calculated, shall not take into consideration (1) any charges on account of establishment and tools and plants other than such as were actually charged direct to the work in case in which the residence was constructed by the Government or (2) in other cases, the estimated amount of such charges:

(v) the Government may, after recording the reason, write off a specified portion of the capital cost of a residence.

(a) When a portion of the residence must be set aside, by the government servant to whom the residence is allotted, for the reception of officials and non-official visitors, visitng him on business, or

(b) When they are satisfied that the capital cost, as determined under the above rules, would be greatly in excess of the proper value of the accommodation provided;

(vi) In assesing the cost or value of the sanitary, water-supply and electric installations and fitting, the Governor may by rules determine what are to be regarded as fittings for this purpose.

(vii) The capital cost of wells and washermen's tanks shall be included in the capital cost of the residence.

(vii) When a work is replaced by work of a more expensive character, the cost of the replacement shall be reduced by the value of the dismantled work before it is added to the capital cost.

(ix) The cost of all structural alterations, additions or repairs to newly purchased or previously abandoned buildings, required for bringing them into use, shall be added to the capital cost

(x) The proportionate capital cost of a portion of a residence set aside for visitor's room shall be calculated in accordance with the following formula:

Proportionate capital cost of the portion set aside for 'he visitor's room

$$= (b + c + d/(a.X)$$

Proportional capital cost to the residential portion

$$= (X - b + c + d/a .X)$$

Total cost of the residence for the purpose of assessment of rent

$$= (X - b + c + d/a .X) + y$$

where:

    a - Total plinth area of the main building

    b - Area of room set aside for the visitors room measured from centre to centre of walls.

    c - Area if portion of verandah or verandahs directly in front of visitor's room and, which is normally utilized by visitors coming to interview the occupant.

    d - Area of bathroom (if any) attached to the visitor's room measured from centre to centre of walls.

    X - Capital cost of building.

a+b+c = plinth area of visitor's rooms portion.

E.9 Rules for calculation of standard rent

The standard rent of residence shall be calculated as follows:

(a) In the case of leased residences, the standard rent shall be the sum paid to the lessor, plus an addition determined under rules which the Governor may make for meeting during the period of lease, such charges for both ordinary and special maintenance and repairs and for capital expenditure on additions or alterations as may be a charge on the Govt. and for the interest on such capital expenditure, as also for municipal and other taxes in the nature of house or property tax payable by the Govt. in respect of the residences.

(b) In the case of residences owned by the Govt. the standard rent shall be calculated on the capital cost of the residence and shall be either-

    (i) a percentage of such capital cost equal to such rate of interest as may, from time to time, be fixed by the Governor plus an addition for municipal and other taxes in the nature of house or property tax payable by the Govt. in respect of the residence and for both ordinary and special maintenance and repairs, such addition being determined under rules which the governor may make,or

    (ii) 6% per annum of such capital cost, which ever is less.

(c) In both cases standard rent shall be expressed as standard for calender month and shall be equal to 1/12th of the annual rent as calculated above, subject to the provision that, in special localities or in respect of special classes of residence, the Govt. may fix a standard rent to cover a period greater than one month but not greater than one year. Where the Govt. takes action under this provision standard rent so fixed shall not be a larger proportion of the annual rent than the proportion which the period of occupation as prescribed above, bears to one year.

(d) For the buildings constructed before 1940, but let out for the first time

after 1-9-1940, the contractual rent determined at the time of first
letting may be taken as the standard rent of the property. However, if it
is still excessive, it can be challenged in the court of law.

(e) For the buildings which are constructed and let out for the first time
after 1940, the contractual rent determined at the time of first letting
may be taken as the standard rent of property. However, if it is still
excessive, it can be challenged in the court of law.

(f) For the buildings which are let on or before 1-9-1940, the contractual
rent of that date is taken as the standard rent of the property.

From the above it is clear that the standard rent is calculated usually at
6% interest on the capital cost of the property. The expenditure on annual and
special maintenance and repairs, municipal and other taxes etc. are also added
while calculating the standard rent. Thus after deducting all outgoings, the
owner must earn or get net 6% interest on the value of his property.

For annual repairs 1.5% of the cost of building, 1% of water supply
works, 1% of the cost of sanitary works and 1.5% of the cost of electric
installations are allowed per annum. In U.P. the limit of additions and
alterations for govt. buildings is

Rs. 500/- or 1% of the capital cost, whichever is less. For Govt. of India
buildings, this limit is 5% of the capital cost. For quadrential and special
repairs 0.6% of the building cost, 3.5% of the cost of water supply works,
3.5% of the cost of sanitary works and 3.5% of the cost of electric works are
taken per annum.

Property and municipal taxes which are actually paid, are to be taken for
calculation purposes.

Example 1: The cost of a new building is Rs.2,50,000.Work out cost of the
building after 10 years,by straight line method and constant percentage
method,if the scrap value is Rs.25,000 assuming the life of building is 50
years.

Solution:

(i) Stright line method:

Cost of building $= 2,50,000$

Annual depreciation $= \dfrac{2,50,000 - 25,000 \text{ (scrap value)}}{50 \text{ (life of building)}}$

$= \text{Rs. } 4500/-$

Cost after 10 years $= $ original cost - no of year $\times$ annual depreciation

$= 2,50,000 - 10 \times 4500$

$= \text{Rs. } 2,05,000/-$

(ii) Constant percentage method:

$$\text{Annual depreciation} = 1 - \left(\frac{25{,}000 \ 1/50}{2{,}50{,}000} \right) = 0.045$$

$$\text{The depreciated cost after 10 years} = 2{,}50{,}000 \left(\frac{25{,}000 \ 10/50}{2{,}50{,}000} \right)^{10/50}$$

$$= \text{Rs. } 1{,}57{,}739.34/\text{-}$$

Example 2: The owner of a building gets an annual rate of interest of Rs.3,500. The future life of the building is estimated as 12 years. But if recommended repairs are carried out immediately at an estimated cost of Rs.30,000, it is estimated to last for atleast 30 years. Assuming the rate of interest as 8% determine whether it is economical to carry out the recommended repairs to the building or leave as it is.

Solution

For comparision the values should be calculated for both the purposes.

(a) With repairs: The present value of annuity of Rs.3500 for 12 + 30 at 8% rate of interest

$$= 3500 \times \frac{1 - (1+0.08)^{-42}}{0.08}$$

$$= \text{Rs.}42023.445$$

(b) Without repairs: The present value of annuity of Rs.3500/- for 12 years at 6 % rate of interest

$$= 3500 \times \frac{1 - (1+0.08)}{0.08}$$

$$= \text{Rs. } 26376.273$$

The difference between both proposal $= 42023.445 - 26376.273$

$$= 15647.172$$

which is less than 30000, so it is not economical.

Example 3: A person owns a land measuring 500 sq m at the rate of Rs.50 per sq m and constructs the building of plinth area of 200 sq m at a cost of Rs.60000. He desires to have 8 % return on the cost of building and 5 % return on the land cost. What rent do you suggest for the property ? (The outgoings shall be suitably assumed and stated).

Solution

Assumptions:

(a) Annual maintenance = 1/2 % of the cost of construction

(b) Municipal taxes & other outgoings= 0.30 of the gross rent

(c) Sinking fund = 4 % and life of the building is 75 yrs = 0.22

Net return per annum:

(i) On the cost of construction= 60000 at 8 % $= \dfrac{60000 \times 8}{100}$

$= Rs.4800/-$

(ii) On the cost of land = 500 x 50 at 5 % $= \dfrac{25000 \times 5}{100}$

$= Rs. 1250/-$

Total return $= Rs.6050/-$

Outgoings:

(a) Annual maintenance $= \dfrac{1}{2} \times \dfrac{60000}{100} = 300/-$

(b) Municipal and other taxes $= \dfrac{30 \times r}{100}$ (r = standard rent)

(c) Sinking fund $= 60000 \times \dfrac{85}{100} \times \dfrac{0.22}{100} = 112.2$

(Scrap value considered 15 %)

Standard rent = net return + outgoings

r $= 6050 + 112.2 + 300 + 0.30 r$

0.7 r $= 6462.2$

r $= Rs.9231.7/annum\ or\ Rs.\ 769.30/month$

Example 4: A shop building was purchased for Rs. 30,000. The building is expected to serve for 10 years at the end of which it is presumed to have a salvage value of Rs.5,000. Determine the amount of depreciation and the book value for each year for the first 5 years after purchase by the sinking fund method, assuming an interest rate of 4 %.

Solution

Total of sinking fund $= \dfrac{(30,000 - 5000)}{10} = 2500$

$$\text{Annual instalment} = \frac{25000 \times 0.04}{(1 + 0.04)^{10} - 1} = 2082.27$$

At the end of	Depreciation for the year	Total depreciation	Book value
1 st year	2082.27	2082.27	27971.83
2 nd year	2082.27 + 83.29	4247.83	27572.269
3 rd year	2082.27 + 86.62	6416.72	23583.2
4 th year	2082.27 + 86.75	8585.75	21414.25
5 th year	2082.27 + 86.76	10754.77	19245.22

Example 5: A lease hold property is to produce a net income of Rs.12000/- per annum for the next 60 yrs.What is the value of the property ? Assume that the landlord desires a return of 6 % on his capital and the sinking fund to replace the capital is also to accumulate at 6 %.

What will be the value of the property if the rate of interest for redemption of capital is 3 % ?

Solution

Years purchase $Y.P = 1/(I_c + I_s)$

Where Ic is the of interest on capital

Is is the coefficient of sinking fund

Coefficient of sinking fund $(I_s) = \dfrac{i}{(1 + i)^{n-1}}$

where i = rate of interest of sinking fund

for 60 yrs $I_s = \dfrac{0.06}{(1 + 0.06)^{60 - 1}} = 0.0019$

Therefore, $Y.P = \dfrac{1}{0.06 + 0.0019} = 16.155$

Value of the property = Net income per annum × Y.P

= Rs.12,000 × 16.155 = Rs.1,93,860/-

When the interest for redemption of capital is 3 % then

co-efficient of sinking fund

$$I_s = \frac{i}{(1 + i)^n} = \frac{0.03}{(1 + 0.03)^{60}} = 0.0061$$

Y.P = $1/(I_C + I_S) = 1/(0.06 + 0.0061) = 15.129$

Value of the property = Rs.12000 × 15.129 =Rs.1,81,548/-

Example 6: What is the value of a plot of land which has been leased out on a ground rent of Rs.1200/- per annum for a period of 19 years unexpired after which the lessor will receive the ground rent of Rs.5000/- per annum in perpetuity. Rate of interest : 6 % (Assume Y.P @ 6 % for 19 years at 11.158)

Solution

(a) For unexpired years:-

Lessor gets an income of Rs.1200/- p.a for a period of 19 yrs.

Year's purchase @ 6 % for 19 years = 11.158

Therefore, Value = Rs. 1200 × 11.158 = Rs.13389.60

(b) For income after 19 years :

Value	=	Net income x Year's purchase in perpetuity deffered for 19 years
Y.P in perpetuity	=	100/6 = 16.667
Deduct Y.P for 19 yrs	=	11.158
Y.P in perpetuity @ 6 % deffered for 19 yrs	= 16.667 - 11.158	
	= 5.509	
Value = Rs.5000 x 5.509	= Rs. 27545	
Value of the plot of land	= (A) + (B)	
Value for next 19 yrs	= Rs. 13389.60	
Value for next 19 yrs	= Rs. 27545.00	
Value of the plot of land	= Rs. 38,684.60	

Example 7: Work out the valuation of cinema house with the following data:

Cost of land for life time period of the house (ie.deffered value)

= Rs.1,20,000

Gross income per year = Rs. 7,50,000

Expenses required per year:

(i) To run the cinema including staff salary,electric charges,municipal taxes including licence free,stationary, printing etc is 30 % of gross income.

(ii) For repairs and maintenance of machinery,plant,equipment, furniture etc.5 % of their capital cost of Rs.9,50,000.

(iii) Sinking fund for the machinery as in (ii) whose life is 25 yrs at 4 % after allowing 10 % scrap value.

(iv) Insurance premium is Rs.10,000 per year.

Solution

(a) Staff salary, electric and printing charges
@ 30 % of gross income = Rs. 2,25,000

(b) For repairs & maintenance of machineries
etc. @ 5% of Rs.9,50,000 = Rs. 47,500

(c) Sinking fund for machineries etc with
25 yrs life @ 4%
= Rs. 9,50,000 × (9/10) = Rs 8,55,000

Sinking fund coefficient for machineries

$$= \frac{0.04}{(1 + 0.04)^{25 - 1}} = 0.024$$

Sinking fund on Rs.8,55,000 = Rs.8,55,000 × 0.024 = Rs. 20,520

(d) Insurance premium per year = Rs. 10,000

(e) Yearly charge for cinema buiding repair @ 2 %
on gross income = Rs. 15,000
Total Rs. 3,18,020

Net income = Gross income - outgoings = Rs.7,50,000 - Rs.3,18,020
= Rs. 4,31,980

Year's purchase for 60 yrs @ 8 % and redemption of capital
@ 4% = 1/(Ic + Is)

Coefficient of sinking fund for 60 yrs ,

$$I_C = \frac{0.04}{(1 + 0.04)^{60 - 1}} = 0.0042$$

$1/(I_C + I_s) = 1/(0.08 + 0.0042) = 11.88$

Capital value = Rs 4,31,980 × 11.88 = Rs. 51,31,922/-

Total valuation = Capital value of house + value of land for 60 yrs
= Rs.51,31,922 + Rs.1,20,000 = Rs. 52,51,922/-

Section F

CONTRACTS

An agreement enforceable by law or by competent authority is "contract". The contract always has two minimum parties. One proposes and other accepts. In absence of any of the above elements of a contract, it becomes void, i.e. without a legal effect.

F. 1 Essentials of Contracts

To form a valid contract, it should have the following particulars:

(a) Contract shall be made by the parties who are competent to contract.

(b) Contract shall be made by free consent of the parties.

(c) Contract should have a definite proposal and its acceptance.

(d) Agreement, the meaning of which shall be certain.

(e) Contract shall be made so that the conditions and objects are lawful.

F. 2 Type of Engineering Contract

Any type of contract can be put in one of the following categories:

(a) Item rate contract

(b) Percentage rate contract

(c) Lump sum contract

(d) Labour contract

(e) Materials supply contract

(f) Piece work contract

(g) Cost plus percentage rate contract

(h) Cost plus fixed fee contract

(i) Cost plus sliding or fluctuating fee scale contract

(j) Target contract k) Scheduled contract

(l) Negotiated contract m) Rate contract

(n) Turn key contract

F. 3 Advantages and Disadvantages of Different Type of Contract

(a) **Item rate or unit price contract**: For this type of contract, contractors quote rates for individual items of work.

Advantages:

(i) It ensures a detailed analysis of cost and it is more scientfic.

(ii) Contractors write their individual rates of separate items, so it is easy to choose a competent contractor.

(iii) Unworkable rate tender can be avoided.

Disadvantages: There is a fair chance to manipulate the rates of individual item.

(b) **Labour contract**: This is a contract where the contractor gives rates for each item of work without including the element of materials which are supplied by the department.

Advantages of Labour contract:

(i) The materials stored by the government is utilized.

(ii) The increase in the cost of work is checked even if there is a rise in the prices of material in the market.

(iii) Sometimes it is difficult to get certain materials in the market which can be avoided.

(iv) A better progress of work can be achieved.

Disadvantages:

(i) There may be delay in obtaining the materials from and by the department.

(ii) Contractor has to be in touch with the department.

(iii) A large area of storage is required to store the materials.

(iv) A constant guarding is required

(v) Responsibilities of department is increased.

(vi) Refund of surplus materials in a good condition is difficult.

(c) **Contracts for Supply of Materials**: In this, the contractor has to specify the rates for supply of the required quantity of materials inclusive of all taxes etc. to the specified stores within the specified time.

Advantages:

(i) If payment of this type of contract is made promptly, contractors take the supply order even at less profit which reduces the cost of materials.

(ii) Department does not worry about the loss, breakage, charges against transit of materials.

Disadvantages:
  (i) Constant control and check of materials in different batches at different time is very difficult.
  (ii) The contractors form a group and try to cheat.

(d) **Engineering and construction contracts:** Sometimes it happens that the owner desires to deal with only one contractor for all services both engineering and construction in connection with the work. Officers are invited from specialised contractors for all planning, design plans, specifications, preparations of estimates and construction services under one contract.

The other types of contract are not dealt in this text and if reader is interested, they can refer any hand book on contracts.

F. 4 Conditions of Contracts

The terms and conditions of contract shall be correct, precise and definite and there shall not be any ambiguity in language or terms and conditions. There are several clauses of contract to govern the character of the work. The important conditions are :

  (a) Amount of security deposit.
  (b) Compensation for delay
  (c) Action when whole of security is forfeited
  (d) Extension of time
  (e) Completion certificate
  (f) Monthly bill
  (g) Payment of bill
  (h) Departmental materials
  (i) Proper execution of work
  (j) Alteration in designs and specifications
  (k) Compensation in case of bad work
  (l) Notice before the work is covered
  (m) Maintenance period
  (n) Labour
  (o) Work on holidays
  (p) Supply of water
  (q) Changes in constitutions
  (r) Arbitration
  (s) Supervision by higher officers
  (t) Sp. terms and conditions etc.

F. 5 Topics Concerned with Contracts

Tender: If a contractor wants to execute certain work or supply some specified articles, he submits a written offer which is called tender. The form in which a contractor submits the offer is supplied by the department.

Tender form: It is a printed standard form of contract giving the following details:

    (a) Standard conditions of contract

    (b) General rules and directions for guidance of contractors

    (c) General description of work

    (d) Estimated cost

    (e) Earnest money

    (f) Security deposit

    (g) Time allowed for the work

    (h) Columns for signature of contractor and witness.

Tender Notice: The notice inviting tender papers is called tender notice and it is very important document. A tender notice should have minimum following things.

    (a) Name of the authority inviting tender

    (b) Particulars of contractors eligible to submit tenders.

    (c) Name of work and its location

    (d) Estimated cost of work

    (e) Price of tender form and other tender documents

    (f) Earnest money

    (g) Time of completion

    (h) Last date of sale of tender paper

    (i) Last date, time limit and place of receipt of tender and time of opening

    (j) Accepting authority

Unbalanced tender: For item rate tender, contractors give the rates for each and every individual item. On the basis of information sometimes a contractor gives high rates for the items which are likely to increase and gives low rates for the items which are likely to go down. If the contractor's anticipation is correct the tender becomes unbalanced and the department looses heavily otherwise the contractor looses.

Tender Notice

    1. Sealed tenders will be received upto .....A.M./P.M. on the ....of 20 .., by the Executive Engineer .... division for the following work:-

Name of work..............Estimated cost Rs............

2. The work must be completely finished to the satisfaction of the Executive Engineer within ....months from the date of the order to commence the work.

3. The tender form with complete sets of blank forms of contracts can be obtained from the office of Executive Engineer......... division at .... every day (except sunday and holiday) from ....A.M. to P.M. at a charge of Rs........per set.

4. Each tender must be accompanied by a deposit of Rs........ as earnest money. Such earnest money may be of the following forms:-
   (i) Cash or Treasury Challan.
   (ii) Post office, bank savings pass-book having the requisite amount in the account, pledged to the Executive Engineer.
   (iii) Deposit receipt of State bank or other approved bank pledged to the Executive Engineer.
   (iv) National plan loan or National Savings Certificate pledged to the Executive Engineer.

5. The tenders will be opened at.......A.M/P.M. on the .....day 20... by the Executive Engineer or his authorised agent at his office at ....

6. Power is reserved to reject any tender or all tenders without assigning any reason or given any explanation.

7. Unless the person whose Tender has been accepted, signs the contract and deposits the security specified within....... days, the earnest money deposited by him will be forfeited and the acceptance of his tender will be withdrawn.

8. The tender rates shall be for the complete work and shall include all quarrying charges, royalty, testing, screening, tools and plants, carriage of materials to site, removal and charges of rejected materials, all taxes, income-tax, octroi charges, materials, labour, etc.

9. The tendered rates will remain valid for a period of three months from the date of opening tenders.

10. The quantities in the bill of quantities are approximate and liable to variation or cancellation for which contractor will not be entitled to any compensation. The quantities of any item or items and the total cost may vary for which ra   shall not be altered.

11. The rates should be quoted in the bill of quantities, legibly both in figures and words.

MUSTER ROLL OF DAILY LABOUR

Division..........District.............Name of work...............................

Estimate to which chargeable............Imprest holder's voucher no................
Gang incharge................

Part I - Nominal Roll

S.No.	Name of labour	Father's name	Design-ation	Days of month	Days wor-ked	Rate of wages	Amt. Ear-ned	To be filled in by the pay-ing officer
Daily total								
Daily Initial								
Inspecting Officer								

Passed for Rs.....................(Rupees...........................)

Signature...........Rank......

Grand total of this muster roll..

Deduct-payments not made as per details transfered to register of arrears.........

Total amount paid in words Rupees................

Date...............   Signature..........Rank......

Part-II

Details of measurements of work done by the labour employed as per this nominal muster roll in cases in which the work is susceptible to measurement.

Depreciation of work (Grouped sub-headwise)	Quantity	Deduct as shown on the last muster roll	Balance

Measurement taken on...............date......   Signature..........

Rank...............

Measurement Book No.......page no.......   Date...............

Liquidated damage: If a contractor fails to complete his works within the time specified in the tender then contractor pays an amount of compensation and the amount is called the *Liquidated damage.*

This amount is also mentioned in the tender.

If the owner uses any part of the completed work before the completion of full work, the liquidated damages for delay is reduced in the proportion of the value of the part to the value of the whole work.

Earnest money: Earnest money is an assurance or guarantee in form of cash on the part of the contractor. A contractor has to deposit this money:

(a) To keep open the offer for consideration

(b) To confirm his intentions to take up the work.

In case where a tenderer fails to start his work the earnest money is returned to the contractor. No interest is paid on earnest money to the contractor. A tender without earnest money is not considered under any circumstances. The earnest money of all the tenders other than the three lowest tenders are returned on application within a week from the date of opening. Earnest money of the second and third lowest tenders also should be returned within 15 days of the acceptance of the tender.

The amount of the earnest money which a contractor should deposit with the tender is mentioned by the department as follows:

(a) For works upto Rs. 5 lakhs - 2.5% of the estimated cost subject to a maximum of Rs 10000/-.

(b) For work more than Rs.5 lakhs - 2% of the estimated cost subject to a maximum of Rs.20000/-.

Security deposit: It is an amount of money to be deposited by the contractor whose tender has been accepted to pay compensation amounting to the part or whole of his security deposit if the work is not carried out according to given specifications or in time limit or conditions of contract.

EXERCISE

1. The arch of a culvert subtends an angle of 120° at the centre. Work out the quantity of arch masonry from the following data:

   Span of arch : 5.00 m; thickness : 50 cm; length of arch : 10m from face to face.

2. Prepare a detailed estimate of a R.C.C. roof slab of 3 meter clear span and 5 meter long.

   The slab is 12 cm thick.

   Main bars : 12mm diameter at 12 cm c to c.

   The alternate bars are bent up.

   Distribution bars : 6 mm diameter at 18 cm c to c.

   The quantity of steel shall be worked from the knowledge of weight of steel/running metre.

   The schedule of bars shall be shown in a tubular form. The dimensions for bearing in the walls, concrete cover etc., shall be suitably assumed.

3. Calculate number of bricks and quantity of cement and sand required for constructing a chimney with the following data :

   Outer diameter, top = 1.40 m

   bottom = 2.40 m

   Inner diameter = 1.00 m (constant)

   Height 16.0 m

4. a) Work out the following :

   (i) Number of cement bags needed for 100 m³ of brick masonry in 1:6 cement mortar;

   (ii) Number of cement bags for 10 sq. m of 25 mm thick cement concrete floor 1:2:4;

   (iii) Quantity of wood work in chowkhat of a door frame 2.10 X 1.20 m size and 7.5 X 10 cm in section.

5. Specify ratio of various ingredients being adopted in respect of the following :

   (i) C.C. slab for roof;

   (ii) cement mortar in brick masonry for partition walls;

   (iii) C.C. for elevated R.C.C.tank;

(iv) C.C for foundation of wall footing.

State number of cement bags needed per unit of each item.

6. Explain with an example, the details you would give for a bar bending schedule in R.C.C. work.

7. (a) What are the methods of working out preliminary estimates for the following :

Hospital, Tube well, road, Hostel of students.

(b) Following are the essential details of a T-beam and slab for the roof of a belt of 6m X 6m. Main reinforcement : Two straight bars of 25 mm and two cranked bars of 20 mm

Top anchor bars : Two of 12 mm

Stirrups : 6 mm at 150 mm c to c at ends upto 2m and @ 300 mm c to c at the central portion.

Cover : 25 mm at sides and bottom, add 50 mm at end supports

Slab :

Main bars : 12 mm @ 120 mm c to c cranked alternately

Distribution bars : 6mm @ 200 mm c to c

Work out the details of bars for the beam and slab giving a bar bending schedule and total weight of steel required.

8. (a) Explain, giving examples, the circumstances which necessitate the preparation of (1) approximate estimate and (2) detailed estimates.

(b) The following essential details refer to an R.C.C. column and its footing.

Footing :

Main bars : 12 mm @ 125 mm c to c. both ways,

Leveling base course : 100 mm thick P.C.C. 1:4:8

Concrete cover above P.C.C. base and sides : 40 mm.

Footing : Square in plan, 2500 mm X 2500 mm.

Sides and depth : 200 mm at ends and 700 mm at, column face.

Column:

side : 500 mm X 500 mm.

Length : 5500 mm above P.C.C. base course.

9. Take a level section, where :

L = length, distance between adjacent cross-sections.

B = constant formation width

S = side slope

C and C_1 = mid heights of two section

M = area of mid section

Derive the Prismoidal correction, $Pc = (V_E - V_P)$ where V_E is the volume by end area method and V_P is the volume by prismoidal method.

(b) Mass diagram details are given below :

S.No.	Distance in m	Volume in cum	
		cutting +	Banking
1	0		
2	100	490	
3	200	927	
4	300	982	
5	380	279	
6	400		31
7	500		226
8	600		654
9	700		1160
10	800		933
11	800		92
12	900	220	
13	1000	420	

Draw the mass diagram with the tabulated values of cuttings and bankings and determine the total volume of cuttings and banking.

10. (a) Calculate the quantity of earth-work in excavating an irrigation canal and forming the banks on either side from the data given below :

Bed width of canal : 4m;

Side slopes of excavation 1 : 1;

Total width of left bank 4m,

Side slopes : 1 1/2 : 1

Top width of right bank 2m;

Height of canal banks above

bed of canal : 3.00 m throughout;

Longitudinal slope of the bed of canal : 1 in 5000.

Station (at 50 m intervals)	0	1	2	3	4	5	6	7	8
R.L. of ground	102.0	102.30	102.48	102.67	101.66	101.25	101.44	101.63	101.42

R.L. of canal bed at the beginning (i.e. at stn. 0) is 100.00

11. (a) Drive the expression for prismoidal formula used in calculating earthwork quantities.

(b) What are prismoidal and curvature correction : Give the expressions used for the same. How are earthwork tables useful in earthwork calculations ?

(c) Work out quantity of earthwork for a canal of length 1,500 m with the following data :

Bed width = 4.5 m

Free board = 0.45 m

Depth = 0.60 m

Bed slope = 1 in 5,000

Side slope cutting = 1 : 1

Filling = 1.5 : 1

Proposed bed level = 137.30 m

Top width (both sides) = 2.0 m

R.L. as given below :

R.D., m G.L

0 137.90

500 137.80

1000 137.60

1500 137.40

12. A hill road is to be constructed in a side long ground in cutting.Calculate the quantities of earthwork in a length of 200 m with the following data:

Ch. m	Depth at centre m	Cross slope of ground
0	0.6	8 : 1
100	1.2	10 : 1
200	1.8	12 : 1

Formation width of road is 8 m and side slope is 1 : 1

13. A road at formation level in cutting is 10m wide and the side slopes are 1 1/2 : 1. The surface of the ground has a uniform side slope of 1 in 6. At adjacent cross-sections 30 m,apart the depths of cutting at the centre line of the road are 2m, 3m, and 4m respectively. Estimate the volume of earthwork in cutting.

14. Determine the quantities of earthwork for the portion of a road between changes 50 and 60 from the following data, lengths being measured with a standard 20 m chain :

Chainage	50	51	52	53	54	55
Ground level	131.1	131.2	130.9	131.2	130.8	130.7
Chainage	56	57	58	59	60	
Ground level	130.6	130.4	129.1	129.1	129.7	

The formation level at chain 50 is 130.0 and the road is in a rising gradient of 1 in 200. The width of formation is 10 m and the side slopes

1½ : 1 in cutting. The lateral slope of the ground may be assumed as level. Calculate also the cost of this earthwork in bank and in cutting at prevailing rates.

15. A portion of a proposed irrigation canal has the following data:

Distance in m	0	50	100	150	200
Ground level		884.80	885.03	884.86	884.50	884.40

Proposed bed level at 0 = 884.00
Bed width = 6m
Full supply depth = 1.0
Free board = 0.5m
Longitudinal slope of canal = 1 in 2500
side slopes in : Cutting = 1 : 1 Embankment = 1.5
Top width of embankment on both sides = 3.0 m
Determine the quantities of earthwork in cutting and in bank.

16. (a) How are lead and lift provided for determining rates for earthwork? Give an example at local rates.

(b) Portion of hill road in straight alignment is to be constructed in side long ground having a uniform lateral slope of 1 in 8. The formation width is to be 20 m with side slopes of 1 : 1 in cutting and 1 1/2 : 1 in embankment.

Assuming the areas in cutting and embankment at each as equal, determine the total volume of earthwork for the portion of road 675 m long, and also its cost at Rs. 60 per m³.

17. Analyze and derive rates for the following items of work. (material and labour part shall be analyzed separately). Assume suitable rates prevailing in your area.

(i) Providing and laying P.C.C. (1:4:8) in foundation trenches including compacting and curing;

(ii) work out quantities of materials (ingredients) required per metre cube concrete of proportion 1:2:4. The basis for your calculation shall be clearly indicated.

18. (a) Prepare rate analysis in the standard form for C.C. for R.C.C. slab or 100 mm thick brick partition wall.

(b) Explain the importance of rate analysis and state the factors affecting the rate analysis.

(c) Work out the rate analysis in a standard form of the following:

(i) R.C.C. work for beams;

(ii) 6 mm thick terrazzo mosaic flooring over 20 mm thick cement concrete;

(iii) Laying 150 mm diameter glazed stoneware pipe.

19. Work out the data rates (analysis of rate) for any two of the following:
    (i) First class brick work in cement mortar 1:6 in superstructure for first floor;
    (ii) 40 mm thick flooring under layers of 30 mm thick c.C. 1:2:4 and top layer of 10 mm thick red oxide cement mortar plaster 1:3;
    (iii) Surface dressing or bituminous painting one coat.

20. (a) What are the items that are considered as overhead expenses in executing a job ?
    (b) Draft detailed specifications for centering and shuttering of an R.C.C. slab and beam work.

21. Workout the unit rates of the following items of work at local rates. Use standard proforma :
    (i) 100 mm thick brick partition wall in cement mortar 1:6;
    (ii) R.C.C. lintel in 1:2:4 cement concrete and 0.9% steel;
    (iii) Providing and fixing porcelain wash basin on cast iron brackets.
    (iv) R.C.C. roofing 15 cm thick with 1.2% reinforcement.
    (v) Asphalt concrete 10 cm thick for a road.
    (vi) Laying 30 cm diameter sewer pipe.

22. (a) What factors are considered in drafting specifications for construction items ? Site the important purposes served by specifications.
    (b) Draft specifications for Ist. class brickwork in C.M. 1:6 or for reinforced concrete work for beams and slabs.
    (c) What are standard specifications ? What are the principles of specifications writing ?
    (d) Draft specifications for woodwork in doors and windows (I class or U.C.R. masonry in foundations.

23. (a) Explain the statement that the site of work influences the rate of an item, also mention the other factors.
    (b) Draft detailed specifications for Manglore tiled roofing including battens or for mosaic flooring.
    (c) Draft detailed specification for cement flooring (I.P.3.) or C.I. pipe for I.S.

24. Write down the detailed specification for the following :
    1. First class brickwork in cement mortar 1:6 for an office building;
    2. Wood work for doors and windows :
    3. Water bound macadam roads.

25. (a) Explain the importance of clarity and precise language in drafting specifications. How do specifications help a contractor in arriving at rates for a tender ?

(b) Draft specifications for the following :

    (i) Terrazzo flooring, (ii) aluminium doors for a shop front.

26. Prepare a material statement for different materials required for each of the following items of work :

    (a) An underground R.C.C. sump 4m X 4m internally and 2m deep to be constructed in 1:2:4 cement concrete with 2% steel. The bottom slab is 250 mm thick laid over a base course of 1:4:8 cement concrete 150 mm thick. The side walls are 250 mm thick at floor slab level and 150 mm thick at top, the inside face being vertical.

    (b) Stone masonry in cement mortar 1:4 for a 600 mm thick segmental arch over a wall opening of span 5m and rise of arch of 1.25m, the thickness of wall being 800 mm.

    (c) First class brick work in cement mortar 1:6 for 170 m².

27. (a) Explain the importance of schedule A and schedule B in contract documents. What information do they contain ? Give the standard form for the schedule B.

    (b) Mention the different types of negotiated contracts. Explain briefly.

28. (a) What do you understand by an unbalanced tender ? Give an example to illustrate the same.

    (b) Explain the purpose of earnest money deposit in contracts.

    (c) Draft a tender notice for the construction of a building for a proposed engineering college, costing Rs 100 lakhs.

29. (a) Give the principal features, advantages and drawbrackets of the following types of contracts:

      (i) Lump-sum;

      (ii) Item rate

    (b) Explain the provisions made in the conditions of contract in respect of the following :

      (i) bad work;

      (ii) Subletting;

      (iii) Arbitration

30. (a) What action will you adopt when :

      (i) A contractor fails to complete his work inspite of repeated reminders.

      (ii) The work done by contractor is of substandard quality.

      (iii) There is a theft in the store under your charge.

31. A hospital building costing rupees 60 lakhs is to be constructed in a metropolitan city by P.W.D. Draft a suitable tender notice to be given in the newspaper.

32. (a) Differentiate between the following :

(i) contract and contractor ;

(ii) Item rate contract and percentage rate contract;

(iii) Tender and quotation;

(iv) Security deposit and earnest money deposit.

(b) Write short notes on the following :

(i) Arbitration;

(ii) Duties of executive engineer;

(iii) Measurement book;

(iv) Mister rool.

33. (a) Discuss the essential requirements of a valid contract.

(b) Draft a typical tender notice to be advertised in leading newspapers, for the construction of a bridge costing Rs. 10,000,000 as estimated. Assume relevant information.

34. (a) What are the conditions affecting the salvage value ? - Explain

(b) Describe the method of depreciation by sinking fund formula.

(c) A structure having a present value of Rs. 1,00,000 (one lakh) shall have a depreciated value of Rs. 10,000 after five years. Find the depreciated value of the above building after three years by sinking, fund method, with a 4% compound interest on the sinking fund.

35. (a) Describe the various methods of valuation and state their advantages and disadvantages.

(b) A building is situated by the side of a main road 10 km from a city on a land of 500 sq. m. The built up portion is 20 m X 15 m. The building is provided with water supply, sanitary and electric fittings and the age of the building is 30 years. Work out the valuation of the property.

36. (a) Distinguish between the following pairs of terms :

(i) Gross yield and net yield,

(ii) cost and value,

(iii) stock account and material at site account, and

(iv) penalty and liquidated damages.

(b) Mention the factors which affect the value of a property. What will be the value of the property if the rate of interest for redemption of capital is 3% ?

37. (a) (i) What annual sinking fund at 4½ % must be invested to produce Re. 1/- at the end of 20 years ?

(ii) Derive the expression you use for the same.

(b) The cost of a pumping set with all accessories is Rs. 20,000/-. Work out the depreciation factor and the depreciated cost at the end of second year, using declining balance (or constant percentage) method.

Take life of the equipment as 8 years and salvage value : Rs . 2,400/-

38. (a) Explain the following pairs of terms :

(i) Free-hold property and lease-hold property;

(ii) Scrap value and salvage value;

(iii) Lessor and lessee.

(b) What is depreciation ? Why do properties suffer depreciation ? Explain briefly the following methods of depreciation :

(i) Straight line method ;

(ii) Sinking fund method.

(c) Explain clearly the term "outgoings". Give the principal outgoings to which land and buildings are subjected to.

39. (a) Explain the terms "value" cost and Price. What are the characteristics of an ideal investment ?

(b) Describe briefly the method of valuation based on life and valuation with reference to cost.

Write explanatory note on the following :

(i) Annuity;

(ii) Use of valuation tables ;

40. (a) explain the methods of valuation for :

(i) Factory building (owner occupied)

(ii) Cinema;

(iii) Five-star hotel;

(iv) Tenanted building.

(b) A building stands on a freehold plot of land measuring 800 sq. The price of land is Rs. 200 per sq.m. The owner gets a gross rent of Rs. 2,000/- per month from the building. The estimated natural life of the building is 15 years but is expected to extend by another 16 years if structural and other repairs costing Rs. 8.000 are immediately carried out. The total outgoings is equal to 25 per cent of the gross rent and yield required is 6 per cent. Find out whether it will be advisable to spend the above cost of repairs from an investment point of view.

Given :

(i) Present value of Re. 1 per annum allowing 6% interest and redemption of capital at 3% as 3.79 for 15 years and Rs. 12.50 for 31 years.

(ii) Present value of Re. 1 receivable at the end of 15 years and 31 years at 3% interest as Rs. 0.6419 and Rs. 0.40 respectively.

41. (a) A motor car was purchased for Rs. 150,000. Assuming its salvage value at the end of 6 years to be Rs. 25,000, determine the amount of depreciation for each year using (i) straight line method, and (ii) constant per cent method.

(b) A newly constructed buildings stands on a plot of land costing Rs. 50,000. The construction cost of the building is Rs. 150,000. The expected net returns are 8 and 10 percent on land and construction costs respectively. The outgoings would be 30 percent of gross return. The annual repairs and sinking fund installment may be assumed at 2 percent of the construction cost. Calculate the monthly rent of the building.

42. (a) Distinguish between :

(i) Cost of construction and valuation.

(ii) Depreciation and sinking fund.

(b) What are the objects of valuation ?

(c) A building constructed on a site measuring 20 m X 30 m is fetching a gross rent of Rs. 2,500 per month. The plinth area of the building is 140 m² and the cost of constructing it is Rs. 2,000 per m² of plinth area. The estimated life of the building is 70 years. Determine the present value of the property based on rental income assuming a net yield of 9%. For sinking fund accumulation, a compound interest rate of 5% may be assumed. Taxes, annual repairs and all other outgoings may be taken as 32% of the gross income, and the cost of land as Rs. 80 per m².

43. (a) Discuss briefly the various items included in the term 'outgoing.

(b) The owner of a property realizes a net annual income of Rs. 15,000 which he invests at 9% interest. At the end of 15 years he carries out certain repairs at a cost of Rs. 20,000. He then gets an offer from a buyer to purchase the property for Rs. 4,50,000. Determine whether this is advantageous to him or to the buyer.

(c) Differentiate

(i) Market value and book value,

(ii) Obsolescence and depreciation,

(iii) Sentimental value and speculative value.

44. (a) How is standard rent of a given premises determined ?

(b) A building is constructed at a cost of Rs. 2,50,000 on a land purchased at Rs. 50000. The owner of the property expects a return of 9% on the cost of construction and 8% on the cost of land. The building is estimated to have a future life of 60 years at the end of which it requires Rs. 3,25,000 for constructing a new building in its place.

Determine the standard rent of the property. Given :

  (i) Rate of interest for sinking fund at 6%;

 (ii) Annual repairs at 1½ % of the cost of construction;

(iii) All other outgoings : 28% of the net income of the property;

(iv) Scrap value at the end of the useful life of the building as 10% of its present cost.

98 Construction Planning and Technology

Chapter II

Section A

INTRODUCTION

A. 1 Need For Planning

When a new project has to be undertaken for which one does not have any previous experience, it is necessary to visualise all the operations of the project; arrange these operations in the sequence, acquire the know-how and means necessary to perform them and feel convinced that the method thought out for performing each operation is the most economical. Such confidence is achieved through systematic planning.
systematic planning.

When a project is in progress, it grows in size, complexity and technology. For example, in the construction of a plant several specialists-such as structural, constructional, mechanical,electrical, heating and ventilating, sanitary, safety, production and system engineers-may be involved. Each is responsible for his own specialised technology. Each has his own view points and thinks differently about the project scheduling. Some methods must be found to facilitate communication so that each is working towards the same set of project objectives. The emphasis that project manager puts on time, cost, quality and goods may be different from that of his colleagues. It is only by using a project plan that the project managers can convey to his team his emphasis on these variables.

A project can be efficiently planned to make optimal use of the limited resources available. Resources may include labour, capital, equipment etc.

A. 2 Historical Development Of Planning Methods

Practically every construction project is sufficiently complex than its breakdown and its interrelationships must be recorded on paper apart from the memory of the planner. Therefore, when the plan is formulated, some type of 'paper model' of the project is developed to communicate results of the plan to others to serve as a basic for evaluating progress and controlling the works and to take effective decisions.

A large number of management techniques or tools were developed in order to assist management in the construction field. The most widely used techniques are

    (a) Bar chart or Gantt chart

    (b) Critical path planning

Bar charts failed to provide suitable project model because the relationship dependancy among different operations were not indicated. The division of a project into its component tasks can be very gross and the bars representing the tasks can be freely overlapped in terms of time.

These limitations of bar charts forced to improve on the method. As a result, new techniques like *Critical path method* (in 1956)and *Program Evaluation & Review Technique* (in 1957) were developed. These methods were developd to such an extent that projects with thousands of activities, as well as smaller projects with fewer activities can be planned and controlled.

Other tools used to achieve the timely completion, economy & control of a project include the following:

    (a) **Labour Schedule:** The purpose of labour schedule is to know well in advance the type of labour and their numbers required from time to time so that steps required for their recruitment are taken in advance.

    (b) **Material Schedule:** The purpose of the material schedule is to know the requirement of various construction materials well ahead of the start of the project so that at no stage during execution of project, work is held up for want of matetials.

    (c) **Equipment Schedule:** The purpose of equipment schedule is to show the seperate sizes and types of equipment required on rent, lease or outright purchase.

    (d) **Expenditure Schedule:** The purpose of the expenditure schedule is to show the income & expenditure to ensure that the financial position during a period is kept within the allotment for the year.

The applications of these tools are based on bar charts.

A. 3 Bar Charts

The GANTT chart or bar chart was developed by Henry L Gantt in 1919, in connection with military requirements of World War I. The bar chart is essentially a graphic representation of programme of procurement of materials and time bound schedule of completion of various elemants of the project. Gantt chart can be the only schedulig tool used for preparing a schedule or it can be used as a format for the presentation of schedules developed by the use of more sophisticated techniques.

The scheduled programme of various activities are indicated on the left hand side and proposed duration of execution is represented by a number of bars as shown in Fig. A1.

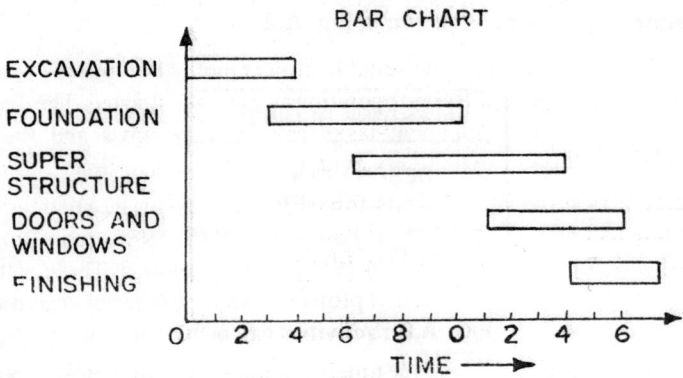

Fig. A.1. Bar chart

The main advantages of the Bar charts are

    (a) The Bar chart is simple to prepare & interpret.

    (b) Each item of work or item is shown seperately.

    (c) Modifications to the chart can be carried out easily.

    The Bar charts have the following disadvantages

    (a) It is difficult to prepare the bar chart for large projects if there are numerous and complex relationships between the tasks that make up the schedule.

    (b) It does not show the interdependancy among different activities of the project.

Bar charts were later modified to cope up with Critical path method networks. The following Bar charts were introduced later:

    1. Consecutive – job bar charts

    2. Cross connected bar charts

3. Cascade bar charts
4. Sequenced or float-linked bar charts.

These bar charts are discussed later on with CPM network.

A. 4 Milestone Charts

On a large complex project there are hundreds and even thousands of operations. Managers cannot keep track of all the details of each task. They are satisfied by keeping tabs on certain tasks or MILESTONES. Milestones can be indicated on the bar charts using any symbol desired.

The milestone event is refered to either the start of an event or completion of an event. A milestone network can be developed so that a comprehensive chart of important events is available for the top management. The illustration of milestone is shown in Fig. A.2.

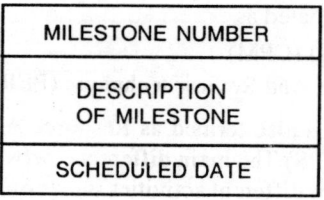

Fig. A.2. Milestone symbol

Example 1: A project manager has identified the following major events of milestones in a Highway Bridge project.

1. Excavation & Backfill
2. Piling
3. Abutments
4. Steel Girders

The milestone chart with deadlines of the milestones is shown in Fig. A.3.

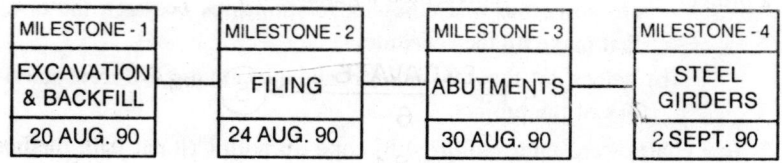

Fig. A.3. Milestone chart.

If some additional information is required at a lower level of management, additional milestones can be added.

Section B

CRITICAL PATH METHOD (CPM)

B.1 Network Representaton Of A Project

On account of the shortcomings of the bar chart discussed in Section A, new techniques based on network analysis were developed in late sixties. These new techniques are designated as :

(i) Critical Path Method (CPM)
(ii) Program Evaluation And Review technique (PERT)

In some cases CPM is also termed as Resource Allocation For Multi-project Scheduling(RAMPS).The main difference between CPM and PERT is that the time estimate for different activities is deterministic in CPM and it is probabilistic in PERT.Therefore,the term CPM is applied to project with known magnitude of time whereas PERT is applied to projects under research.

A network is a diagramatic representation of a project or plan of a project that shows the correct sequence and relationship of activities and events required to achieve the end objectives.In a network,each arrow represents an activity. Each circle or node represents an event (Refer Fig. B.1).
one arrow to the other.Each circle or node represents an event.(Refer Fig 1)

on site	on paper
a job or activity	is represented by an arrow
an event	is represented by a node
a project	is represented by a network

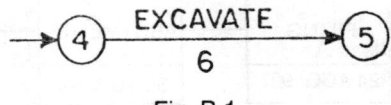

Fig. B.1

The number on the arrow is the duration in units of time of the activity.The length of the arrow has no significance;it merely represents the passage of time in the direction of arrowhead.

The start of all the activities leaving a node depends on the completion of all the activities entering that node. Hence the event represented by any node is not achieved untill all the activities entering that node have been completed.

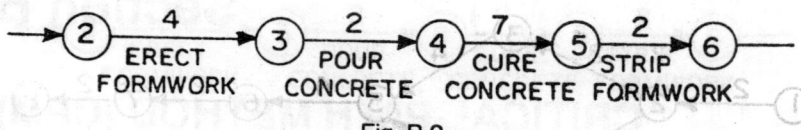

Fig. B.2

In case of Fig. B.2, it can be inferred from the sequence of activities that the concrete cannot be poured until the formwork is completely erected.

Each activity can be represented by the activity description or by a capital letter or using i-j convention. In i-j convention, the number of the event at the tail of the arrow is known as the i-number and j is the number at the head of the arrow. Using i-j convention, the activities in Fig. B.2 are identified as follows:

2-3	**Erect Formwork**
3-4	**Pour Concrete**
4-5	**Cure Concrete**
5-6	**Strip Formwork**

Example 1: Consider the simple example of erection of portal frame as shown in Fig. B.3, in which the operations and their time estimates (in days) can be the following:

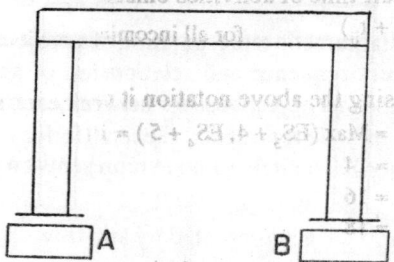

Fig. B.3. Portal frame

Activity	Duration (days)
Clear site	2
Prepare base A	2
Prepare base B	3
Erect column A	4
Erect column B	5
Erect beam	4
Plumb & line structure	2
Grout bases	2

As we can see, the preparation of base A and base B can be done concurrently after clearing the site. Also, erection of column A depends only on base A and not on base B. At the same time, the erection of the beam can only follow the erection of columns. All these arguments lead to the following representation of the network.

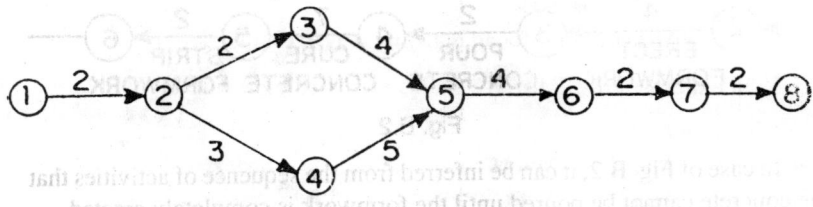

Fig. B.4.

With the help of this example, we can define some terms used in network analysis.

1. **Dummy Activity:** A dummy activity is an artificial activity, represented on the network by a dotted line, which indicates that an activity (or activities) following the dummy activity cannot be started until the activity or the activities preceding the dummy activity are completed. It has zero duration.

2. **Earliest start time (EST):** This is the earliest time at which an event can occur, that is, the earliest time by which all activities leading to that event can be completed. If t is the duration of an activity (i,j) and if ES is the earliest start time of activities emanating from node j, then

$$ES_j = Max (ES_i + t_{ij})$$ for all incoming activities (i,j), j > 1

$$ES_1 = 0$$

In example 1, using the above notation it can be verified that

$ES_1 = 0$ $ES_5 = Max (ES_3 + 4, ES_4 + 5) = 10$
$ES_2 = 2$ $ES_6 = 14$
$ES_3 = 4$ $ES_7 = 16$
$ES_4 = 5$ $ES_8 = 18$

The earliest start time of every node is entered inside a rectangle at every node (Fig. B.5).

3. **Earliest completion time (ECT):** This is the earliest time at which an activity can be completed if it is started at its EST and is completed within its duration.

$$ECT = EST + t_{ij}$$

where EST is the earliest start time of activity (i,j)

t_{ij} is the duration of the activity (i,j)

(ECT is used in resource allocation & resource levelling).

4. **Latest completion time (LCT):** This is the latest time at which an activity can be completed without delaying the project completion. If LC is the latest completion time of the activities i coming into node i, then

LC$_i$ = ES$_i$ for last node

LC$_i$ = Min (LC$_j$ - t$_{ij}$) for all outgoing activities (i,j)

In example 1, LCT for few nodes are calculated below.

LC$_8$ = ES$_8$ = 18
LC$_7$ = Min (LC$_8$ - 2) = 16
LC$_2$ = Min (LC$_3$ - 2 , LC$_4$ - 3) = 2

LCT for every node is entered inside a rectangle at the corresponding node (Fig. B.5).

5. **Latest start time :** This is the latest time at which an activity can be started if it is to be completed by its latest completion time.

LST = LCT - t$_{ij}$

where LCT is the latest completion time of activity (i,j)

t$_{ij}$ is the duration of the activity (i,j)

6. **Total Float (TF):** Total float of an activity (i,j) is the amount of time by which the start of an activity can be delayed without delaying the project's duration.

TF$_{ij}$ = LC$_{ij}$ - EC$_{ij}$
= LST$_{ij}$ - EST$_{ij}$
= LC$_j$ - ES$_i$ - t$_{ij}$

7. **Free Float (FF):** Free Float of an activity (i,j) is the amount of time by which the start of the activity may be delayed without delaying the start of any succeeding activity.

FF$_{ij}$ = ES$_j$ - ES$_i$ - t$_{ij}$

8. **Independent Float (IF) :** The part of the float which remains unaffected by utilization of float by the preceding activities and does not affect the succeeding activities is called Independent float.

IF + EST (Head event) – LST (Tail event) – duration

9. **Interfering Float (InF) :** The portion of the total float which affects the start of the subsequent activities is known as Interfering float.

InF + total float – Free float.

10. **Critical Activity:** The duration of the project depends on certain activities called critical activities. If a critical activity is delayed, the entire project is delayed. If a project is to be accelarated, one or more of the critical activities are to be accelarated. The path joining these activities is called CRITICAL PATH. The sum of the durations of the activities along the critical path determines the duration of the project.

An activity (i,j) is said to be critical if,

$$ES_i = LC_i$$

$$ES_j = LC_j$$

$$LC_j - LC_i = ES_j - ES_i = t_{ij}$$

It can be verified that for a critical activity, Total Float is zero.

Example 2: The network in Fig. B5 represents a section of work being undertaken by a contractor. Duration (in weeks) are shown alongside each arrow. We are interested in calculating EST, LST, ECT & LCT for this network.

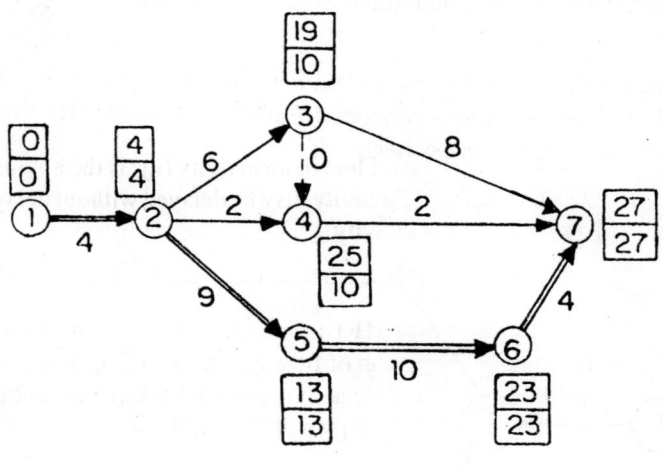

Fig. B.5

Activity $(3,4)$ is a dummy activity and hence it is shown with a dotted line. The Table B.1 shows EST, LST, ECT, LCT, Total float & Free float for all activities. EST & LCT of each node are shown in rectangles at each node.

Table B.1

Job (i,j)	Duration t ij	Earliest start time	comple- -tion time	Latest start time	comple -tion time	Total float	Free float
(1,2)	4	0	4	0	4	0	0
(2,3)	6	4	10	13	19	9	0
(2,5)	9	4	13	4	13	0	0
(2,4)	2	4	6	23	25	19	4
(3,7)	8	10	18	19	27	9	9
(3,4)	0	10	10	25	25	15	0
(4,7)	2	10	12	25	27	15	15
(5,6)	10	13	23	13	23	0	0
(6,7)	4	23	27	23	27	0	0

The CRITICAL PATH (1-2-5-6-7) is shown in Fig. B.5 with double lines.

Therefore, the duration of the work is 27 weeks.

Example 3: Fig. B.6 depicts a CPM network and Fig. B.7, Fig. B.8, and Fig. B.9 shows the Cross connected Bar chart, Cascade Bar Chart, and Sequenced Bar Chart respectively.

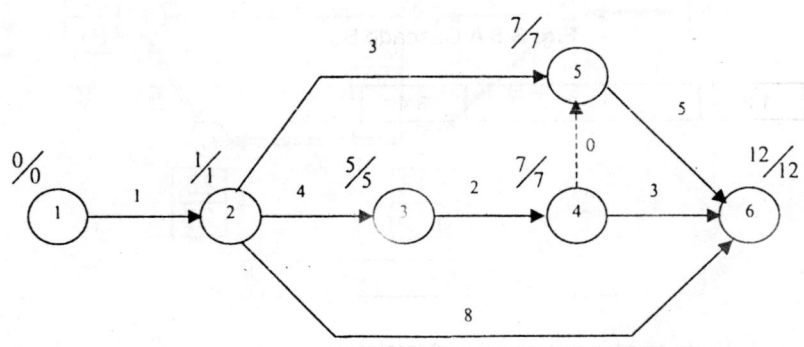

Fig. B.6 CPM

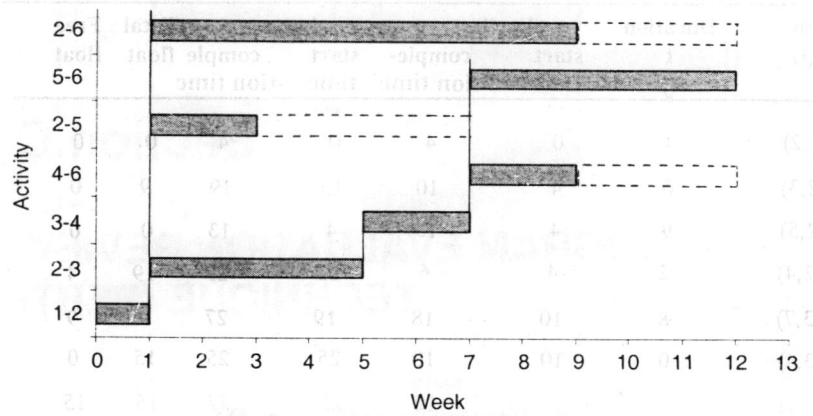

Fig. B.7 A cross connected Bar Chart.

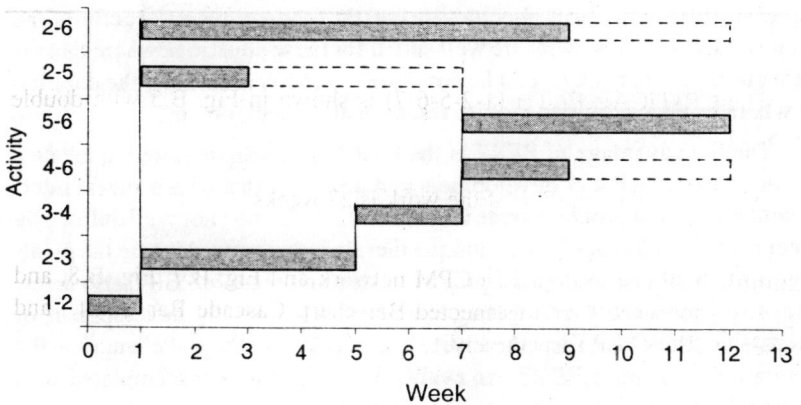

Fig. B.8 A Cascade Bar Chart.

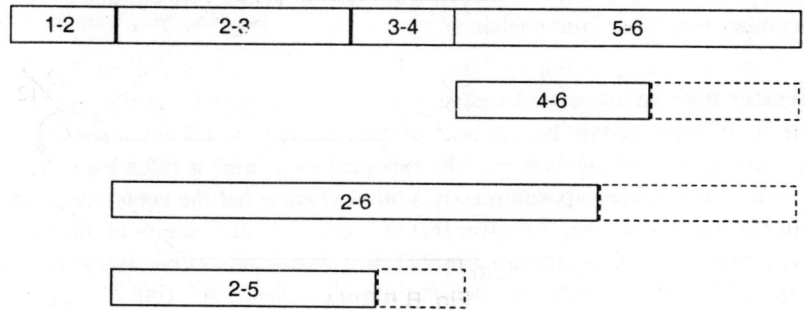

Fig. B.9 Sequenced or Float – Linked Bar Chart.

Section C

PROGRAM EVALUATION REVIEW TECHNIQUE (PERT)

Planning with CPM requires reasonably accurate knowledge of time and cost for each activity; CPM network model is essentially deterministic. In many situations,however,the duration of an activity cannot be accurately forecast.PERT introduces uncertainity into the time estimates for activity and project durations.It is therefore well suited for those situations where there is either insufficient background information to specify accurately the duration or where project activities require research and development.

The big advantage of PERT is the kind of planning required to create a major network.Network development and critical path analysis reveal inter-dependencies and problem areas that are neither obvious nor well defined by other planning methods.The technique therefore determines where the greatest effort should be made for a project to stay on schedule.PERT also determines the probability of meeting specified deadlines by development of alternative plans.PERT has the ability to evaluate the effect of changes in the program.For example,PERT can evaluate the effect of a contemplated shift of resources from the less critical activities to the activities identified as probable bottle necks.PERT also allows a large amount of sophisticated data to be presented in a well organised diagram from which both contractor and customer can make joint decisions.

PERT uses an activity duration called the expected mean time ($E(T_x)$) together with an associated measure of the x uncertainity of this activity duration.This uncertainity may be expressed as the standard deviation (σ) or the variance (V) of the duration.The expected mean time is intended to be a time estimate having approximately a 50 % chance that the actual duration will exceed it.From this it is clear that the formal determination of such activity data necessiates assuming probability distribution curve as shown in Fig. C.1 for the activity completion times. To ensure tailoring of this assumed distribution curve to the circumstances of each individual activity, three Engineering time estimates are made and embedded within the theoretical curve (see Fig. C.1). These three estimates of the activity's durations enable the expected **mean time, as** well as the standard deviation and the variance to be derived mathematically.

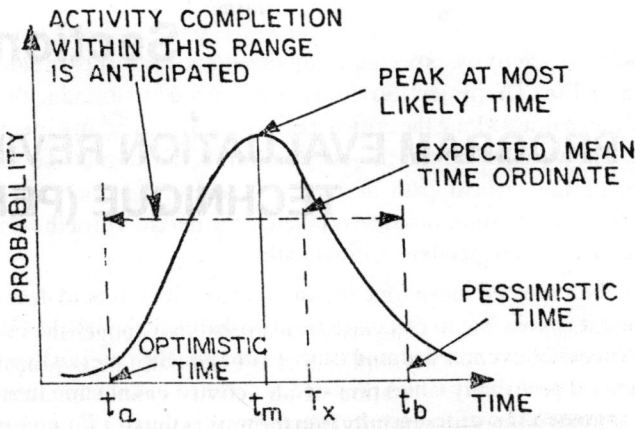

Fig. C.1 The graph between probability & time

The optimistic time (t_a) is an estimate of the minimum time required for an activity if it is completed in ideal conditions.

The most likely time (t_m) is based on experience and judgement,being the time required if the activity is repeated a number of times under essentially the same conditions.

The pessimistic time (t_b) is an estimate of the maximum time required if everything goes wrong.It may take account of an initial failure or delay,but should not be influenced by major hazards unless these are inherent in the activity.

The calculation of these three estimates forces the planner to take an overall view of the particular difficulties involved in the activity.They tend to offset the effects on the planner's judgement of known target dates.

The expected mean time is derived from the equation

$$E(T_x) = \frac{t_a + 4t_m + t_b}{6}$$

The standard deviation(the statiscal measure of uncertainity being the spread of the distribution curve about its mean value)is given by

$$\sigma_T = \frac{t_b - t_a}{6}$$

The variance is defined as the square of the standard deviation

$$V_T = \sigma_T^2 = \left[\frac{t_b - t_a}{6} \right]^2$$

By adopting activity expected mean times, the critical path calculations proceed as before. The project duration is determined by summing the activity expected mean times along the critical path and will thus be an expected mean duration. The variance of the project duration is the sum of the individual variances of the critical path activities. If more than one critical path exists, the project duration variance is taken as the maximum of those summed along the various independent critical paths.

Once the expected mean time for an event and its standard deviation are determined, it is possible to calculate from probability theory the chances of meeting a specific event scheduled time. To do this, the practical approach is to use standard probability tables prepared for normal distribution functions. To use this approach, the diiference between the scheduled (T_s) and expected mean times (T_x) for the event is scaled down to the standard curve by computing a factor Z where

$$Z = \frac{T_s - T_x}{\sigma_T}$$

Using this computed value of Z, a direct entry into table II.2 (Appendix-II) gives the probability of meeting the scheduled time T_S.

Example 1: (Construction of a pipeline)

The table below gives activities in a pipeline construction and their time estimates. The expected mean time T_x, standard deviation and variance are calculated and they are shown in Table C.1.

Table C.1

Activity	Description	t_a	t_m	t_b	T_x	σ_{Tx}	V_T
(1,2)	Move to site	8	10	12	10	0.67	0.45
(1,3)	Obtain pipes	12	14	16	14	0.67	0.45
(1,6)	Obtain valves	6	10	20	11	2.33	5.43
(2,4)	Layout pipeline	4	6	8	6	0.67	0.45
(3,5)	Dummy	0	0	0	0	0	0

(Contd.)

Activity	Description	t_a	t_m	t_b	T_x	σ_{Tx}	V_T
(3,6)	Cut specials	7	9	17	10	1.67	2.79
(4,5)	Dig trench	10	15	26	16	2.67	7.13
(5,6)	Prepare valve chambers	8	13	18	13	1.67	2.79
(5,7)	Lay pipes	18	20	46	24	4.67	21.81
(6,8)	Fit valves	4	6	20	8	2.67	7.13
(7,8)	Concrete anchors	3	6	9	6	1.00	1.00
(8,9)	Dummy	0	0	0	0	0	0
(8,10)	Backfill	8	8	20	10	2.00	4.00
(9,10)	Test pipeline	2	3	4	3	0.33	0.11

Fig. C.2 shows the network for the above table with T_x as the duration shown on the arrows.

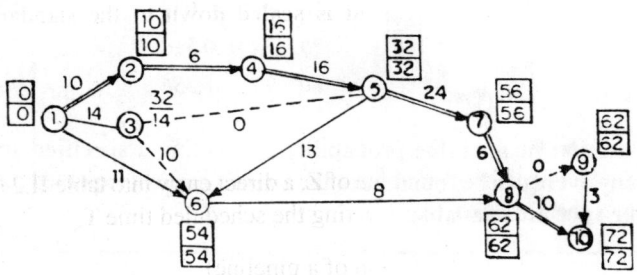

Fig. C.2. Network of pipeline problem

The PERT network calculations for the earliest expected starting time and latest expected completion time are shown in Fig. C.2. These times are calculated in exactly same manner as those of CPM. Critical path (1-2-4-5-7-8-10) is shown with double lines in Fig. C.2.

The variance of an expected mean time of any event or node is the sum of the variances of all those activities along the most time consuming path (in terms of expected mean time) leading to that node. Thus, for example, for event 6, its variance is computed from the chain (1-2-4-5-6) as this path is the most time consuming path leading to event 6.

$$V_{T6} = 0.45 + 0.45 + 7.13 + 2.79 = 10.82 \text{ T}$$

Referring to Fig. C.2, it is observed that event 10, and hence the project itself, has a finish time of 72 days and a variance of 34.84 (and therefore a standard deviation of 5.9). From this data, it is possible to compute readily the approximate probabilities of meeting specified or selected target dates for completion of the project. For example, we are interested in meeting the target date of 75 days for the completion of the project i.e, we are interested in finding the probability of completing the project in 75 days.

Therefore, $Z = \dfrac{75 - 72}{5.9} = 0.51$

Z has standard normal distribution. From Table II.2 this probability is found to be 0.69. Similarly, the approximate probabilities of scheduled project duration for completion of the project can be computed readily as shown in Table C.2.

Table C.2: Probability of meeting scheduled project duration

Expected mean time (T_x)	Standard deviation σ_x	Target or scheduled T_S	$Z = \dfrac{T_s - T_x}{\sigma_T}$	Probabilty of meeting target (T_S) from table 4
72	5.9	65	-1.186	0.12
		70	-0.039	0.36
		75	0.510	0.69
		80	1.356	0.91

In a similar manner, the probability of meeting a specified scheduled time for any event can be found, once the event's earliest expected mean time and its variance are available.

Section D

OPTIMAL SCHEDULING

We have discussed the problem of representing a project by a network and the significance of the critical path.Critical path identifies those activities (critical activities) which determine the project duration and these activities are to be paid attention if the duration of the project is to be reduced.If the duration of the project is to be cut down,there will be corresponding change in the cost of the project.

D.1 Cost-Duration Relatonships

A project has two types of costs—Direct cost and Indirect cost. Direct costs are those which increase with decrease in project duration. For example, labour. If a project duration is to be reduced, more labour has to be employed. Indirect costs are those which increase with increase in duration or vice versa. Administrative, Overhead, Depreciation, Insurance costs, etc. are the examples of this. The total project cost would be the sum of the Direct and Indirect costs (Refer Fig. D.1). We are interested in finding the optimum duration with minimum cost.

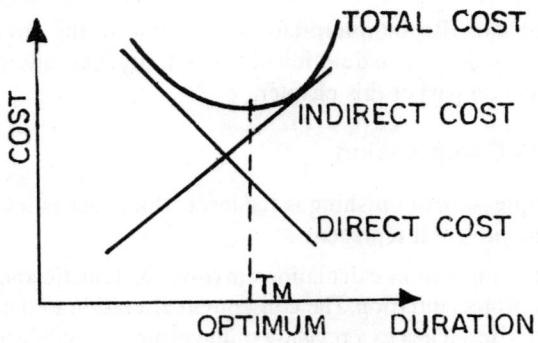

Fig. D.1 Cost-duration relationship

We have assumed Cost-Duration relationship as linear.In general,this is non-linear and it is approximated by a piecewise broken line for simpler computations.

Fig. D.2 shows the relationship between the direct cost and the time taken for its completion. The cost of the activity is at a minimum if the work is completed in what is thought to be 'normal' time (T_N). The minimum cost corresponding to this **maximum duration is called 'normal cost'**(C_N).It is usually possible on any individual activity to continue to reduce its duration by additional labour,overtime working,etc untill the point is reached where it just cannot be done any more quickly and this duration is called 'crash duration' (T_C).The corresponding cost will be maximum and it is called 'crash cost'(C_C).Refer Fig. D.2 for more details.

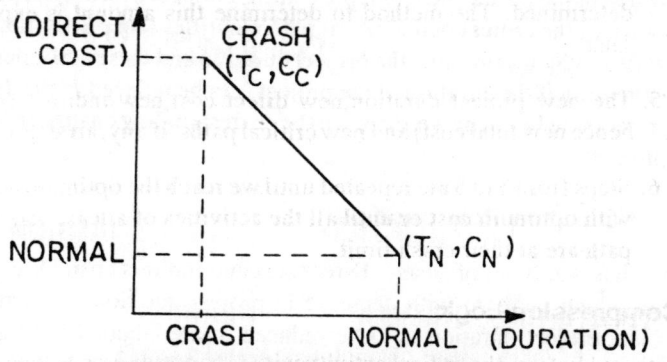

Fig. D.2 Direct cost-duration relationship

The slope of the straight line in Fig. D.2 can be expressed as

$$\text{Slope} = \frac{\text{Crash cost }(C_C) - \text{Normal cost }(C_N)}{\text{Normal duration}(T_N) - \text{Crash duration}(T_C)}$$

The slope can also be interpreted as increase in the direct cost of an activity for unit reduction in duration of that activity.This interpretation will be used in the later part of this chapter.

D. 2 Network Compression

Network compression or crushing is the term which means reduction in the duration of the project it represents.

Network compression calculations involve systematic and progressive reductions in project duration.The consequent alteration to the durations of individual activities leads to a revision of the project schedule.At each stage of the network compression calculations,a logical analysis must be made in accordance with the following rules.

1. The activities along the critical path are listed. (As mentioned be-

fore, one or more critical activities have to be compressed for reducing the duration of the project).

2. We delete the critical activities with zero potential for compression; these include those activities whose normal and crash durations are identical and those which have already reached their crash duration in previous stages.

3. We select the critical activity with smallest slope for compression (smallest slope is interpreted as cheapest compression). If there is more than one critical path, duration of activities along all the critical paths must be reduced simultaneously by the same amount.

4. The amount by which this activity can be compressed is to be determined. The method to determine this amount is explained later.

5. The new project duration, new direct cost, new indirect cost (& hence new total cost) and new critical paths, if any, are determined.

6. Steps from 1 to 5 are repeated until we reach the optimum duration with optimum cost or until all the activities of atleast one critical path are at their crash limit.

D. 3 Compression Logic

As mentioned before, the critical activity with minimum slope is selected for compression. The amount by which it is to be compressed is to be selected carefully. The amount of compression should not be so much that the stage when one or more critical paths are developed is surpassed. One method is to compress or crush the activity selected by one unit at each stage and hence identify the new critical paths.

The other method is we check the Free Floats (FF) of the non-critical activities. We reduce the duration of the selected critical activity (with minimum slope) say (i, j) by one unit and find the non-critical activities whose FF is changing. This step is called TEST STEP. If (p, q) & (s, t) are the non-critical activities whose FF is changing, then

$$F = Min (FF_{pq}, FF_{st})$$

where F is called Free-Float limit (FF Limit) of (i, j).

The Crash Limit (C) of selected critical activity (i, j) is defined as

C= Present Duration of (i, j) - Crash Duration of (i, j)

Therefore, Compression Limit which is the duration by which the activity (i, j) can be crushed is

K = Compression Limit = Min (F,C)

Hence, if we compress critical activity (i,j) by K units,then no new critical path will develop in between. All these steps are illustrated in Example 1.

Remark 1: If in the test step,there is no activity whose FF changes, then FF limit is infinite.

Remark 2: If there are two selected critical activities along with two critical paths with crash limits C_1 and C_2, then common crash limit is Min (C_1, C_2).

Example 1: The following table gives the details of a network which represents a part of project of a contractor.

Activity	Normal		Crash		Slope
(i,j)	Duration(T_N) (days)	Cost(C_N) (Rs.)	Duration (T_C) (days)	Cost(C_C) (Rs.)	$\dfrac{C_C - C_N}{T_N - T_C}$
(1,2)	6	100	4	120	10
(2,3)	9	200	5	280	20
(2,4)	3	80	2	110	30
(3,4)	0	0	0	0	0
(3,5)	7	150	5	180	15
(4,6)	8	250	3	375	25
(4,7)	2	120	1	170	50
(5,8)	1	100	1	100	0
(6,8)	4	180	3	200	20
(7,8)	5	130	2	220	30

Direct Cost = Rs 1310

Assume the indirect cost of the project is Rs. 20/day. We are interested in finding

(a) Optimum schedule of the project.

(b) Minimum duration of the project with corresponding cost.

Solution: The cost slope of each activity is calculated and entered in the table.As the duration of the activity (3,4) is zero,it is a dummy activity.Activity (5,8) cannot be crushed further as it is at its crash limit.

Fig. D.3 shows the network with all the activities at their normal durations. Star (*) over the duration of the activity (5, 8) indicates that it is at its crash duration. The earliest start time and latest completion time for each node are calculated and are indicated at each node. (1-2-3-4-6-8) is the critical path and hence the normal duration of the project is 27 days. FF of all non-critical activities are shown at each arrow.

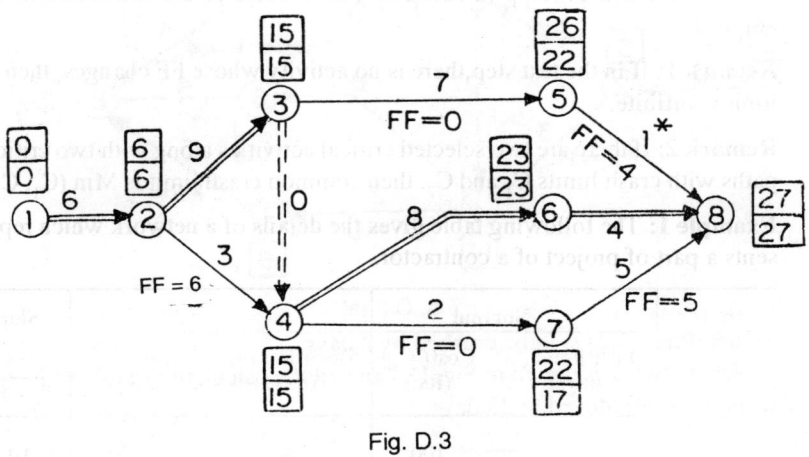

Fig. D.3

Table D.1 gives the summary of the Fig. D.3.

Table D.1

Duration	Cost		
(days)	Direct	Indirect	Total
27	Rs.1310	27 × 20=Rs.540	Rs.1850

I Compression

We find (1, 2) has the minimum slope (of 10) among all the critical activities (because (5, 8) cannot be crushed further). To find the amount of compression of activity (1, 2), we perform the test step.

We reduce the duration of (1,2) by one unit and calculate the FF.Since FF needs only EST of each node, only EST is calculated for each node.All details are shown in Fig. D.3(a).

We find no FF is changing.Therefore, (1,2) has infinite FF limit.

Crash limit of(1,2)=Present duration of(1,2)-crash duration of (1,2)

$$C = 6 - 4 = 2$$

Hence, Compression Limit K of (1,2) = 2

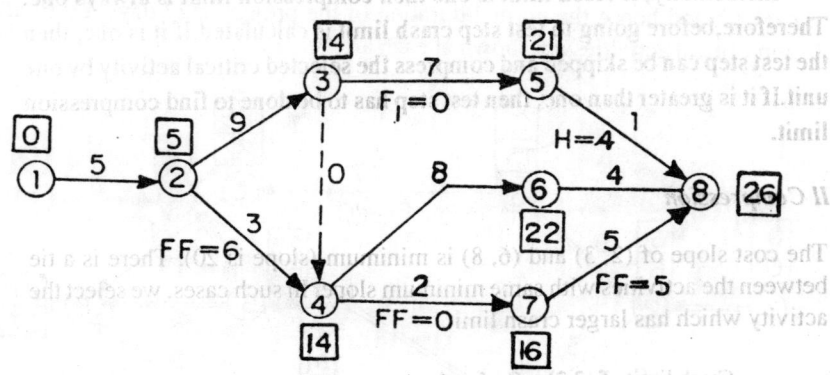

Fig. D.3 (a)

Therefore, (1,2) is to be crushed by 2 days.

New network is drawn in Fig. D.4 and all the calculations are repeated. The duration is reduced to 25 days.

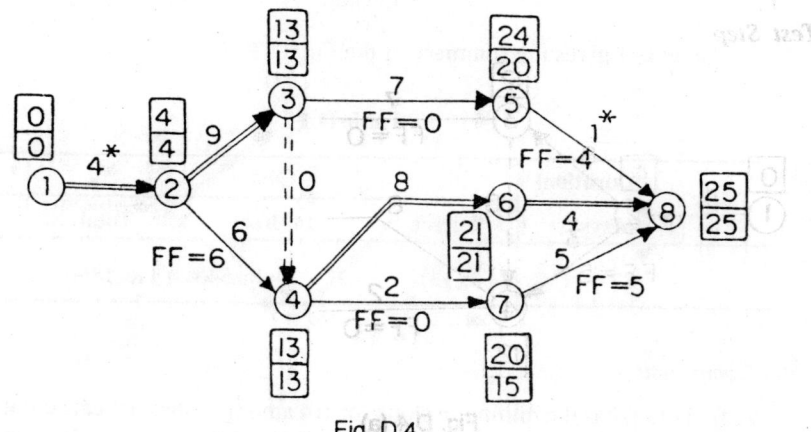

Fig. D.4

Summary of Fig. D.4 is shown in Table D.2.

Table D.2

Duration	Cost		
(days)	Direct	Indirect	Total
25	Rs. 1310	25×20	
	$+ 10 \times 2$	= Rs.500	Rs.1830
	= Rs.1330		

Incidentally, if crash limit is one then compression limit is always one. Therefore, before going to test step crash limit is calculated. If it is one, then the test step can be skipped and compress the selected critical activity by one unit. If it is greater than one, then test step has to be done to find compression limit.

II Compression

The cost slope of (2, 3) and (6, 8) is minimum (slope is 20). There is a tie between the activities with same minimum slope. In such cases, we select the activity which has larger crash limit.

> Crash limit of (2,3) = 9 - 5 = 4
>
> Crash limit of (6,8) = 4 - 3 = 1

Hence, select (2,3) for compression. As crash limit is not equal to one, we have to go through test step Fig. D.4(a).

Test Step

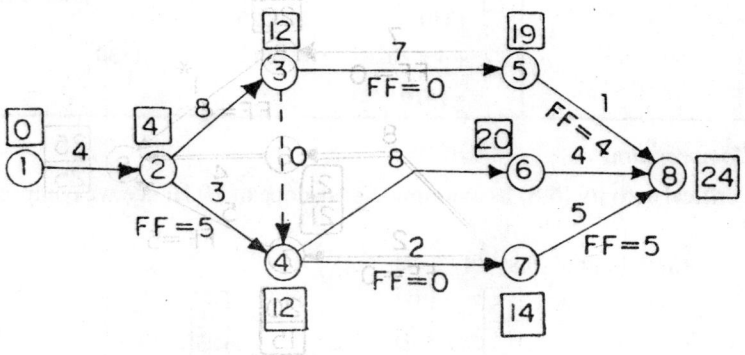

Fig. D.4 (a)

We find FF of (2, 4) is changed from 6 in Fig. D.4 to 5 in Fig. D.4(a). Therefore,

> FF limit of (2, 3) = 6

Hence, Compression limit of (2, 3) = Min (6, 4) = 4

Therefore, compress (2, 3) by 4 units and all necessary calculations of network are shown in Fig. D.5.

The duration of the project is reduced to 21 days. Table D.2 gives the summary of schedule for 21 days.

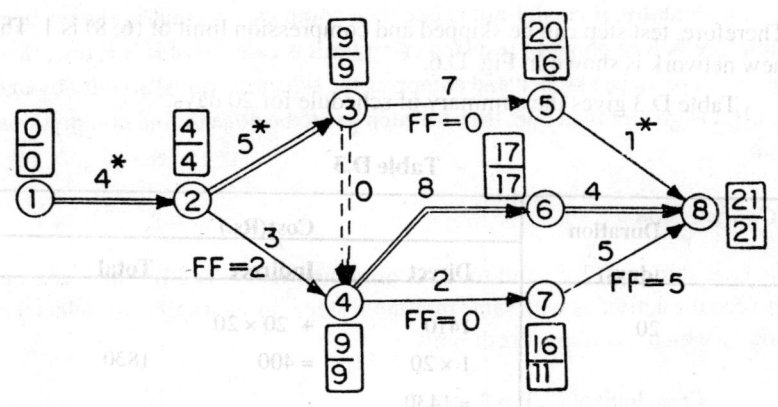

Fig. D.5

Table D.

Duration	Cost(Rs.)		
(days)	Direct	Indirect	Total
21	1330 + 4 x 20 = 1410	21 x 20 = 420	1830

III Compression

The critical activity (6,8) has minimum cost slope of 20.Hence,we compress (6,8).

Crash limit of (6,8) = 4 - 3 = 1

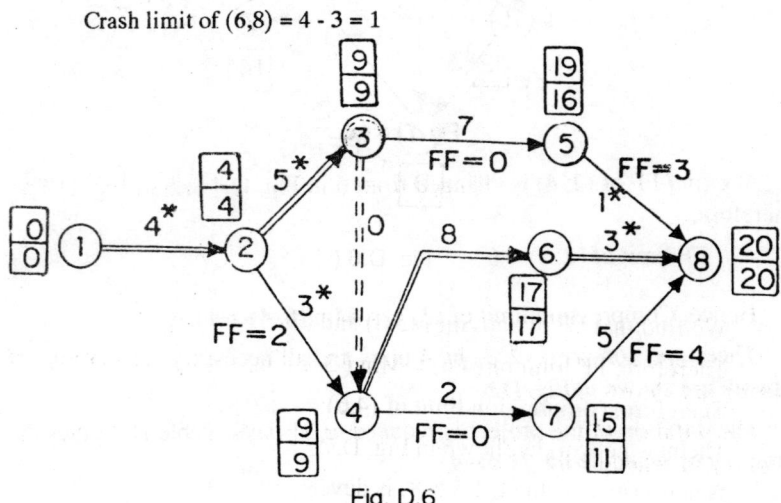

Fig. D.6

Therefore, test step can be skipped and compression limit of (6, 8) is 1. The new network is shown in Fig. D.6.

Table D.3 gives the summary of schedule for 20 days.

Table D.3

Duration	Cost(Rs.)		
(days)	**Direct**	**Indirect**	**Total**
20	1410 1×20 $= 1430$	$+ 20 \times 20$ $= 400$	1830

IV Compression

At this stage (4,6) is the only critical activity which can be crushed along the critical path (1-2-3-4-6-8).

Crash limit of (4,6) = 8 - 3 = 5

We go for test step.

Test step

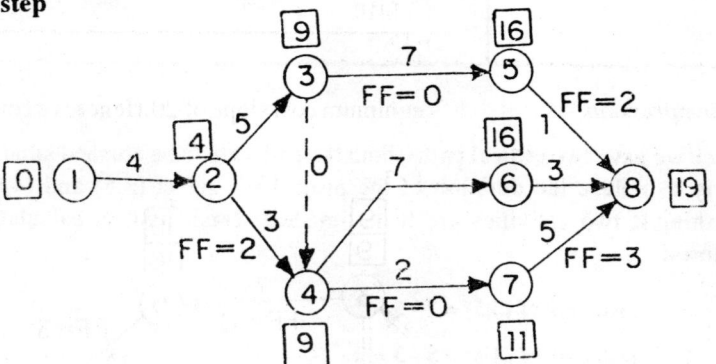

Fig. D.6 (a)

We find that FF of activities (5,8) and (7,8) are changed.

Therefore, FF limit of (4,6) = Min (3,4) = 3

Therefore, compression limit of (4,6) = 3 days

The new network is shown in Fig. D.7

A new critical path (1-2-3-5-8) is developed.

Table D.4 gives the summary for the schedule of 17 days.

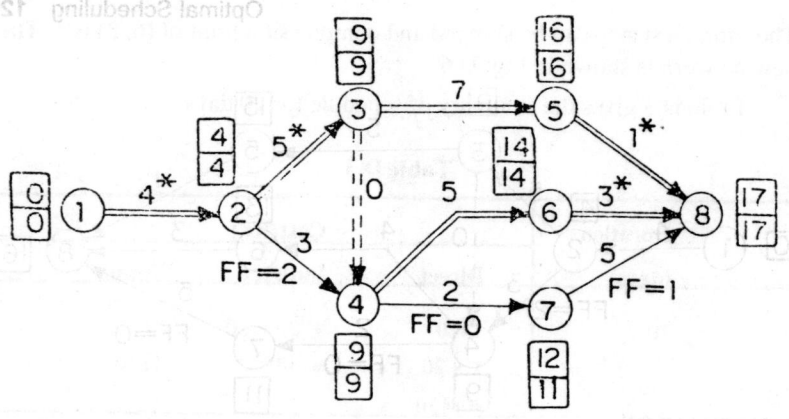

Fig. D.7

Table D.4

Duration	Cost(Rs.)		
(days)	**Direct**	**Indirect**	**Total**
17	1430 +	17 × 20	
	25 × 3	= 340	1845
	= 1505		

V Compression

Since we have two critical paths, both the paths are to be crushed simultaneously to reduce the duration of the project.We select (3,5) and (4,6) for crushing.If two activities are to be crushed, crash limit is calculated as follows:

Crash limit of (3,5) = 7 - 5 = 2

Crash limit of (4,6) = 5 - 3 = 2

Therefore, common crash limit = Min (2,2) = 2

Hence we go for test step.

Test Step

FF limit of (3,5) & (4,6) = 1

Compression limit = Min (1,2) = 1

Crush (3,5) & (4,6) by one unit.

Details are shown in Fig. D.8

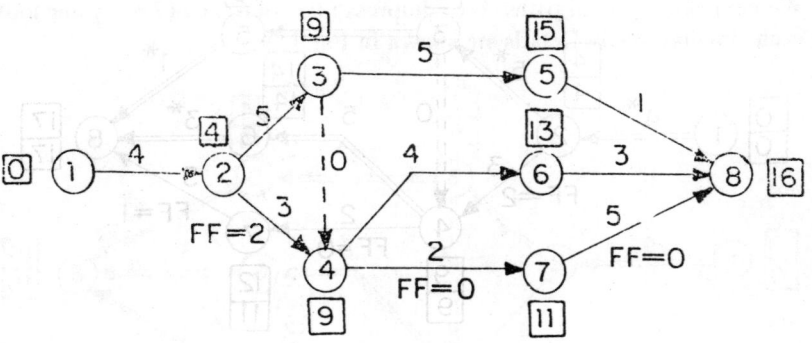

Fig. D.7 (a)

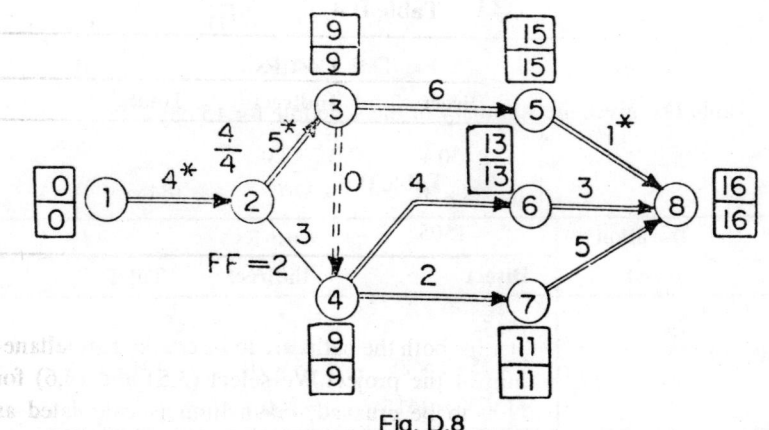

Fig. D.8

Table D.5 gives the summary for the schedule of 16 days.

Table D.5

Duration	Cost (Rs.)		
(days)	Direct	Indirect	Total
16	1505 +	16 × 20	
	1 × 15+	= 320	1865
	1 × 25 +		
	= 1545		

VI Compression

We have three critical paths. We compress (3,5),(4,6) and (7,8) by one unit each simultaneously. Details are shown in Fig. D.9

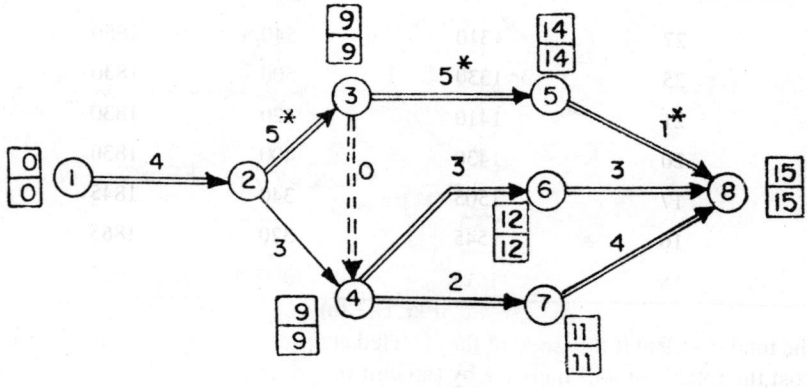

Fig. D.9

Table D.6 gives the summary of the schedule for 15 days.

Table D.6

Duration	Cost(Rs.)		
(days)	Direct	Indirect	Total
15	1545+		
	$1 \times 15 + 1 \times 25$	15×20	1915
	$+1 \times 30 = 1615$	$= 300$	

We find, all the activities along the critical path (1-2-3-5-8) are at their crash duration. Hence, the duration of the project cannot be reduced further. The schedule in Fig. D.9 is called "CRASH SCHEDULE". The minimum duration of the project is 15 days.

The results are summarised in Table D.7 below.

From this table, we see that optimum total cost is Rs. 1830 and it remains constant betwen 20 & 25 days. Therefore, optimum duration is 20 days.

NOTE: The optimum duration and optimum cost of a project can also be found with another method. In this case, we reduce the duration by one unit at each stage. At each stage, the activity with minimum slope is selected and its slope is compared with the indirect cost of the project. If the slope is less than the indirect cost, it implies that unit reduction in project's duration will reduce

Table D.7

Duration (days)	Direct Cost (Rs.)	Indirect Cost (Rs.)	Total Cost (Rs.)
27	1310	540	1850
25	1330	500	1830
21	1410	420	1830
20	1430	400	1830
17	1505	340	1845
16	1545	320	1865
15	1615	300	1915

the total cost.But if the slope of the selected activity is more than the indirect cost,the total cost will increase by the unit reduction in duration.Therefore, optimum duration with optimum total cost will be that resulted from the previous unit reduction.This can be verified with the previous example.In the IV compression,activity (4,6) is selected for compression with a slope of 25,greater than the indirect cost of Rs.20.Hence,the total cost for 17 days has increased to Rs.1845 from Rs.1830 for 20 days.

D. 4 Activity Decompression

Activities previously compressed can often be decompressed (i.e,their durations are extended) with consequent savings in cost,in conjunction with other activities now more heavily compressed, to ensure an overall compression.This combination of decompression and overcompression can lead to the correct combination of activities yielding the cheapest effective cost.This situation is made possible when an expensive activity is decompressed and a relatively inexpensive activity is overcompressed.

For example, consider the network shown in Fig. D.10. The durations are shown with the arrows.

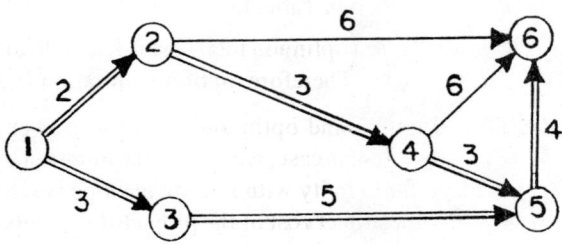

Fig. D.10

The critical paths are (1-2-4-5-6) and (1-3-5-6) and hence the duration of the project is 12 units. The durations of the non-critical activities (2,6) and (4,6) can be extended upto maximum of 10 and 7 units respectively unaffecting the duration of the project. The direct cost of the project reduces and the indirect cost remains same thereby reducing the total cost.

This process of decompressing the non-critical activities without affecting the total duration helps in a smoother resource allocaton(discussed later).

Section E

UPDATING THE NETWORK

As construction proceeds,diversions from the established plan and schedule inevitably occur.Unforeseen job circumstances result in changes of activity durations,activity delays and changes in project logic.As such deviations occur and accumulate, the true job status diverges further and further from that indicated by the programmed plan and schedule.At intervals,therefore,it becomes necessary to incorporate the changes and deviations into the working operational program if it is to continue to provide realistic management guidance.This is accomplished by a procedure called "network updating".The basic objective is to reschedule the work yet to be done using the current project status as a starting point for the redetermination.

The usual causes of time delay on construction work include :

1. Incorrect estimates of activity durations.
2. Unforeseen weather conditions or site hazards.
3. Unpredictable delays in delivery of materials.
4. Strikes or other labour troubles.
5. Unexpected site conditions.
6. Extras or deductions in works quantities.

Information for the periodic reviewing of site operations is collected in activity status reports; the simplest of it is illustrated in Table E.1. Each report should cover every relevant activity in progress, date of start or date of finish, during the review period. This information may be transferred to a copy of the network diagram.

The introduction of CPM into the building industry made many project managers to really know where the project stands at any desired time.To obtain this result,however a formal system of reporting is essential.One such report is shown in Table 1 of previous section.Various systems of progress report is shown in Table E.1. Various systems of progress reporting can be devised, each suited to the problems of the particular construction work.

Table E.1. Activity Status Report

Critical Path Program-Report on Activity Status

Works Section: _____   Date: _____

| Activity | | Started ? | Scheduled | | Expected | Reasons |
No	Descri-ption	Not started ? Finished ?	start date	finish date	or actual finish date	for any delay, if any

Example 1: Consider the network shown in Fig. E.1 which represents a part of a project.

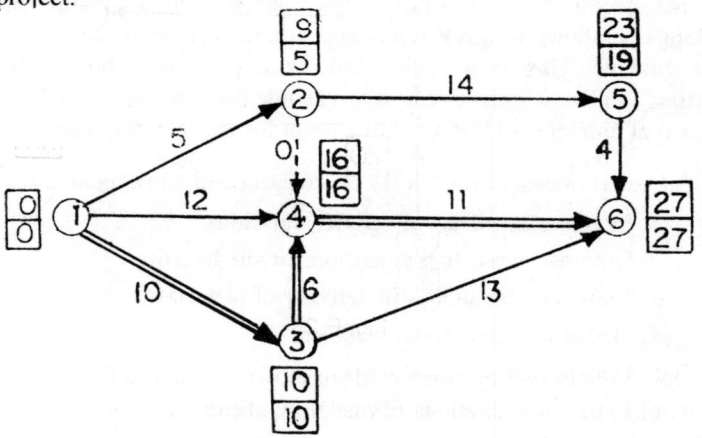

Fig. E.1

The work report for this project at time 20 days is given in the Table E.2.

Table E.2

Activity	Status	Time worked	Time to complete
(1,2)	Finished	8	0
(1,3)	Finished	9	0
(1,4)	Finished	12	0
(2,5)	Working	12	6

(Contd.)

Activity	Status	Time worked	Time to complete
(3,4)	Finished	5	0
(3,6)	Not started	0	13
(4,6)	Working	6	5
(5,6)	Not started	0	4

Fig. E.1 shows the EST and LST for each event and the critical path is (1-3-4-6) and the completion time is 27 days. The stock of the situation is taken at the end of the 20 days and is shown in Table E.2. It can be verified that activities have taken more time or less time than previously scheduled. The corresponding network is shown in Fig. E.2.

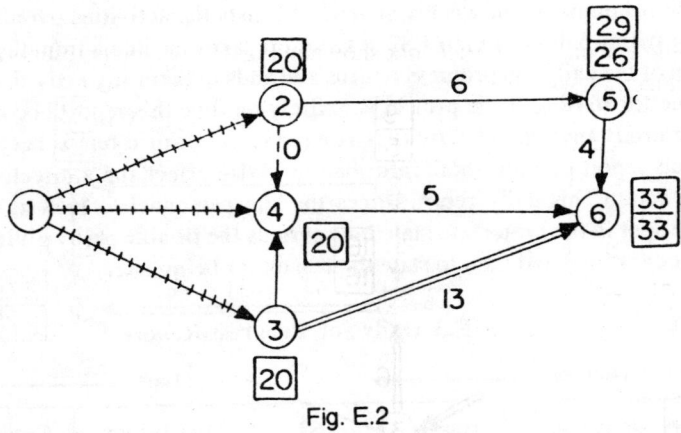

Fig. E.2

The activities that are completed are shown by hatched lines. Nodes 2, 3 & 4 will have 20 as earliest start time. Now, the duration of the project becomes 33 days.

Regular reviewing of site operations may be made at any appropriate time. Usually, weekly or two-weekly periods will be sufficient; but on large fast moving jobs daily review of the major operations is required.However,no definite rule exists about how often a network update should be made.

E. 2 Progress Reports

Table E.3 shows a daily job progress report. It is filled in no later than the evening of the day to which it refers so that it is available to the office the next morning.

This daily report is simple to use and requires very little time to compile.It is divided into three sections.The first shows the activities completed today, the second shows the activities started today whereas the third section lists those operations that are now five days before their LST.The first two sections,which merely list the activities finished or begun,enable keeping upto date records of planning.It is quite obvious that,where the days before LST or LFT is a positive number and there are no critical days lost,this part of the project is well under control and indeed ahead of schedule.The third section of the report has two purposes.First,it serves as a warning that activities are due to start shortly,for as soon as an operation first appears here it has only 5 days to go before its LST will expire.Second, it forces the site personnel to look ahead atleast 5 days, and hence focusses their attention on preparations for starting an activity.

The second report is shown in Table 4 and is a combined daily critical operation summary and weekly statement.It lists the activities overdue for start or within 5 days of their LST.It therefore takes the items from the third section of the daily job progress returns and adds to them any activities now overdue for starting.If the project is entirely on time,this report becomes a "nil"return.If any critical activity is late (or a near critical one is becoming late),this report provides both information to that effect and a directive for taking action.This daily return forces the site personnel to look at every trouble spot in the project, to state how serious the trouble is (by giving the critical days lost),and then to state the reason for being late.

Table E.3. Daily Job Progress Report

Project Day Number ———————— Date ————————

Activity	Description of operation	Total Float	LST LFT	Days before LST/LFT	Critical Days lost
		Operation Completed Today			
		Operations Started Today			
	Operations 5 days before latest start date				

Signature:

Table E.4. Weekly Job Progress Report

Activity	Operation Description	Days before LST	Critical Days lost	Reason for Being behind Schedule	Action taken to Recover lost critical days

E. 3 Cost Records

One of the basic reports for the control of costs on site is the Weekly Cost Record. This gives a complete record of the quantities of work which have been carried out in the week previous to the date of the report, together with the lump sum total costs of the labour and plant which have been incurred in respect of these operations. Each of the figures of quantity, cost and unit cost are prepared on the basis of the work carried out during the week, the total amount of work carried out in previous weeks, the total work carried out to date and an estimated total quantity of the work which has to be carried out in order to complete the operation. Table E.5 shows an example of a typical weekly cost record with two sample operations filled in.

This weekly cost record would not be prepared for every operation being carried out on a site during the week. Only major operations involving large quantities and considerable cost would be chosen, since the preparation of cost records of this nature involve a great deal of time and cost.

At monthly intervals, it is desirable to prepare a cost statement in somewhat more detail than the weekly cost records. In addition to information regarding the cost to date and the value of work carried out, it is highly desirable that estimate of the cost of the work yet to be completed, should be made. This may vary from the original estimate, and should be prepared in the light of information which has been gained on the work so far. A suitable arrangement of a cost report for the monthly interval is shown in Table E.6. Since this is the type of report which is prepared with more accuracy than the intermediate weekly reports, there is an opportunity of inserting more detailed information than on the previous records, discussed.

Table E.5. Weekly Cost Record

XYZ Contracting Ltd   Week Ending............

Code	Operation	Unit	Quantity — This week (1)	Previous total (2)	Total to date (3)	Esti overall total (4)	Cost (Rs.) (1)	Cost (Rs.) (2)	Cost (Rs.) (3)	Cost (Rs.) (4)	Unit Cost (Rs.) Week Ending (1)	(2)	(3)	(4)
.1	Excavation to pumphouse	m³	750	1200	1950	7840	200	350	550	1910	0.26	0.29	0.28	0.25
2	Concreting	m²	56	130	186	780	21	68	89	390	0.38	0.52	0.48	0.5

Table E.6. Monthly Cost Summary

CVP Contracting Ltd.   Contract Snooker-B

Code	Operation	Unit	Quantity — Initial EST	To Date	EST Final Total	Cost — To Complete	Initial EST	Cost To Date	Valuation To Date	Profit (+) Or Loss (–) to Date	EST Final Cost	Final Valuation	Final Profit + Loss

Section F

RESOURCE ALLOCATION

F. 1 Introduction

The completion of a construction project at maximum efficiency of time and cost requires the judicious scheduling and allocaton of available resources.Manpower,equipment and materials are important project resources that require close attention.The supply and availability of these resources can seldom be taken for granted because of seasonal shortages,labour disputes, equipment breakdowns,competing demands,delayed delivaries and many other uncertainities.Nevertheless,if time schedules and cost budgets are to be met, the work must be supplied with the necessary men,equipment and materials when and as they are needed on the job site. This section discusses methods and procedures involved with the management of these resources.The project manager has to identify and schedule the future job needs so that the most efficient employment is made of the resources available.He must establish what resources will be needed, when they must be on site and the quantities required.He should have detailed compilation of resource requirements.If it appears that there will be adequate numbers of resources available to satisfy these projected requirements, the work presumably goes according to established schedule and no adjustment of the job completion date is required.If the resource requirement discloses that demand will exceed the supply, remedial measures to combat inadequate resource supply can include diverting resources from non-critical to critical activities or resorting to some method of expediting the critical activities or decrushing the non-critical activities and thereby reducing the resource demand.If there are conflicts among project activities for the same resource items,rescheduling the non-critical activities will often solve the problem.

F. 2 Allocation of Unlimited Resources

To start with, we assume that the number of resources available are unlimited (which may not be true in most practical cases).The resource requirement of each activity is assumed to be readily available from the original estimate. The resource allocation can be done based on EST of an activity or LST of an activity or in between Resource allocation based on EST assumes that all activities start as early as possible (i.e. at their EST). Resource allocation based on LST assumes all activities start as late as possible (i.e. at their LST). Let us consider the following example for illustration.

Example 1: The table below gives the activities representing a section of a work being undertaken by a subcontractor, their durations and resources required per day. We are interested in making the resource allocation charts.

Activity	Duration (days)	Resources (per day)	Activity	Duration (days)	Resources (per day)
(1,2)	4	2	(3,7)	8	4
(2,3)	6	3	(5,6)	10	2
(2,5)	9	4	(6,7)	4	2
(2,4)	2	4	(4,7)	2	1
(3,4)	3	3			

Solution

The network is shown below. EST and LST of each node are calculated.

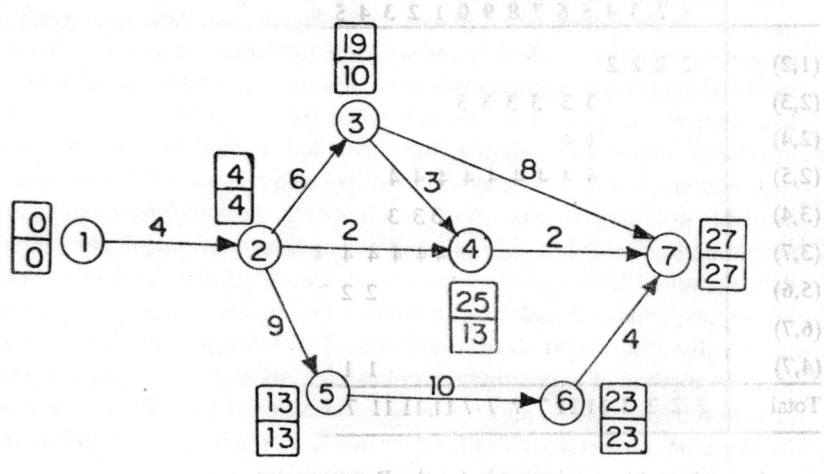

Fig. F.1.

Table below gives the total float, EST and LST of all the activities.

$$\text{EST of } (i,j) = ES_i$$

$$\text{LST of } (i,j) = LC_j - t_{ij}$$

where LC_j is latest completion time of node j

ES_i is earliest starting time of node i

t_{ij} is duration of activity (i,j)

(i,j)	Resources	EST	LST
(1,2)	2	0	0
(2,3)	3	4	13
(2,4)	4	4	23
(2,5)	4	4	4
(3,4)	3	10	22
(3,7)	4	10	19
(4,7)	1	13	25
(5,6)	2	13	13
(6,7)	2	23	23

Table below gives total resources required per day calculated based on EST.

Activity	Days with Resource allocation based on EST																										
	1	2	3	4	5	6	7	8	9	0	1	2	3	4	5	6	7	8	9	0	1	2	3	4	5	6	7
(1,2)	2	2	2	2																							
(2,3)					3	3	3	3	3	3																	
(2,4)					4	4																					
(2,5)					4	4	4	4	4	4	4	4	4														
(3,4)											3	3	3														
(3,7)											4	4	4	4	4	4	4	4	4								
(5,6)														2	2	2	2	2	2	2	2	2					
(6,7)																							2	2	2	2	
(4,7)														1	1												
Total	2	2	2	2	11	11	7	7	7	7	11	11	11	7	7	6	6	6	2	2	2	2	2	2	2	2	2

A similar table can be made for the Resource Allocation based on LST (left to the reader). Figs. F.2 & F.3 give the resource allocation charts.

F.3 Resource Levelling

Comparison of two schedules in Figs. F.2 & F.3 shows that the earliest start schedule for all resources requires a heavy initial rate of investment which

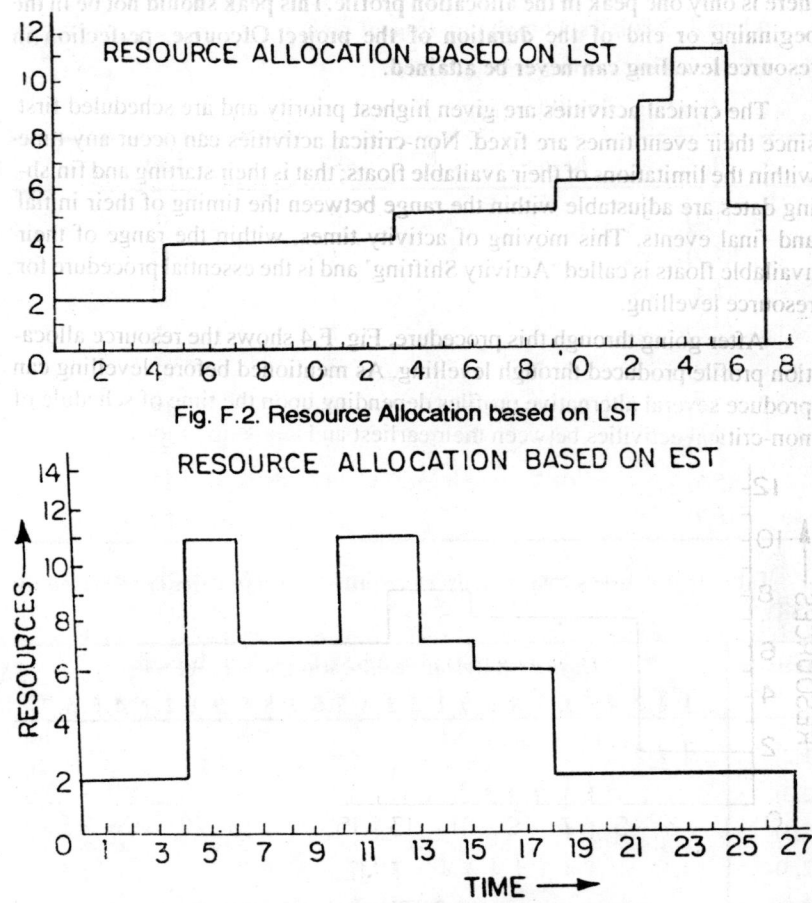

Fig. F.2. Resource Allocation based on LST

Fig. F.3. Resource allocation based on EST

decreases considerably later in the project. The latest start schedule in Fig. F.2 on the other hand, permits a small initial rate of investment but requires a considerable increase later. It is obvious that a compromise between two schedules (earliest & latest start) for the project described in the preceeding section should result in a more constant work force over the whole project duration. The development of such a compromise schedule is known as "RESOURCE LEVELLING".

The variation in the resource demand result in recurring hiring and laying off resources on a short term basis. This is troublesome, inefficient, expensive and scarcely conducive to attracting and keeping top workers. Hence, in resource levelling attempts are made to build up the resource pool step by step to a peak level and the pool is allowed to drop step by step. It is also seen that there is only one peak in the allocation profile. This peak should not be in the beginning or end of the duration of the project. Ofcourse, perfection in resource levelling can never be attained.

The critical activities are given highest priority and are scheduled first since their event times are fixed. Non-critical activities can occur any time within the limitations of their available floats; that is their starting and finishing dates are adjustable within the range between the timing of their initial and final events. This moving of activity times, within the range of their available floats is called 'Activity Shifting' and is the essential procedure for resource levelling.

After going through this procedure, Fig. F.4 shows the resource allocation profile produced through levelling. As mentioned before, levelling can produce several alternative profiles depending upon the time of schedule of non-critical activities between their earliest and latest start times.

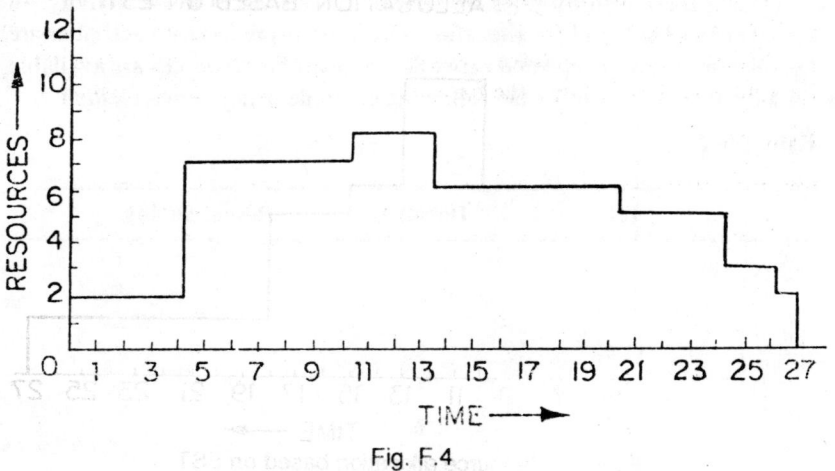

Fig. F.4

F.4 Allocation of Limited Resources

In the previous section, we have discussed earliest and latest start time solutions assuming no restriction on the number of resources available. In this section, we shall discuss the methods to allocate resources that are fixed in number. As the number of resources is fixed and the project has to be completed using this fixed number of resources, we may have to extend the duration of the project.

There are two methods to allocate the resources.

(a) Series method: In this method, the resources are allocated to an activity for its entire duration. This method is applicable under the assumption that the number of resources available is not less than the number of resources required by a single activity.(Otherwise that activity will never be allocated any resources).

There are certain rules to allocate resources to the activities on priority basis.The priority rules that are generally used are as follows:

(1) The activity with least total float is given the top priority.

(2) In case of tie for the least total float,the activity with largest number of days is given the priority.(Resource days of an activity is the product of number of resources required per day and its duration).

(3) If there is a tie for largest number of resource days,activity requiring largest number of resources per unit time is allocated.

(4) If there is a tie,activity with lower (i,j) value is given priority.(If tie between (1,2) and (1,3) then (1,2) is prefered).

Using these priority rules,we allocate the resources to the activity.An activity can be selected for allocation only if all its predecessor activities are finished.We draw the updated network whenever the resources are available for allocation.Let us solve the following example using Series method.

Example 2:

(i,j)	Duration	Resources/day
(1,2)	4	5
(1,3)	2	6
(1,4)	2	4
(2,3)	3	5
(2,4)	2	3
(2,5)	2	4
(3,5)	3	3
(4,5)	3	3

The objective is to allocate 9 resources using Series method.

Solution: Table F.1 (at the end of this example) gives the details of

(a) TF at each stage

(b) Activities and their priorities.

(c) Number of resources allocated at each day.

STEP 1: (Refer Fig. F.5)

(1,2),(1,3) & (1,4) are the activities with TF as 0,5 & 5 and priorities as 1,2 & 3 respectively.(We have used rule 2 in breaking the tie between (1,3)

& (1,4)).Activity (1,2) is allocated 5 resources for its duration of 4 days and we have 4 resources left. As (1, 3) requires 6 resources, we allocate 4 resources to (1, 4) for 2 days. This brings to the end of 2 days.

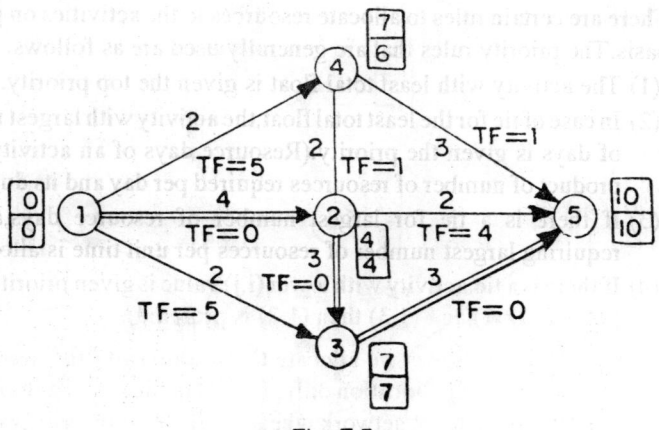

Fig. F.5

Note: On the 3rd day, 4 resources are available.But,we don't have any activity for allocation.

STEP 2: (Refer Fig. F.6)

We draw the updated network with the EST of nodes 1 & 2 after 4 days. The probable candidates for allocation are (1, 3), (2, 3), (2, 4) & (2, 5) with TF 1, 0, 1 and 4 respectively.

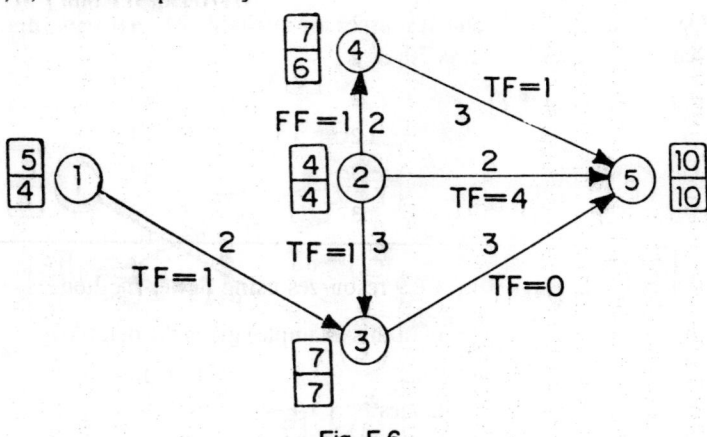

Fig. F.6

(2, 3) is given the top priority and 5 resources are allocated for 3 days. (1, 3) gets the 2nd priority but sufficient number of resources are not available (2, 4) gets the third priority and 3 resources for 2 days are allocated. (2,

5) has the 4th priority but only one resource is left. This brings to the end of 6th day.

In step 3,we draw the updated network with EST of 6.

STEP 3: (Refer Fig. F.7)

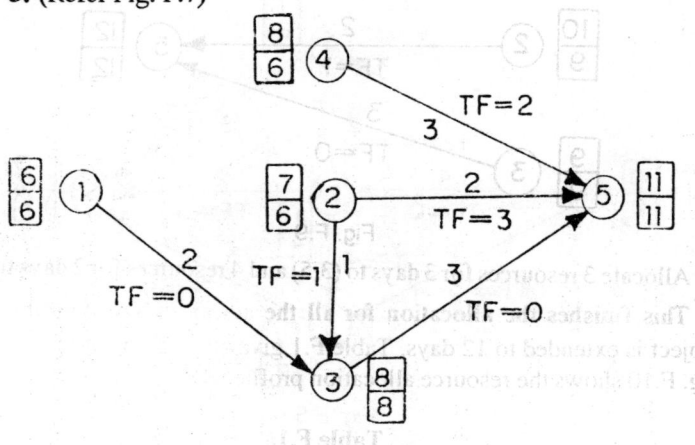

Fig. F.7

The activity (2,3) has only one more day.(1,3),(4,5) and (2,5) are the activities for allocaton.After finding the total floats,we find (1,3) gets 1st priority;(4,5) gets 2nd priority and (2,5) gets 3rd priority.On 7th day,only 4 resources left,therefore (1,3) cannot be allocated.Hence,3 resources are allocated to (4,5) for 3 days.

On 8th day, we find some resources are available.Therefore,we draw the updated network with EST as 7th day.

STEP 4: (Refer Fig. F.8)

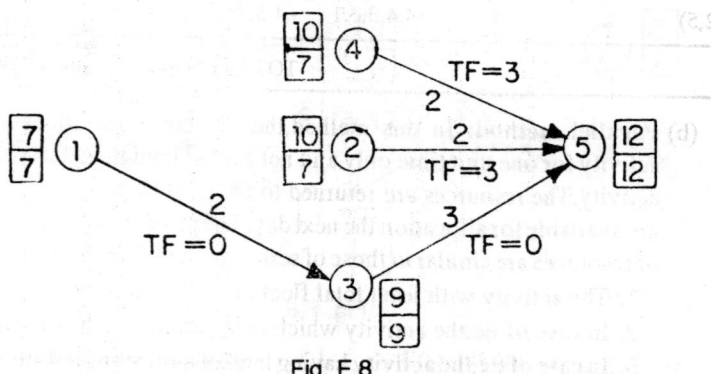

Fig. F.8

Activities for allocation are (1,3),(4,5) and (2,5).Activity (1,3) gets the 1st priority,(4,5) 2nd priority (for largest resource days) and (2,5) 3rd priority. 6 resources are allocated to (1,3) for 2 days.No further allocaton is possible.This brings to the end of 9th day.

STEP 5: (Refer Fig. F.9)

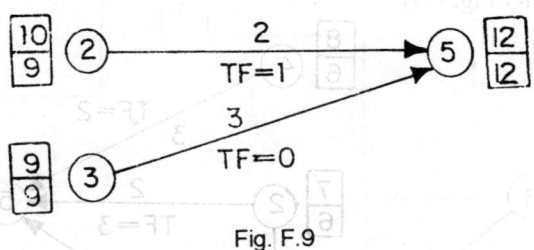

Fig. F.9

Allocate 3 resources for 3 days to (3,5) and 4 resources for 2 days to (2,5).

This finishes the allocation for all the activities.The duration of the project is extended to 12 days. Table F.1 gives the details of the allocation. Fig. F.10 shows the resource allocation profile.

Table F.1.

(i,j)	Duration	Resources	TF	Priority	1 2 3 4 5 6 7 8 9 0 1 2
(1,2)	4	5	0	1	5 5 5 5
(1,3)	2	6	5,1,0,0	2,2,1,1	6 6
(1,4)	2	4	5	3	4 4
(2,3)	3	5	0,0,1	-,1	5 5 5
(2,4)	2	3	1,1	-,3	3 3
(3,5)	3	3	0,0,0,0,0	-,-,-,-,1	3 3 3
(4,5)	3	3	1,1,2,3	-,-,2,2	3 3 3
(2,5)	2	4	4,4,3,3,1	-,4,3,3,2	4 4
				TOTAL	9 9 5 5 8 8 8 9 9 7 7 3

(b) **Parallel method:** In this method,the resources are allocated to an activity for one unit time only and not for the complete duration of the activity.The resources are returned to the pool at the end of a day and are available for allocation the next day.The priority rules for allocation of resources are similar to those of series method.They are listed below:

1. The activity with least total float gets the top priority.
2. In case of tie,the activity which is in progress is given priority.
3. In case of tie,the activity having largest number of resource days is preferred.

(4) If there is a tie, activity requiring largest number of resources per day is preferred.

(5) In case of tie, activity with lower (i,j) value is allocated.

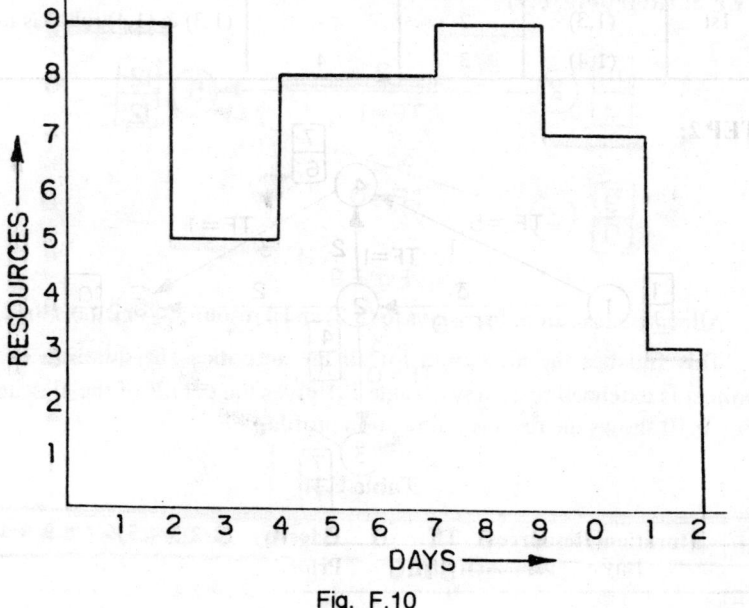

Fig. F.10

Example 2: Let us solve Example 1 using parallel method.

Solution:

STEP 1

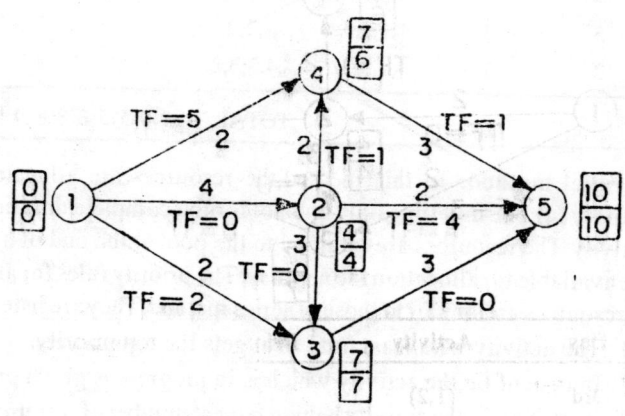

Day	Activity	Priority	Resources	
1st	(1,2)	1	5	For breaking the tie between
	(1,3)	2	-	(1,3) & (1,4),rule 3 is used.
	(1,4)	3	4	

STEP 2:

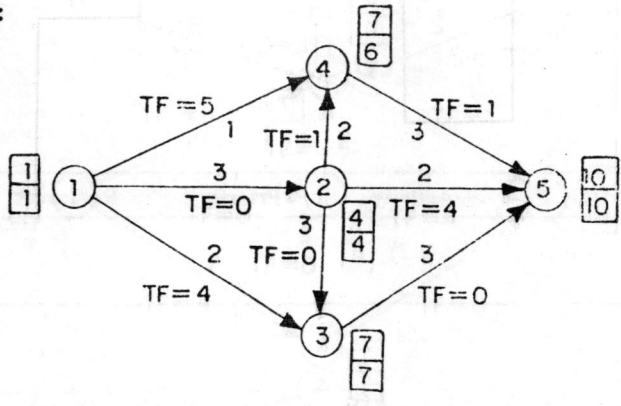

Day	Activity	Priority	Resources
2nd	(1,2)	1	5
	(1,3)	2	-
	(1,4)	3	4

STEP 3

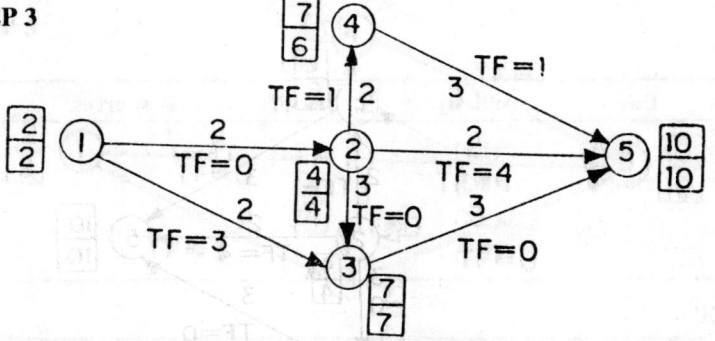

Day	Activity	Priority	Resources
3rd	(1,2)	1	5
	(1,3)	2	-

STEP 4

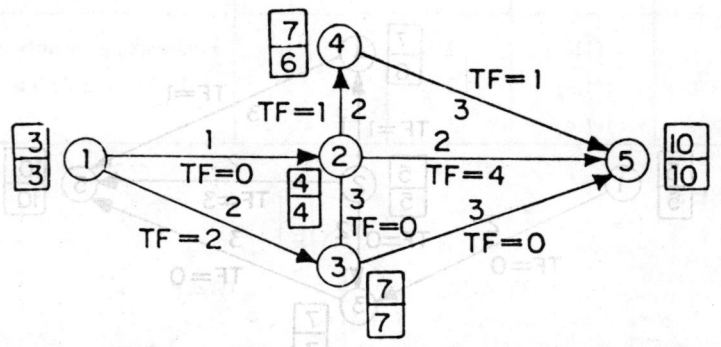

Day	Activity	Priority	Resources
4th	(1,2)	1	5
	(1,3)	2	

STEP 5

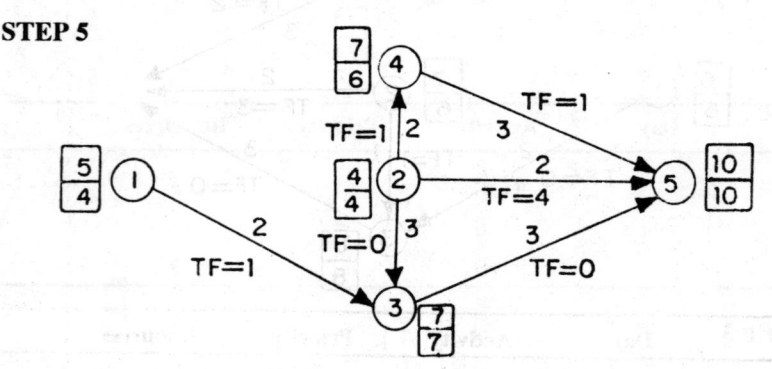

Day	Activity	Priority	Resources
	(1,3)	2	-
5th	(2,3)	1	5
	(2,4)	3	3
	(2,5)	4	-

STEP 6

Day	Activity	Priority	Resources	
	(1,3)	2	-	Priority rule 2 is used
6th	(2,3)	1	5	to break the tie between
	(2,4)	3	4	(1,3) & (2,3).
	(2,5	4	-	

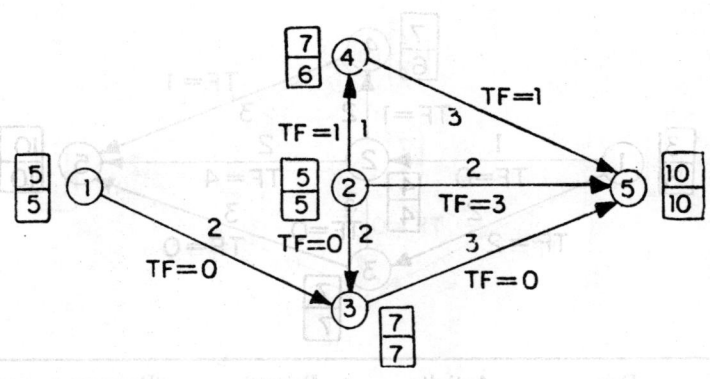

STEP 7

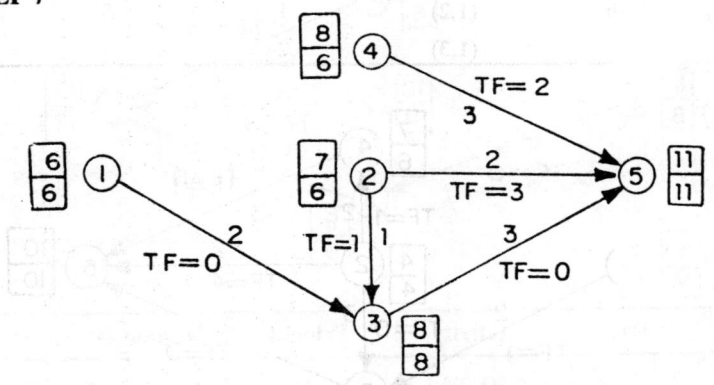

	Day	Activity	Priority	Resources
		(1,3)	1	6
	7th	(2,3)	2	-
		(2,5)	4	-
		(4,5)	3	3

STEP 8

Day	Activity	Priority	Resources	
	(1,3)	2	-	Rule 3 is used to break the tie
8th	(2,3)	1	5	between (1,3) & (2,3).
	(2,5)	4	-	
	(4,5)	3	3	

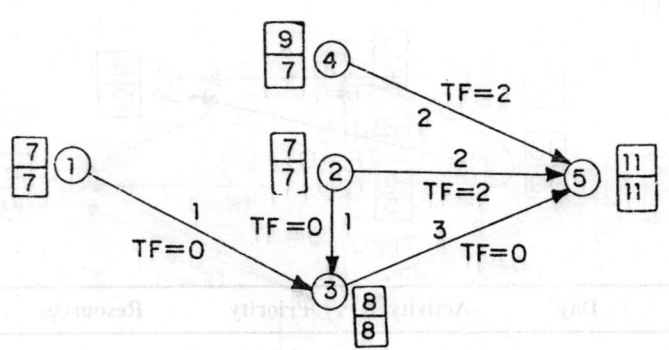

STEP 9

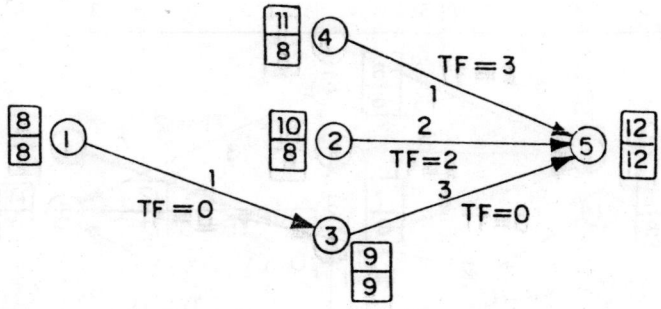

Day	Activity	Priority	Resources
	(1,3)	1	6
9th	(2,5)	2	-
	(4,5)	3	3

STEP 10

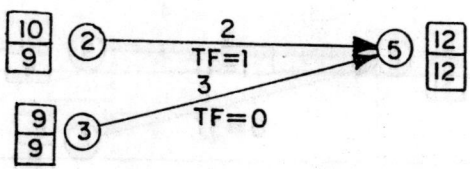

Day	Activity	Priority	Resources
10th	(2,5)	2	4
	(3,5)	1	3

STEP 11

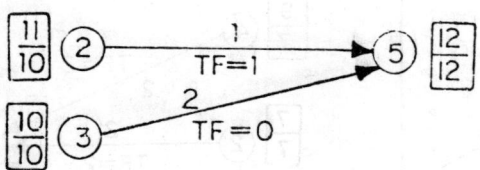

Day	Activity	Priority	Resources
11th	(2,5)	2	4
	(3,5)	1	3

STEP 12

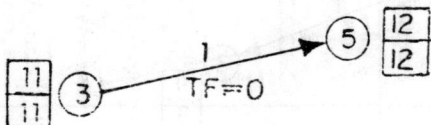

Table F.2 gives the details of the steps in parallel method. Fig. F.11 gives the resource allocation profile. We can see that the resource allocation profile in parallel method is slightly different from the one of series method.

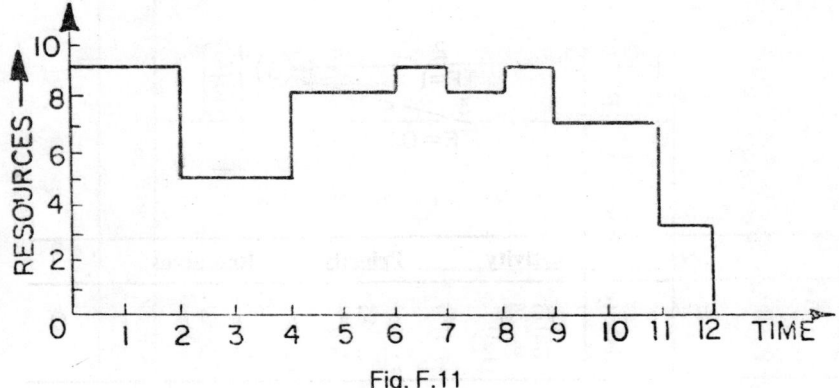

Fig. F.11

Table F.2

Activity	Duration	Resources	Total Float	Priority	Days											
					1	2	3	4	5	6	7	8	9	10	11	12
(1,2)	4	5	0,0,0,0	1,1,1,1	5	5	5	5								
(1,3)	2	6	5,4,3,2,1,0,0,0,0	2,2,2,2,2,2,1,2,1						6	6					
(1,4)	2	4	5,5	3,3					5	5						
(2,3)	3	5	0,0,0,0,0,1,0	-,-,-,-,1,1,2,1					5	5	5					
(2,4)	2	3	1,1,1,1,1	-,-,-,-,3,3						3	3					
(2,5)	2	4	4,4,4,4,3,3,2,2,1	-,-,-,-,4,4,4,4,2,2									4	4		
(3,5)	3	3	0,0,0,0,0,0,0,0,0,0	-,-,-,-,-,-,1,1,1							3	3	3			
(4,5)	3	3	1,1,1,1,1,1,2,2,3	-,-,-,-,-,3,3,3										3	3	3
Total					9	9	5	5	8	8	9	9	8	8	7	7

MULTI-RESOURCE ALLOCATION

Last section we discussed about the resource levelling considering the single resource. In most of the civil engg. projects, we require more than one resource. The resources could be man power like engineers, masons etc. or it also could be the different equipments. Because of a large no. of resources required for civil engg projects resource levelling is very important and can be used effectively not only to save the money but also to complete the project in time. To understand the procedure let us take an example. (All the rules of allocating resources are same as we have for single resource). The activity can be executed only when all the resource are available for the activity.

Example: Let us consider the finish work of a building. After completing the brickwork & cementing, the following activities are required

			Resources		
		Duration	Beldars	Carpenters	Plumbers
(A)	Door, window shutter fixing	3	2	2	0
(B)	Glazing	5	2	0	0
(C)	White washing	3	4	0	0
(D)	Colour washing and painting	3	2	0	0
(E)	Sanitry fitting	3	4	0	2
(F)	Fitting of door and windows	2	2	2	0
(G)	Site cleaning	2	6	0	0

The network to the example is given in Fig. F.12.

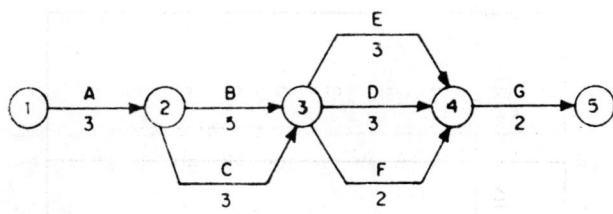

Fig. F.12

Step 1: For first day, we have only one activity so we allocate all the resources to A i.e. 2 belders and 2 carpenters for 3 days.

Step 2: After 3 days the activities B and C are to be started. Activity B is more critical so we allocate 2 belders for activity B. Activity C can be started immediately or can be delayed by 2 days.

Step 3: After 8 days we have to start activities E.D and F. Activities D and E should be started immediately but F can be delayed by one day.

Step 4: After 11th day we start activity G and complete the project in 13 days.

See the table F.3 for the resource allocation.

Series Method
With unlimited resources
A. Earliest start time

Table F.3

Belders = b, Carpenters = c, Plumbers = p

Activity	Duration	EST/ LST	Allocation												
			1	2	3	4	5	6	7	8	9	10	11	12	13
			2b	2b	2b										
A	3	%	2c	2c	2c										
B	5	3/3				2b	2b	2b	2b	2b					
C	3	3/5				4b	4b	4b							
D	3	8/8									2b	2b	2b		
E	3	8/8									4b	4b	4b		
											2p	2p	2p		
F	2	8/9									2b	2b			
											2c	2c			
G	2	11/11												6b	6b
Total			2b	2b	2b	6b	6b	6b	2b	2b	8b	8b	6b	6b	6b
											2p	2p	2p		
			2c	2c	2c						2c	2cs			

B. With latest start time

Activity	Duration	EST	Allocation												
			1	2	3	4	5	6	7	8	9	10	11	12	13
A	3	%	2b 2b 2b 2c 2c 2c												
B	5	3/3				2b	2b	2b	2b	2b					
C	3	3/5				4b	4b	4b							
D	3	8/8									2b	2b	2b		
E	3	8/8									4b 4b 4b 2p 2p 2p				
F	2	8/9									2b 2b 2c 2c				
G	2	11/11												6b	6b
Total			2b 2c	2b 2c	2b 2c	6b	6b	6b	2b	2b	8b 2p 2c	8b 2p 2c	6b 2p	6b	6b

If we draw resource levelling curve we find that B is smoother than A.

Series method (with limited resources) :-

The no of resources require for whole project are

Belders $= 3X2 + 2X5 + 4X3 + 2X3 + 2X2 + 6X2$
$= 6 + 10 + 12 + 6 + 12 + 4 + 12$
$= 62$

Carpenters $= 2X3 + 2X2 = 10$
Plumbers $= 3X2 = 6$

Resources/day required = 62/13 + $\begin{cases} 0 \text{ if resource/day is an integer} \\ 1 \text{ if resources/day is not integer} \end{cases}$

(belders) = 4.7 + 1 = 5

(carpenters) = 10/13 + $\begin{cases} 0 \\ 1 \end{cases}$ = 1

(plumber) = 6/13 + $\begin{cases} 0 \\ 1 \end{cases}$ = 1

So no of resources available = 6,2,2

(Resources/day required can not be taken less than the resources required for an activities)

Step 1: For activity A, allocate belders and carpenters for 3 days

Step 2: Fourth day activities B and C have to be started. We need 6 belders for B & C. So we can start both activities. Activity C can be started 2 days late also

Step 3: 9th day. Activities D,E & F can be started. We have only 6 belders and required no is 8. So we start activities D & E which are more critical.

Step 4: On 12th day we start activity F

Step 5: On 14th day we start activity G. Total duration of project is = 15 days

Table F.4

Activity	Duration	EST/ LST	Allocation 1 2 3 4 5 6 7 8 9 10 11 12 13
A	3	%	2b 2b 2b 2c 2c 2c (days 1–3)
B	5	3/3	2b 2b 2b 2b 2b (days 4–8)
C	3	3/5	4b 4b 4b (days 6–8)
D	3	8/8	2b 2b 2b (days 9–11)
E	3	8/8	4b 4b 4b (days 9–11) 2p 2p 2p (days 9–11)
F	2	8/9	2b 2b (days 12–13) 2c 2c (days 12–13)
G	2	11/11	6b 6b
Total			2b 2b 2b 2b 2b 6b 6b 6b 6b 6b 6b 2b 2b 6b 6b 2c 2c 2c 2p 2p 2p 2c 2c

With earliest start time, readers can make the allocation table. Only change will be the activity C can be started on the 4th day itself.

With limited resources project duration is increased by 2 days. Cost wise nothing can be said until direct and indirect costs are calculated.

PARALLEL METHOD

In this method, as mentioned earlier the resources are allocated after each day (unit time). The resources are allocated to most critical activity each day.

(a) With unlimited resources: If the no. of resource is not limited, there will not be any difference in allocation table for this example because without seeing the criticality of the activity we allocate the resources each day. The only difference while allocating will be the activities C and F will be allocated the resources when they become the critical activities.

(b) With limited resources (6 belders, 2 carpenters, 2 plumbers)

Step 1:   First day, only candidate activity is A. Allocate required no. of resources.

Second day → Only activity is A, Allocate.

Third day → Only activity is A, Allocate.

Step 2:   Fourth day → Activities B and C can be started. Activity B is critical but not C. So allocate 2b to activity B.

Fifth day → Activity B and C both are critical. So allocate 2b to B and 4b to C.

Seventh and Eighth day also, allocate 2b to B and 4b to C.

Ninth day → Activities D, E and F can be started. Activities D and E are critical but not F, so allocate the resource to D & E.

10th day → Activities D, E and F are critical. We need 8 belders but we have only 6. So allocate to those activities which are in progress so resources are allocated to D & E.

11th day → Activity F become more critical than D and E so the top priority will go to F. Alocate 2b to F. Then allocate 4b to activity E. In this case both activities D and E are equally critical, so we can allocate the resources to any activity. We try to allocate the resources to the activity which require more resources and the resources are available

12th day → Activity E is over. The resources can be allocated to D and F which are equally critical.

13 and 14th day → Only activity left is G. so we allocate the resources to G.

Column 3 of table F.5 shows the float for each activities for example. For example activity F, which can be started on 9th day, float is 1 on 9th day. On 10th day float becomes 0, on 11th day it becomes 1 (very critical) and on 12th day it becomes 0.

Table 5

Activity	Duration	Float	Allocation														
			1	2	3	4	5	6	7	8	9	10	11	12	13	14	15
A	3	0	2b 2c	2b 2c	2b 2c												
B	5	0				2b	2b	2b	2b	2b							
C	3	2,1,0						4b	4b	4b							
D	3	0,0,0,-1									2b	2b		2b			
E	3	0,0,0									4b 2p	4b 2p	4b 2p				
F	2	1,0,-1,0											2b 2c	2b 2c			
G	2	0													6b	6b	
Total			2b 2c	2b 2c	2b 2c	2b	2b	6b	6b	6b	6b 2p	6b 2p	6b 2p 2c	4b 2c	6b	6b	

Section G

CASH AS A RESOURCE

Of all the resources that are required for carrying out a project, money or cash is the most vital resource, which is required for harnessing all the other resources. So the availability of cash, at different stages of execution of the project can have a considerable bearing on the final project schedule.

The availability of cash, like the availability of labour, equipment or space can be a constraint upon the achievement of a project. There is however, one factor, which in general sets it apart from labour - namely that of time. Consider the following statement : "The labour resource ceiling for the duration of the projet (or for some duration of the project) is 10 men. This implies that 10 men are available throughout the specified period ; they can be used first on one activity and later on others. Cash, however is exhausted once it is used; "Rs.2 lakh is available "means, that in fact, that once Rs. 2 lakh is used up, it is no longer available for further allocation. Furthermore, it need not be present until it is required; it may be in a bank or some other store, available at call but not generating cost. The Rs. 2 lack may be on deposit in a bank generating interest, only generating cost, when it is actually used. The difference between these two types of resources is recognised as "NON-STORABLE" and 'STORABLE' resources, labour being typically non-storable, cash being typically storable.

The storable and exhaustible nature of cash means that a histogram is not the appropriate measure and display of cash. Consider two activities, A and B, where B follows A and A involves an outlay of Rs. 6000 while B involves an outlay of Rs. 4000. A histogram such as Fig. G.1 clearly does not represent the situation. By time t assuming that the cash is engaged at the start of each activity, Rs. (6000 + 4000) = 10000 have been committed, and a more useful diagram could be given (Fig. G.2). If the cash is committed in some other way; for example linearly with time, Fig. G.2 would be different, but the accumulative nature of the process would remain same.

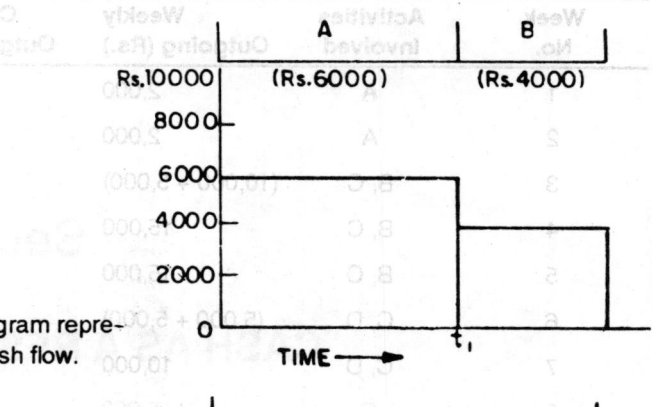

Fig. G.1. Histogram representation of cash flow.

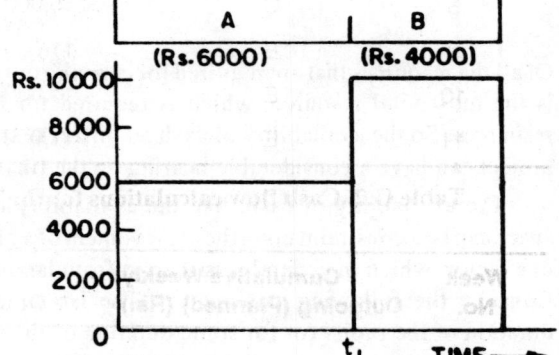

Fig. G.2. Cash flow diagram.

G.1 Cash Flow Diagram

The direct costs involved with a network are assumed to be as shown in Table G.1, G.2 shows the cash flow calculations. Table G.3 is a cash flow diagram showing how the accumulated cash flows out of the organisation.

Activity	Duration (weeks)	E.S.T.	E.F.T.	Cost (Rs.)	Effective Weekly cost
A	2	0	2	4,000	2,000
B	3	2	5	30,000	10,000
C	6	2	8	30,000	5,000
D	2	5	7	10,000	5,000
E	3	8	11	33,000	11,000

Table G.1. Direct costs involved with the network diagram for G.3. casting of foundation of a building

Week No.	Activities Involved	Weekly Outgoing (Rs.)	Cumulative Outgoing (Rs.)
1	A	2,000	2,000
2	A	2,000	4,000
3	B, C	(10,000 + 5,000)	19,000
4	B, C	15,000	34,000
5	B, C	15,000	49,000
6	C, D	(5,000 + 5,000)	59,000
7	C, D	10,000	69,000
8	C	5,000	74,000
9	E	11,000	85,000
10	E	11,000	96,000
11	E	11,000	1,07,000

Table G.2. Cash flow calculations for the network

Week No.	Cumulative Weekly Outgoing (Planned) (Rs.)	Cumulative Weekly Outgoing (Actual) (Rs.)*
1	2,000	3,000
2	4,000	6,000
3	19,000	25,000
4	34,000	45,000
5	49,000	60,000
6	59,000	75,000
7	69,000	85,000
8	74,000	95,000
9	85,000	1,10,000
10	96,000	1,30,000
11	1,07,000	1,50,000

* Slightly higher spend rate than planned is assumed.

Cash flow diagram (tabulated)

The sum Rs. 107000 is the total expenditure on the project: it is not a particularly useful figure for control purposes since it is only reached at the conclusion of the project, and deviations revealed at this time are of historical interest only : no correcting action can be taken. A more useful measure is the outflow of cash with time, allowing comparision of the actual outflow with a planned outflow, assuming that the flow of cash is constant with time across each activity (i.e. the 'spend rate' is constant). Refer G.3b from the data the each activity (i.e. the 'spend rate' is constant). From Table G.1, G.2 and G.3, the cumulative cash flow can be calculated, which is shown in Fig. 4.3

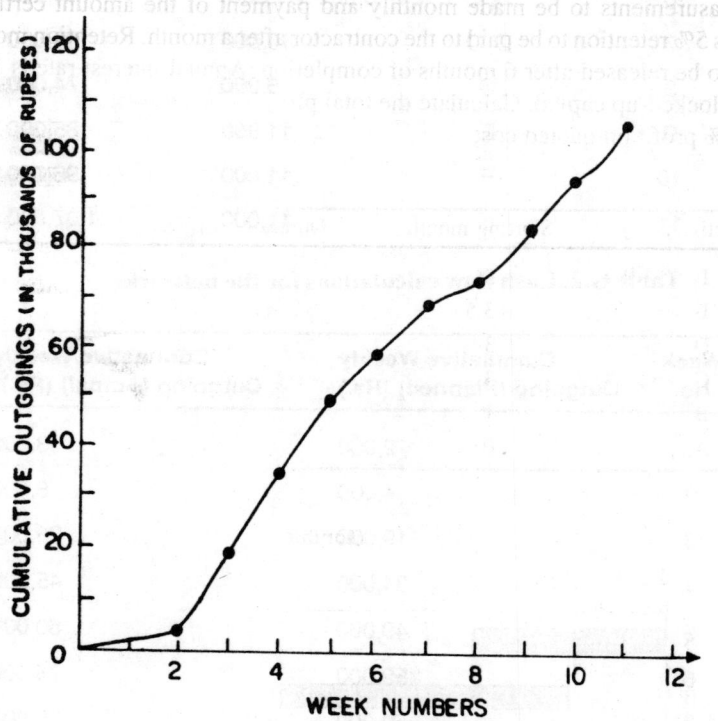

Fig. G.3. Cumulative cash flow diagram

G.2 Cost Control

The cash flow calculations above show the planned situation. To be able to exert control, it is necessary to compare the actual situation with the planed one. It is not "sufficient " however, to record that at week 10 the actual spending were 10,00,00, the planned outgoings being Rs. 87000. This apparent over_expenditure have come about by weekly costs being in excess of planned costs - for example, the actual costs of activity C could have been Rs. 8000 a week instead of Rs. 5000 a week - or more work could have been done than anticipated. To resolve the above difficulty, three figures are required : (1) Planned Expenditure, (2) Actual Expenditure and (3) Value of work completed. The cash flow diagram, the 'value of work completed (VWC) figure is obtained from the initial budgeted figures, i.e. for example, activity A, budgeted at Rs. 1000 a week for 3 weeks have a VWC of $1000 \times 3 = 3000$.

Example: Table shows a contractor's project's activities along with starting and finishing time and the budgeted cost. The conditions of contract allow measurements to be made monthly and payment of the amount certified less 5% retention to be paid to the contractor after a month. Retention money is to be released after 6 months of completion. Annual interest rate is 15% on locked-up capital. Calculate the total profit of the contractor if he earns 10% profit on quoted cost.

Activity	Starting month	Duration (month)	Cost/month
F	5	1	200000
E	3.5	1.5	40000
D	3	1	150000
C	3	1.5	120000
B	1	3	40000
A	0	2	45000

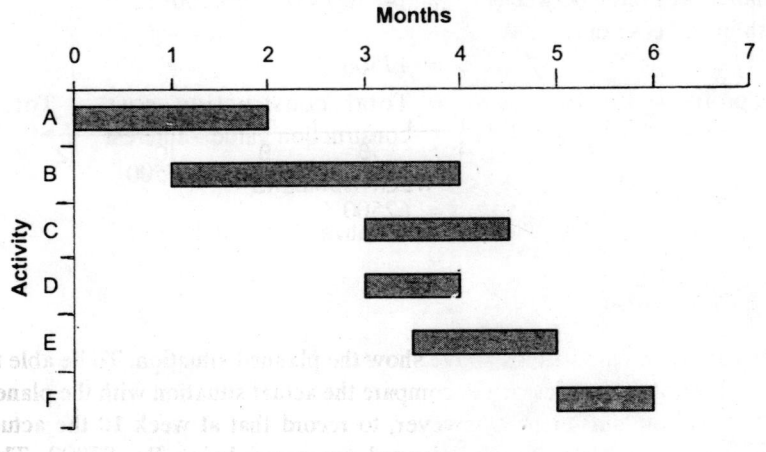

Monthly value	45000	85000	40000	330000	100000	200000	
Cumulative value	45000	130000	170000	500000	600000	800000	
Cumulative profit	4500	13000	17000	50000	60000	80000	
Cumulative cost	40500	117000	153000	450000	540000	720000	
Cumulative value less retention	42750	123500	161500	475000	570000	760000	
Cumulative money received after end of the month	0	42750	123500	161500	475000	570000	760000

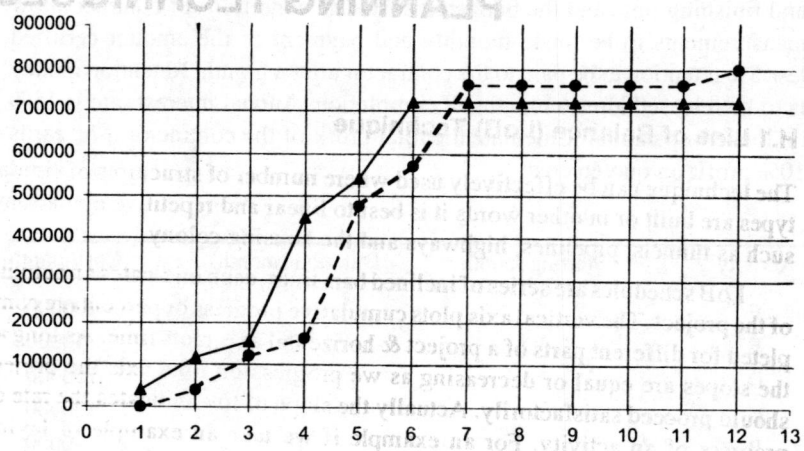

800000 after six months

Calculation of Interest and net profit

Assume the interest rate is 15%.

Finance area (area between cash in and cash out)	= 14*100000*1*15/100*12
	= 17500
Net profit	= Total construction cost – Total construction value – Interest
	= 800000 – 720000 – 17500
	= 62500

Section H

OTHER CONSTRUCTION PLANNING TECHNIQUES

H.1 Line of Balance (LoB) Technique

The technique can be effectively used where number of structures of similar types are built or in other words it is best to linear and repetitive operations, such as tunnels, pipelines, highways and the housing colony.

LoB schedules are series of inclined bars lines, each inidicates an activity of the project. The vertical axis plots cumulative progress or percentage completed for different parts of a project & horizontal axis plots time. As long as the slopes are equal or decreasing as we progress on time axis the project should proceed satisfactorily. Actually the slope of line indicates the rate of progress of an activity. For an example if we take an example of laying pipeline, the different activities can be shown by line of balance chart. (Fig. H.1)

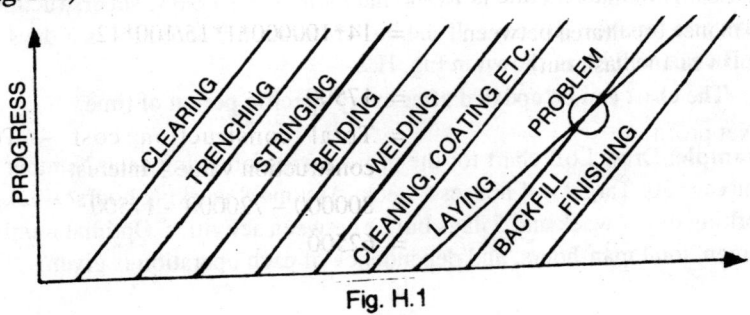

Fig. H.1

There are four major phases of LoB Techniques

    1. **Objective** : Objective of the project in terms of
        (a) number of units/time period
        (b) Scheduled completion time/date
    2. **Program** → determination of limiting and/or principal steps.
    3. **Program** → Progress

(i) It is an inventory of the stock or progress of the events for all limiting and/or principal steps identified in phase two. The progress can be depicted by a bar-chart (discussed in previous section.)

4. Comparison of project progress with the objective.

The LoB technique is fairly simple to understand. With the help of this technique resource allocation could be optimum & it takes the fluctuations of produtivity into considerations.

To draw the lines of balance for any project for the repeating units, the following points should be taken into considerations.

(a) Prepared gang size

(b) Realisable hand over rate

(c) Requisite working duration for each activity.

For example: Assume that the standard output for one mason with 2 labourers is $2m^3$ brickwork in 8 hours work (1 day). Let the brickwork for one house is required 103.2 m^3. So house requires 52 masons and 104 labourers, giving 1248 hours in a working day. If 25 masons and 75 labourers are recruited continuously in a working week, the realisable handover rate, to construct 7 houses, will be 2.77 houses per week, if 24 masons and 48 labourers are employed continuously in a working week. Each house requires, 4.34 working days (or 0.723 working week). The line of balance for 10 sets will span over $9 \times 0.723 = 6.507$ and for 40 sets it will span over $39 \times 0.723 = 28.2$ working weeks. The Lob can be constructed. Let us take an example of constructing a housing colony where you have to construct 7 houses. Assume for each house foundation take 10 days, superstructure 5 days and finishes 101 days. If the buffer between 2 activities is 5 days the LoB chart will be as shown in Fig. H.2.

The chart can be updated after every specific period of time.

Example: Draw LoB chart for the project with following data: Number of houses = 30; Target = 4 houses / week. Assume 8 working hours/ week, 5 working days / week and 5 days buffer between activities. Optimal number of men, total man-hours, and dependency of each operation is given:

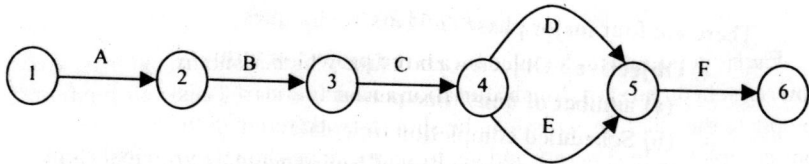

Sl. No.	Operation	Man-hours	Optimal number of Men per operation
1.	A	120	3
2.	B	290	6
3.	C	250	4
4.	D	40	3
5.	E	30	2
6.	F	220	5

H. 2 Matrix Schedules

The method is suitable for the projects where the operations are of repeatitive nature. The technique is quite useful for high rise buildings with successive floors repeating the same plan. It does not mean that it can not be used where the floors have different plan. To understand the technique refer Fig. H.3a The horizontal rows on the schedule correspond to floors within the building, starting with basement levels at the bottom and moving upto the top floor, The vertical columns correspond to the operations to be performed on each floor. They are listed from left-to-right and list the opeations in chronological order. The operations are given at the top of the columns.

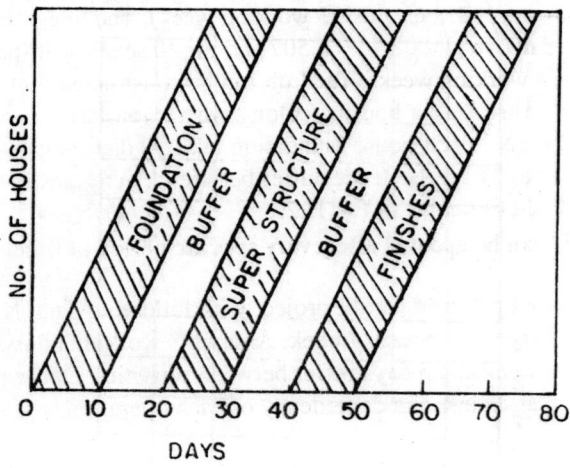

Fig. H.2.

Each operation is scheduled by a box like which is subdivided in 6 parts showing scheduled start, finish, dur-ation and actual start, finish and duration. To update the chart, progress can be shown by different colours (the operations completed, in progress and yet to start can be shown by three different colours).

The technique is easy to understand and need no explanation. It is easy and fast to read the chart. The logical interrelationships among operations are very obivious. With proper planning the work can be given to many subcontractors and it can be shown by the technique effectively. It is compact, one page schedule.

H. 3 GRAPHICAL EVALUATION AND REVIEW TECHNIQUE

H.3.1 Introduction

There are situations where CPM and PERT networks and the network concepts developed so far cannot faithfully represent the actual interrelationsips. In both CPM and PERT networks, an activity can start only after the previous activities, on which it depends, have been completed. This is known as finish-start (FS) relationship. However in situations where different relationships are encountered, for example : The start of a work may depend upon start (and not finish) of another activity. It is not possible to represent such

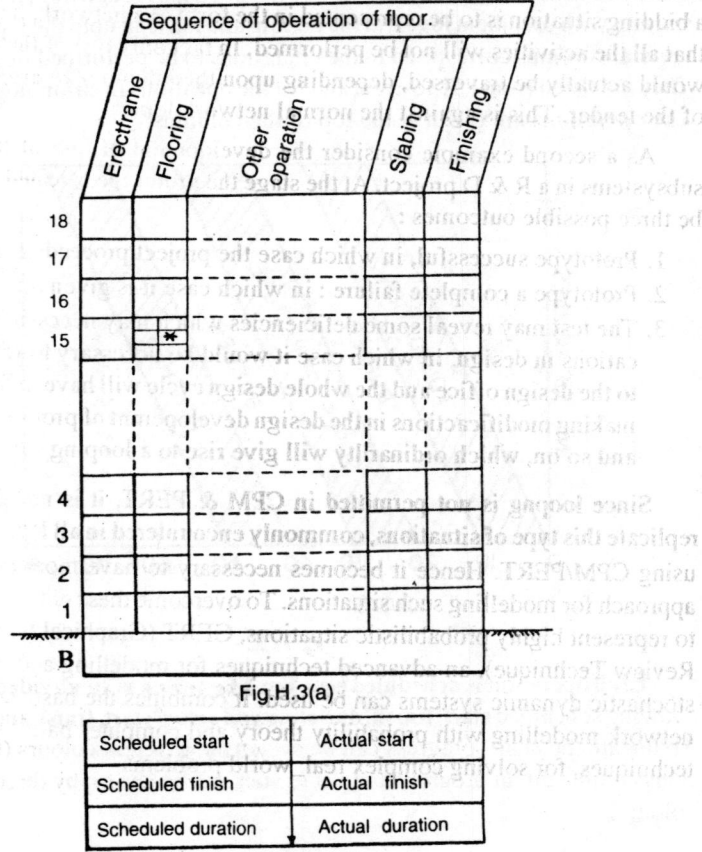

Fig.H.3(a)

Scheduled start	Actual start
Scheduled finish	Actual finish
Scheduled duration	Actual duration

Fig.H.3(b)

Start-to Start (SS), finish-to-finish (FF) relationships with the network logic discussed in previous sections. CPM & PERT networks are deterministic in nature i.e. there is definiteness about the various components jobs/event taking place. PERT networks deals with uncertainty; but the uncertainty relates to the actual performance times. There is no uncertainty about the actual occurance of the activities or events.

In situations, therefore, where there is uncertainty about the occurance of some or all the activities or events in the network, viz, development of new products (R&D), bidding situation etc., ordinary network logic cannot be used.

As an example consider a Civil Engineering bidding situation. Let a contracting company bid for construction work. There would be uncertainty about who would win the contract. There are two possibilities: either the company bags the contract or loses it to one of the competitors. In the former case, a large number of activities will have to take place up to job completion. If the contract is lost, nothing further will be required in this regard. If such a bidding situation is to be represented in the form of a network, it is obvious that all the activities will not be performed. In fact only one of the two paths would actually be traversed, depending upon the outcome (award/rejection) of the tender. This is against the normal network logic.

As a second example consider the development of one of the crucial subsystems in a R & D project. At the stage the prototype is tested, there can be three possible outcomes :

1. Prototype successful, in which case the project proceeds further.
2. Prototype a complete failure : in which case it is given up.
3. The test may reveal some deficiencies which may necessitate modifications in design, in which case it would be necessary to refer it back to the design office and the whole design cycle will have to be repeated making modificactions in the design developemnt of prototype, testing and so on, which ordinarilty will give rise to a looping situations :

Since loopng is not permitted in CPM & PERT, it is not possible to replicate this type of situations, commonly encountered in all R & D projects using CPM/PERT. Hence it becomes necessary to have more generalised approach for modelling such situations. To overcome these deficiencies, and to represent highly probabilistic situations, GERT (Graphical Evaluation & Review Technique), an advanced techniques for modelling and analysis of stochastic dynamic systems can be used. It combines the basic concepts of network modelling with probability theory and computer based simulation techniques, for solving complex real world problems.

H.3.2 Basic Concepts of GERT

GERT networks, like CPM/PERT use arrows and nodes. GERT networks are more probabilistic in nature. It is not necessary in such networks, for all activities or nodes to get completed. It is possible that some of the paths may not get traversed at all. The components of GERT networks are directed branches and logic nodes. A directed branch has two nodes: one from which it emanates and the other in which it terminates. A node may be realized with the completion of one or more of the incoming branches.

Logical nodes: A node in a stochastic network consists of inouts (receiving, contributive) side and an output (emitting, distributive) side. There are in general three logical relations on the output side. They are :

1. Input side :

Name	Characteristic
EXCLUSIVE_OR	The realization of any branch leading into the node causes the node to be realized; however, one and only one of the branches leading into this node can be realized at a given time.
INCLUSIVE_OR	The realization of any branch leading into the node (inclusive_or) causes the node to be realised. The time of realization is the smallest of the competing times of the activities leading into the INCLUSIVE_OR node.
AND	The node will be realized only if all the branches leading into the node are realized. The time of realization is the largest of the completion time of the activities leading into the AND node.

2. Output side:

Name	Characteristic
1. DETERMINISTIC	All branches emanating from the node are taken if the node is realized, i.e., all branches emanating from this node have a definite parameter.
2. PROBABILISTIC	Exactly one branch emanting from the node is taken if the node is realized.

For notational convenience, the input and the output symbols and their combinations are shown in Fig. H.4. GERT nodes can be realized more than once. Nodes can also be characterized by their function in the network. A GERT analyst can specify a node as : (1) Source Node, (2) Sink Node, (3) A Statistics Node, (4) A Mark Node. Activities emanating from a source node are started at time zero. A sink node is a node that indicates that the network may be realized when it is realized. Sink nodes have input sides only. There may be multiple sink nodes in a GERT network. A statistics node is one on which statistics are maintained. All sink nodes are automatically made statistical nodes. A mark node establishes a reference time and permits the calculation of the time it takes to go between two nodes of the network.

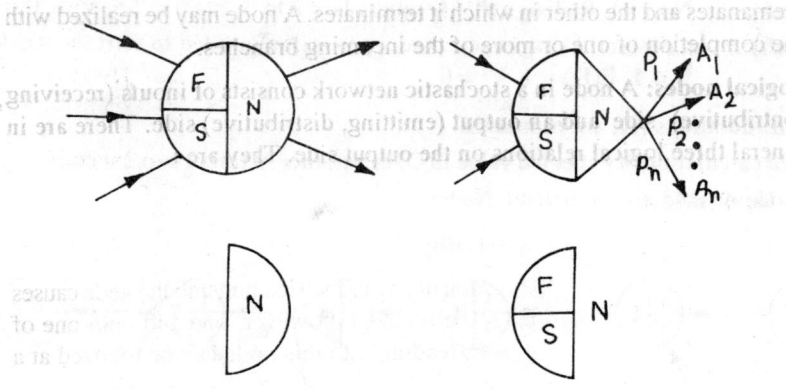

Fig. H.4. Nodes

H.3.3 Advantages of GERT

1. GERT allows the planner to choose from any of the 11 types of distributions, viz, constant, normal, uniform, beta, gamma, poisson, triangular, log etc.
2. GERT permits multiple sink nodes
3. Looping is allowed, Hence multiple realization of nodes is possible.
4. Activities can get repeated.
5. GERT uses probabilistic nodes and also a combination of deterministic and probabilistic nodes.
6. Some or all of the nodes or activities may not be traversed at all in GERT network.
7. It is applicable in many diverse fields

Example: Consider the following activities.

Activity Description

(1,2)   Constant → Contract is given to draw the drawings for the building
         D = 0.5, E(D) = 0.5, $V^2 = 0$.

(2,3)   Normal → Drawings submitted D = 2 to 3, E(D) = 2.5 $V^2 = 1/36$.

(3,4) Normal, Drawings evaluated by the committee. D = 0.75 to 1.25
E(D) = 1, V^2 = 1/144

(4,2) Constant, not acceptable, given back to contractor, D= 0, E(D) = o,
V^2 = 0

(4,5) Normal, Accepted, Design the members of the structure D = 2 to
3, E(D) = 2.5, V^2 = 1/36

(5.5) Normal, Redisign the members D = 2 to 3, E(D) = 2.5, V^2 = 1/36

(5,6) Constant, Designs accepted, permission granted to start the work
D = 1, E(D) = 1, V^2 = 0.

The durations (D) are in months

For calculation of expected value E(D) & variance V refer chap 2 section (C).

Node 4 and 5 are probalistic Nodes

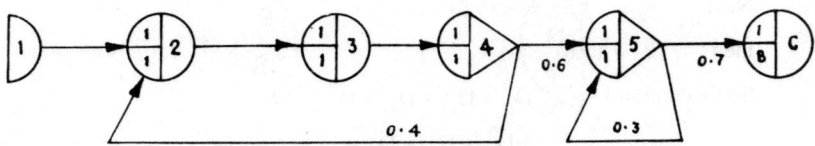

Fig H.5 GERT Network

At node 4 the possibility of branching activities are 0.4 & 0.6. It implies

for a random number x_1

if $0 < x_1 <= 0.6$ then it branches to activity (4, 5) otherwise to (4, 2)

Similarly at node 5, for a random number x_2

if $0 < x_2 < 0.7$ then it branches to activity (5, 6) otherwise to (5, 5)

If n denotes a normal deviate, then

D_{23} = $E(D_{23})$ + n.(D_{23})

= 2.5 + n/6

D_{34} = 1 + n/12

D_{45} = 1 + n/12

D_{55} = 2.5 + n/6

Generate random samples for the activities and for x_1 and x_2.

Select at random a position in the tableof random numbers at the end of
the section. Proceed row-wise or column wise. Group the numbers according

to the digit given in probability.

For example If you take column and row 1 number is 07018. Proceed column or row wise. Either 07018, 31172, 12572 or 07018, 34482, 27420 etc.

Random nos will be 0, 0.7, 0, .1, .8, .3, .1 etc

or 0, .7, 0, .8, .3, .4 etc.

Total duration of the project

(1) $\qquad D = D_{12} + D_{23} + D_{34} + D_{45} + D_{56}$

$\qquad\qquad = 0.5 + 2.55 + 0.96 + 2.26 + 1$

$\qquad\qquad = 7.27$ Months

(2) $\qquad x_1 = (4,2), x_2 = (5,5)$

Next $\qquad x_1 = (4,5), x_2 = (5,5)$

then $\qquad x_2 = (5,6)$

So total duration = $\qquad D_{12} + D_{23} + D_{34} + D_{42}$

$\qquad\qquad\qquad + D_{23} + D_{34} + D_{45} + D_{55}$

$\qquad\qquad\qquad + D_{55} + D_{56}$

$\qquad = \ 0.5 + 2.65 + 0.97 + 0 + 2.54 + 1.05$

$\qquad\quad + 2.65 + 2.44 + 2.45 + 1$

$\qquad = \ 16.25$

(3) Duraton X_1 (4,5) & X_2 (5,5)

$\qquad\qquad$ Then (5,5)

$\qquad\qquad$ Then (5,6)

Total Duration $\quad = \quad D_{12} + D_{23} + D_{34} + D_{45}$

$\qquad\qquad\qquad\qquad D_{55} + D_{56}$

$\qquad\qquad\quad = \quad 0.5 + 2.34 + 1.05 + 2.65$

$\qquad\qquad\qquad +2.50 + 2.37 + 1$

$\qquad\qquad\quad = \quad 10.04$ Months

This way we can find out the distribution of total duration D. From the distribution find out E(D) & V(D) & then the probability to complete the project in a particular duration

For example for three simulation

D $\qquad = 7.27, 16.25$ & 10.04

$$E(D) = (7.27 + 16.25 + 10.04) / 3 = 11.18$$

$$V(D)^2 = E(D^2) - (E(D))^2 = 19$$

$$= 139.239 - 124.99$$

$$= 14.24$$

$$V(D) = 3.77$$

The probability to complete the project between 5 to 10 months

$P(5 <= D <= 10) = 1/3$ there is only one value between 5 to 10 out of three.

S.N	Normal	D_{23} Normal deviate	D_{34} Normal deviate	D_{45} normal deviate		D_{55} deviate	Rudem activity number	Rudem actitvity number
1	0.31	2.55	-0.51	0.96	-1.45	2.26	-0.35	2.44	0.0 (4,5)	0.1 (5,6)
2	0.90	2.65	-0.36	0.97	0.33	2.56	-0.28	2.45	0.7 (2,2)	0.7 (5,5)
3	0.22	2.54	0.58	1.05	0.87	2.65	-0.02	2.50	0.0 (4,5)	0.9 (5,5)
4	-1.00	2.34	0.53	1.05	-1.90	2.18	-0.77	2.37	0.1 (4,5)	0.1 (5,6)
5	-0.12	2.48	-0.43	0.96	0.69	2.62	0.75	2.63	0.8 (4,2)	0.8 (5,5)
6	0.01	2.5	0.37	1.03	-0.36	2.44	0.68	2.61	0.3 (4,5)	0.7 (5,5)
7	0.16	2.52	-0.83	0.93	-1.88	2.19	0.89	2.65	0.1 (4,5)	0.5 (5,6)
8	1.31	2.72	-0.82	0.93	-0.36	2.44	0.36	2.56	0.1 (4,5)	0.0 (5,6)
9	0.33	2.44	-0.26	0.98	-1.73	2.21	0.06	2.5	1 0.7(4,2)	0.7 (5,50
10	0.38	2.56	0.42	1.03	-1.39	2.27	-0.22	2.46	0.2 (4,5)	0.1 (5,6)

Table
Random Normal Deviate

0.31	-0.51	-1.45	-0.35
0.90	-0.36	0.33	-0.28
0.22	0.58	0.87	-0.02
-1.00	0.53	-1.90	-0.77
-0.12	-0.43	0.69	0.75
0.01	0.37	-0.36	0.68
0.16	-0.83	-1.88	0.89
1.31	-0.82	-0.36	0.36
-0.38	-0.26	-1.73	0.06
0.38	0.42	-1.39	-0.22
1.07	2.26	-1.68	-0.04
-1.65	-1.29	-1.03	0.06
1.02	-0.67	-1.11	0.08
0.06	1.43	-0.46	-0.62
0.47	-1.84	0.69	-1.07
0.10	1.00	-0.54	0.61
-0.71	0.04	0.63	-0.26
-0.94	-0.94	0.56	-0.09
0.29	0.62	-1.09	1.84
0.57	0.54	-0.21	0.09
0.24	0.19	-0.67	3.04

Table
Random Numbers

07018	31172	12572	23968	55216	85366	56223	09300	94564	18172
34482	42158	40128	48436	30254	50029	19016	56837	05206	33851
27420	97534	89707	97453	90836	78964	00704	85734	21776	85764
73904	89123	19271	15792	72675	62175	48746	56084	54029	22296
19193	99621	66899	12351	72438	99839	24228	32079	53517	18558
13623	76165	43195	50205	75736	77473	07268	31330	07337	55901
17918	75071	91057	46829	47992	26797	64423	42379	91676	75127

Section I

ASSEMBLY LINE PROCESS

In this section we are concerned with the work that men do along the lines and not with machines or equipments. We shall be dealing almost wholly with assembly work and not with actual fabrication work. We shall also be talking about the use of the technique in Civil Engineering projects.

Line production saves handling because products move by conveyors or by some other meachnical means. It eliminates a great deal of paperwork because matrial and production workers are not required to be instructed, Lines can handle minor but not the major variations in productions so it is normally used where a process is repeated for a long time.

The sequential diagram for assembly line scheduling can be a basic network, CPM network, PERT diagram or precedence diagram. The precedence diagram is mostly used. First of all, the precedence network have been discussed and then the asembly line process.

Precedence networks are relatively new. They are more similar to CPM than to PERT. In appearance, they look much like event - oriented PERT networks. The connections between work items are zero-time logical connections, so that there are no dummies as such. Some authorities claim that the precedence network is simpler to use and the network is less complex in appearance.

I.1 Precedence Logic

One reason for the greater simplicity of precedence networks is that a work item can be connected from either its start or its finish. This allows a start - finish logic presentation without breaking the work item down. Fig. I.1 illustrates the three relationships - start-to-start, end-to-start, and start-to-end.

(a) The start of B depends on the start of A (a start-to-start relation ship).

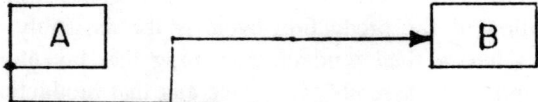

(b) Starting of B depends on finishing A (an end-to-end relationship).

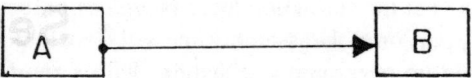

(c) The start of B depends on finishing A (a start-to-end relationship).

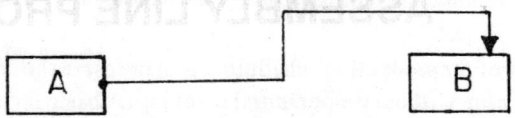

Fig. I.1 (a,b,c). Precedence Relationships

Although the network is simpler in appearance, deeper thought must be given to the reading and interpretation of the network. Fig. I.2 & I.3 model the same situtation through CPM and precedence, respectively. The superiority of the precedence in this case is demonstrated by the relative simplicity of its diagram. Two sets of four activities are each reduced to one, and no dummies are required.

Activity identification appears in the left end of each rectangle, the description in the middle, and the duration in the right end. The left edge of the rectangle represents the presented on a node the activity is called nodal.

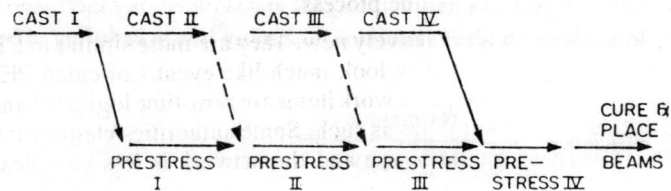

Fig. I.2

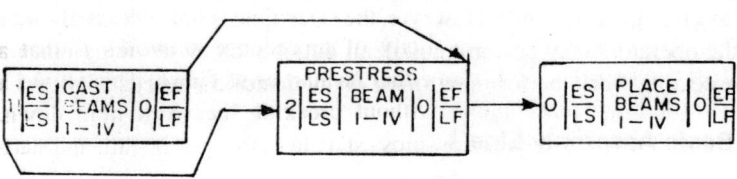

Fig. I.3

I.2 Assembly Line scheduling

The culmination of the production cycle is the assembly of parts and subassemblies into a final product. Assuming that inventory levels are suitable to support the assembly line pace and that production quotas are correctly established, the assembly line scheduling functions as a separate entity. Assembly line scheduling is a balancing of factors to produce an even flow of production at the substation level as well as in the final product. A bottleneck at any portion of the assembly line will serve to slow down the final product and build up over supply at stations with a short cycle time. The problems of non continuous assembly lines are obvious. Accordingly the goal of the assembly line scheduling is to evenly balance the various factors involved.

The value of preproduction scheduling is manifest in the costly problems which can develop without proper initial planning. A basic factor in assembly line scheduling is the theoretical cycle time.

Theoretical cycle time,

$$= \frac{\text{Productive time per hours}}{\text{units to be produced per hour}} \times F$$

where F is a factor considering incentive factors.

The theoretical minimum number of assembly lines or stations required to support a specific cycle is as follows :

$$N\,(\text{minimum}) = \frac{\text{Sum of all activity times}}{\text{cycle time}}$$

The station figure is a minimum figure and is inevitably increased by the inability to achieve an ideal situation. Assembly line scheduling efficiency is :

$$\text{Efficiency} = \frac{N\,(\text{minimum})}{N\,(\text{Actual})}$$

Slack time is the value of 1 minus efficiency and is generally to be carefully reviewed if it exceeds a cycle time. In other words, slack should always be a positive value. However, the slack time is not necessarily wanted as the operators may perform collateral duties such as servicing machines, reinspecting work, performing minor maintenance duties etc.

I.3 Basic Assembly Line

The backbone of any significant assembly line scheduling or balancing procedure is the logical sequence which must be followed in assembling one production unit.

The assembly of an item consists of a series of jobs called ACTS. These can be more or less complex, depending upon the unit to be manufactured. The Precedence diagram can be a summarization of various subassembly collections or can include the detailed portion of simpler unit. For example, a bolt cannot be placed untill the drill hole has been laid out, chasis drilled and reamed, and the basic material has been punched. Cycle time is independent of total assembly time and is the function of the production rate required. Accordingly, the summation of assembly times for one unit could be perhaps one day, while the cycle time for a single unit might be one hour. Accorigly if the plant were on 30 hours production and requird a daily rate of 30 units, one approach would be to provide 30 parallel assembly lines, another would be to break the work stations into a large number of concurrent activities within one or several lines, Fig. I.4 indicates parallel and serial assembly lines. Another approach is the feeder line, which can bring subassembled units into the main assembly line.

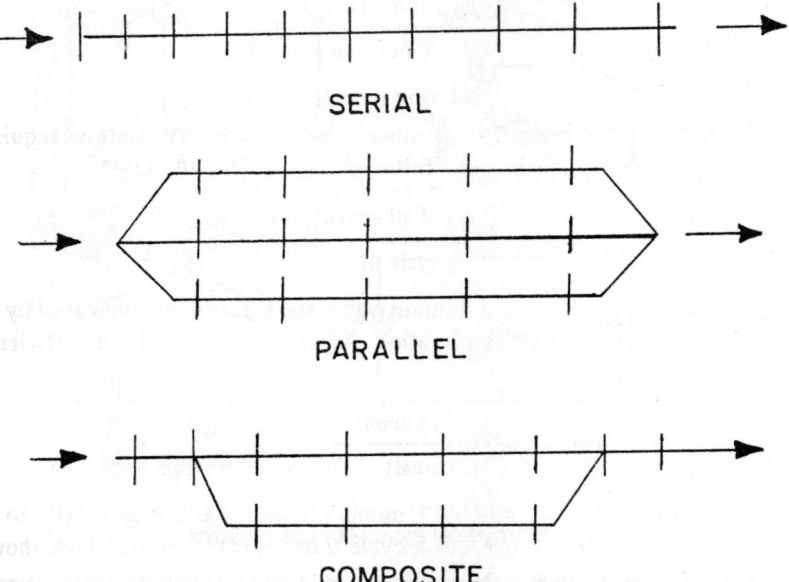

SERIAL

PARALLEL

COMPOSITE

Fig. I.4. Functional assembly line forms

3.4 Assembly Line Layout

The physical configuration of the assembly line affects the degree of balancing which can be achieved. In the case of an existing line, routine line balancing does not usually consider major rearrangement. Rearrangements of tools, small equipments and fixtures are usually feasible and should be considered in diagramming. However, major fixed conveyors, large equipment such as machine tools, and other major installations are not usually considered for relocation. Nevertheless, in preparing the precedence diagram, although it should be considered to indicate the impact upon balancing of layout fixity.

Conversely, where the assembly line is for a new building or product, the precedence diagam may well have a substantial effect upon the selection of the physical layout. In either case, a planned view of the assembly line, whether existing or proposed, drawn to scale can be invaluable in considering the assembly line scheduling routines. Fig. I.5. shows five basic assembly line flow patterns.

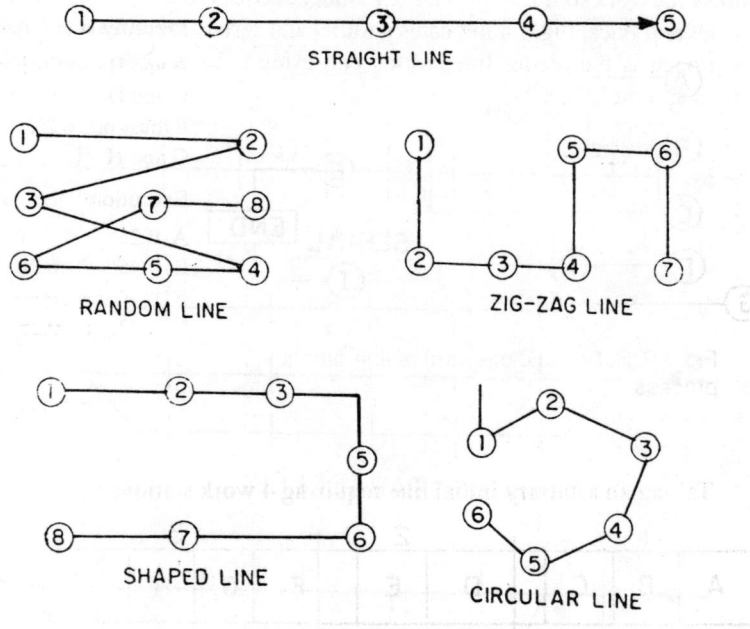

Fig. I.5. Physical assembly line patterns

One of the basic steps in the development of production layout is the flow process chart. The precedence chart can be substituted for this flow process chart, or a flow process chart can be developed directly from the precedence chart. Utilizing these two and knowing the usual funtional and physical configuration, a flow diagram can be prepared on functional, departmental or equipment lay out bases. These would involve analysis of travel time within the plant. Naturally, optimal assembly line schedling minimizes internal travel time.

I.5 Assembly Line Balancing

The traditional approaches to line balancing or assembly line scheduling are intiative, Fig. I.6 shows a small precedence sequence for an assembly line. The cycle time for this assembly operation is to be one time unit. Establishing stations along a strict sequential basis, four work stations are required. The first three have reasonable slack periods, but the fourth station is 75 percent slack. By making some rearrangements on a trial basis, three work stations are scheduled, with only a very small (5 percent) slack at one station and no slack at the other two.

Defining the line which is to be balanced.

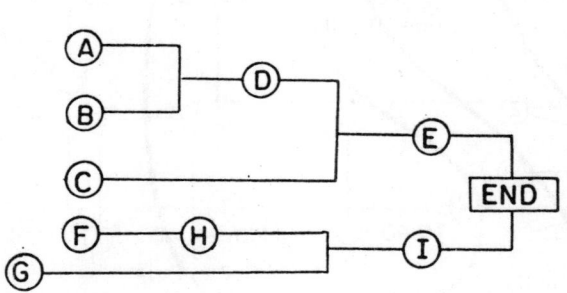

Precedence relations :

A and B must precede D
C and D must precede E
F must precede H
G and H must precede I

Execution times :

A	0.25	F	0.20
B	0.45	G	0.30
C	0.25	H	0.40
D	0.53	I	0.25
E	0.35		

Fig. I.6. Schematic diagram of line balancing process

Cycle time = 1.0

Taking an arbitrary initial line requiring 4 work stations :

1			2		3			4
A	B	C	D	E	F	G	H	I	
0.25	0.45	0.25	0.5	0.35	0.2	0.3	0.4	0.25

And making job shifts which preserve all precedence relations

F and G to station 1   B to station 2

E and I to station 3

When balanced, the line requires only 3 work stations

Although this approach is effective in a very simple situation, it cannot cope with the practical line balancing situation. Another approach is shown in Fig. I.7. In this case, balancing is done directly on the diagram. This is a practical approach and simplifies the problem of maintaining sequential requirements while balancing. The approach is to group stations into minimum. A number of solutions can be attempted rapidly in this manner.

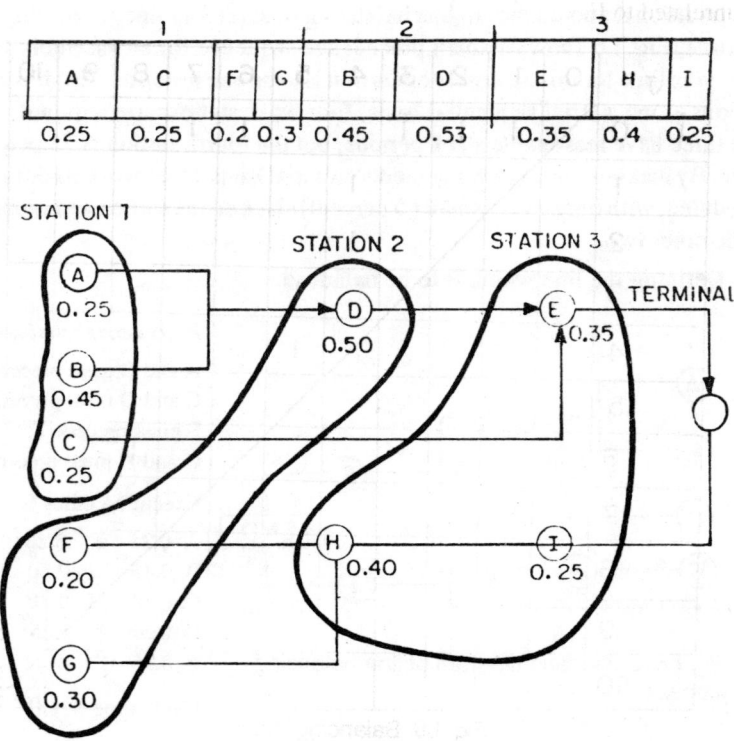

Fig. I.7. Balancing directly on precedence diagram

The matrix approach to assembly line scheduling is a rigorous and disclined one. Fig. I.8 shows the balancing matrix for the precedence network in Fig. I.6. The first step is to convert the act letter into number. These numbers must represent the precedence relationship. With j numbers being greater than i numbers, matrix is then constructed indicating the immediate successors for the various i acts. This is similar to the matrix solution for networks, but different in that the acts are similar to a precedence diagram rather than to events in a CPM or PERT network. In the vertical column, the i or beginning acts are noted. The j acts across the horizontal axes represent succeeding acts. The succeeding acts immediately following each i act are noted with a "1". For instance, 5 separate acts immediately follow the starting act 0. The starting and terminal acts serve only to tie the beginning and completion of the network together. Solution is shown in matrix form. This approach is to schedule generally in order of i, but to level this sequence in order to balance. Accordingly, act 0 must be scheduled first following this, the acts are scheduled in order until act 5. Since no act is less than 20. Accordingly, at this point, a search is made for a balancing act. F is tried since it is unrelated to E.

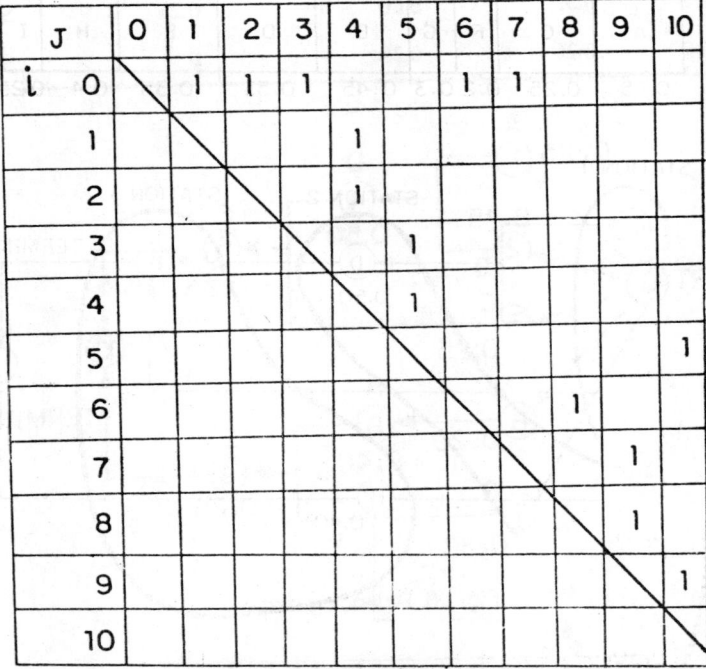

Fig. I.8. Balancing matrix

Act Selected	Duration	Station (before) time remaining	Station (after) time remaining	Station
- 0	0	1.0	-	-
A 1	0.25	1.0	0.75	S-1
B 2	0.45	0.75	0.30	S-1
C 3	0.25	0.30	0.05	S-1
D 4	0.50	1.00	0.50	S-2
F 6	0.20	0.50	0.30	S-2
G 7	0.30	0.30	0	S-2

(Contd.)

Act Selected	Duration	Station (before) time remaining	Station (after) time remaining	Station
H 8	0.40	1.00	0.60	S-3
E 5	0.35	0.60	0.25	S-3
I 9	0.25	0.25	0	S-3

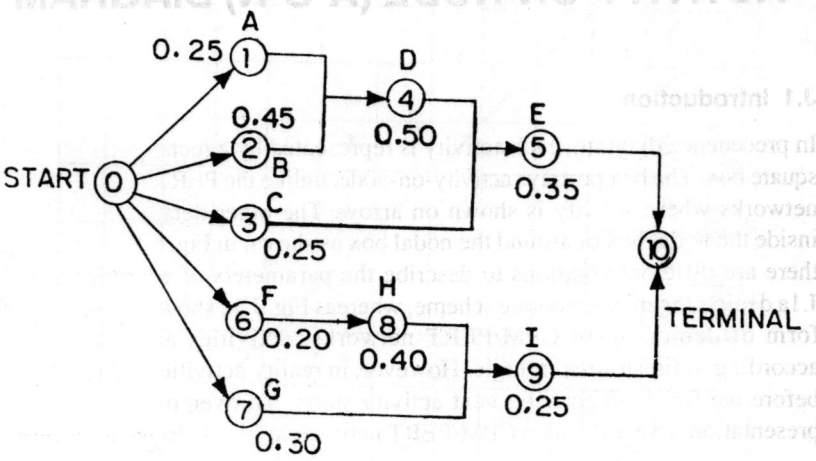

Fig. I.8. Balancing matrix

ACT	i \ j	0	1	2	3	4	5	6	7	8	9	10	DURATION	
—	0		1	1	1			1	1				0	
A	1					1							0.25	1
B	2					1							0.45	1
C	3						1						0.25	1
D	4						1						0.50	2
E	5											1	0.35	3
F	6								1				0.20	2
G	7									1			0.30	2
H	8									1			0.40	3
I	9											1	0.25	3
—	10													

Section J
PRECEDENCE DIAGRAMS (PN) OR ACTIVITY-ON-NODE (A-O-N) DIAGRAM

J.1 Introduction

In precedence diagram, each activity is represented by a rectangular box or square box. The box portrays activity-on-node, unlike the PERT/CPM/GERT networks where activity is shown on arrow. The other details are shown inside the nodal box or around the nodal box as shown in Fig. J.1. However, there are different notations to describe the parameters of activities. Fig. J.1a depicts the most elaborate scheme, whereas Fig. J.1c shows the simplest form of depiction. In CPM/PERT networks, activities are connected according to finish-to-start logic. However, in reality activities overlap and before we finish an activity, next activity starts. To overcome such mispresentation, a very detailed CPM/PERT network has to be drawn for a very

ID #	Duration		Cost
SS	Description		SF
FS			FF
EST	LST	EFT	LFT
TF	FF	InF	IF

Fig J.1a

EST	Duration ID #	EFT
EST	Description	EST
Activity no.	TF	Activity no.
LST		LFT

Fig J.1b

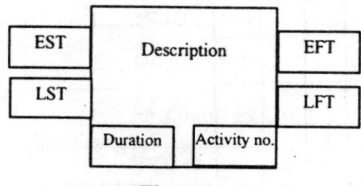

Fig J.1c

Legend

ID # :	Identity number
SS :	Start to start
SF :	Start to finish
FS :	Finish to start
FF :	Finish to finish
EST :	Earliest start time
LST :	Latest start time
EFT :	Earliest finish time
LFT :	Latest finish time
TF :	Total float
FF :	Free float
InF :	Interfering float
IF :	Independent float

Fig J.1

small project. Such detailed network may become unmanageable and whole purpose of project planning is defeated. Another advantage of AON diagrams is to incorporate the delay during execution, which is not simple in case of CPM/ PERT.

J.2 Fundamentals

The procedure for drawing AON diagram is similar to that of AOA diagram. Though the procedure and parameters are same in case AON, model presentation is different. As mentioned earlier, AON depicts activities by rectangular/square boxes, which are connected by lines. These lines show the dependency among the activities. The length or shape of the line has no significance and generally move from left to right. Arrowheads are also not essential to show. Crossing of lines should be avoided or proper symbols (as in electrical circuits) should be used to show the crossings, if unavoidable. The precedence logic among the activities is of mainly two types:

(i) Job dependency: It shows the sequence in which the activities progress. In general, the activity start after the preceding activity (ies) is (are) completed.

(ii) Construction constraints: It generally arises due to restrains on start and completion of dependent activities due to construction process at site. They are of following types:

(a) Start-to-start: It shows the delay from the start of an activity to the start of another dependent activity (Fig. J.2a).

(b) Start-to-finish: It shows the delay from the start of an activity to the finish of another dependent activity (Fig. J.2b).

(c) Finish-to-start: It shows the delay from the finish of an activity to the start of another dependent activity (Fig. J.2c).

(d) Finish-to-finish: It shows the delay from the finish of an activity to the finish of another dependent activity (Fig. J.2d).

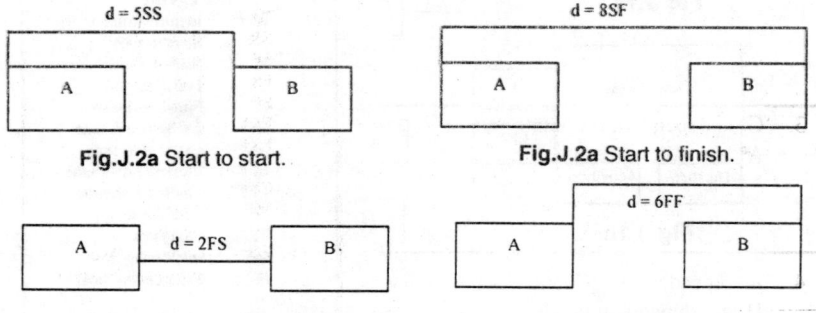

Fig.J.2a Start to start.   Fig.J.2a Start to finish.

Fig.J.2a Finish to start.   Fig.J.2a Finish to finish.

Fig J.2 Dependency of activities

J.3 Calculations on a AON network

Numbering of nodes and calculations are more or less similar to that of AOA networks. However, delays are incorporated during calculation of EFT, EST, etc. of dependent activities. Some representations and calculations are shown in Table J.1. Calculations are done normally in two phases:

1. **Forward pass:**
   i) Assign 0/1/ date as the early start time (EST) of first activity.
   ii) Calculate the early finish time for the activity (EFT = EST + duration)
   iii) Assign EST to each succeeding activities. It is the maximum of EFTs of all preceding activities.
   iv) Repeat steps (ii) and (iii) for each activity.

2. **Backward pass:**
   i) Proceed from right to left. Determine the EFT of last activity and that will be the latest finish time for last activity.
   ii) Calculate the earliest finish time for the last activity (LST = LFT – duration)
   iii) Assign LFT to each preceding activities. It is the minimum of LSTs of all preceding activities.
   iv) Repeat steps (ii) and (iii) for each activity.

Float calculations are more or less same as that of AOA network. However, in AON network, delay has to be considered.

Table J.1 Logic and Calculations on Precedence Diagram

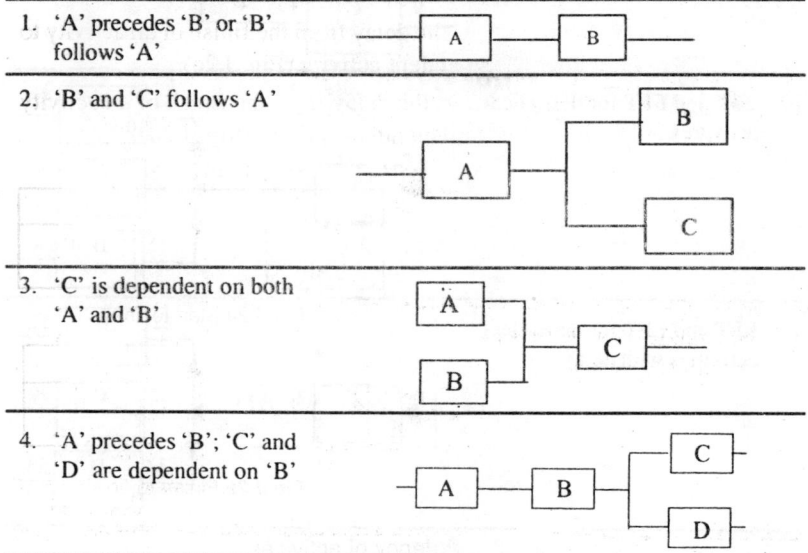

1. 'A' precedes 'B' or 'B' follows 'A'	
2. 'B' and 'C' follows 'A'	
3. 'C' is dependent on both 'A' and 'B'	
4. 'A' precedes 'B'; 'C' and 'D' are dependent on 'B'	

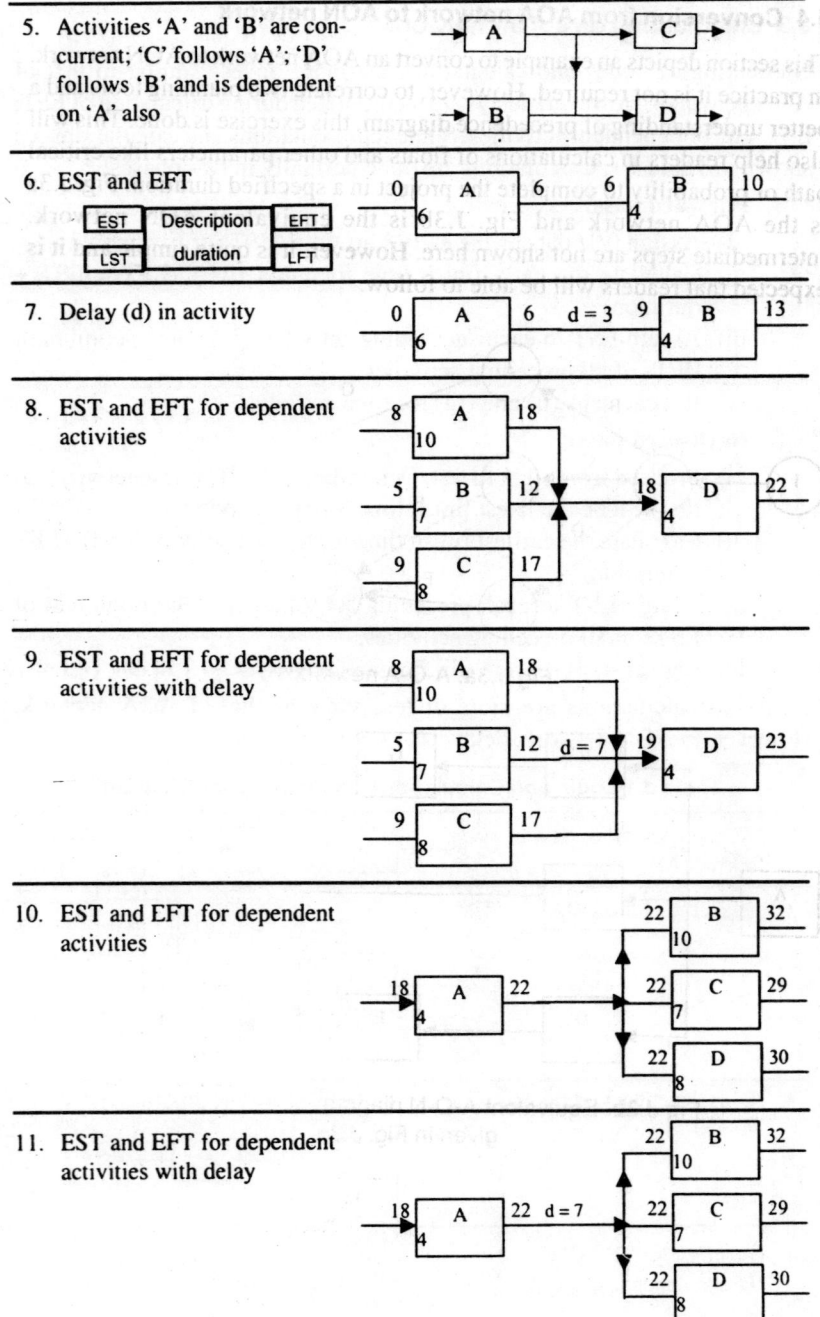

5. Activities 'A' and 'B' are concurrent; 'C' follows 'A'; 'D' follows 'B' and is dependent on 'A' also

6. EST and EFT

EST	Description	EFT
LST	duration	LFT

7. Delay (d) in activity

8. EST and EFT for dependent activities

9. EST and EFT for dependent activities with delay

10. EST and EFT for dependent activities

11. EST and EFT for dependent activities with delay

J.4 Conversion from AOA network to AON network

This section depicts an example to convert an AOA network to AON network. In practice it is not required. However, to correlate two planning tools and a better understanding of precedence diagram, this exercise is done. This will also help readers in calculations of floats and other parameters like critical path or probability to complete the project in a specified duration. Fig. J.3a is the AOA network and Fig. J.3b is the equivalent AON network. Intermediate steps are not shown here. However, it is quite simple and it is expected that readers will be able to follow.

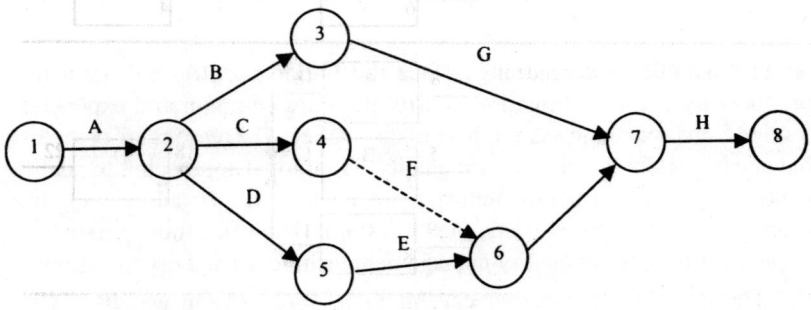

Fig J.3a. A-O-A network

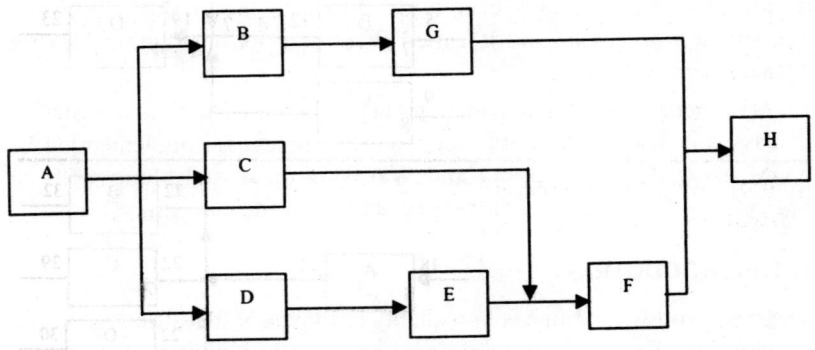

Fig J.3b. Equivalent A-O-N diagram of A-O-A diagram given in Fig. J.3a

Section K

ELECTRONIC - CONSTRUCTION MANAGEMENT

Several troublesome challenges face the Indian construction industry: productivity is down, litigation is up, delays are common and expensive, foreign firms are taking a great share of the market. The reasons are complex, but one important factor is lack of quality. Traditional approaches to quality control in the construction industry are inadequate. To address quality problems and associated costs, the construction industry must pursue and implement innovative quality management techniques and organization.

The advent of new pervasive technologies for managing and displaying digital geographical (natural and/or man-built) data has created a situation highly conducive to other technologies. The proposed work incorporates digital photographs and World Wide Web into the quality control of construction processes and is called Electronic - Construction Quality Management (E-CQM).

The need for quality management in the construction industry is discussed in brief in the initial section. The process for implementing E-CQM using digital mapping and WWW is discussed along with an organizational set-up, need of training, and a real life case study.

K.1 INTRODUCTION

There is a worldwide emphasis on quality in all type of industries, including construction. The quality of works performed by the building professionals has long been a concern. The conditions in developing countries with respect to quality improvement are less favourable, as most developing countries face problems with regard to product quality. The nature of the problem differs depending on the phase of industrial development in the country. However, there are several common factors impeding the improvement in quality, in most developing countries, including shortage of goods, constraints on foreign exchange, incomplete infrastructure, and inadequate knowledge. Despite these challenging circumstances, the need for the quality in developing countries cannot be overemphasized. Improving the quality of construction is particularly urgent in order to save investment of billions of dollars annually. This rapid deterioration has been attributed to a harsh environment, bad construction practice. The introduction of the quality scheme in developing country is sure to face many problems, like:

- There are no local standard specifications.
- The so-called infrastructure, such as professional societies, and accreditation boards, are still non-existent or in early stages of the development.
- Most testing laboratories are not accredited.
- The majority of customers are small owner builders who generally lack a basic knowledge of construction.
- Constructors are not well educated.
- Well-educated human resources/experts are not many in numbers.
- Constructors and quality control managers are in contact, which increases the opportunities of mal-functioning.

Not long after the construction industry began to look for tools that would help productivity improvement, rigorous efforts for computer utilization occurred. Computerized information systems (IS) are widely recognized as an enabler, not only for effective project management, but also for automation of engineering and construction tasks. Moreover, the advent of recent IT has accelerated the adoption of innovative IS in the construction industry. Nevertheless, compared with other industry sectors, current uses of computer applications in construction are not very advanced. In addition, these applications exist independently and have little or no capacity for communications with each other. Apart from these, there are other applications of IT like Computer Integrated Construction Planning Methodology for engineering and construction firms, Computer Imaginary and Visualization in Civil Engineering Education, Digital Imaging in Teaching Construction Process, Integrated approach towards use of computers and IT for construction, Software for integrated analysis and design of RC buildings, etc. However, there is no reference of application of IT in quality control management. Programmes normally choose either the engineering model using quantitative approach or a qualitative approach based on induction from precedent to implement the quality management programme. However, a single approach is not the practical solution of improving the quality of work, safety of the labours, and reduction in cost and time. To train the students and professionals, a critical thinking and innovations in present existing tools are essential. Digital imaging (DI) and World Wide Web (WWW) are two of the significant information technologies and have had a profound impact on several aspects of life. Although, technology substitution is essential, the mature application of a technology involves innovations. The paper deals to take advantage of unique properties of digital imaging, the web, and the synergistic relationship between them in order to improve the quality of construction.

Electronic Construction Quality Management (E-CQM) places emphasis on prevention, and not only correction. Hence, E-CQM needs a fast two-way communication, immediate actions, and complete coordination among the participants involved. The paper discusses the implementation of E-CQM with the help of photographs, DI, and WWW. A comprehension methodology is mentioned in the paper. The scanned photographs of on going construction are put on WWW that can be referred by the experts of the fields after required processing of images. The experts could suggest the control measures and corrective actions could be taken in no time. However, the method is at conceptual level and is implemented for the local construction projects. To implement at the regional, zonal, and national level it may require something more.

K.2 DIGITAL IMAGING (DI) and WORLDWIDE WEB (WWW)

DI offers a powerful medium for extending the utility of photographs in critical visual analysis. The potential of image processing in construction and teaching has been widely recognized. DI permits the adjustment in contrast, colors balance, sharpness, and perspective distortion. The operations reveal unclear or indistinguishable features and highlight features of interest with limitations. Enhancement is important for engineering photographs, which are taken by non-professional photographs, i.e. with no or little knowledge about the subject. It is beneficial as compared to darkroom techniques that are time-consuming and difficult. DI enables annotation of images with text, symbols, guidelines, and transparent overlays. These operations not only highlight the important features but also clarify the relationship among the real phenomena. DI provides to manipulate image content, using operations such as adding, distorting, and rearranging image elements. DI is quite common and no. of software are available which provide the basic techniques of DI. However, depending on the special requirements, processing techniques can be developed.

The demand for better access to and presentation of information has given rise to the need of WWW. The achievement of WWW lies in its ability to present rapidly a variety and quality of information to meet the needs of different types of users. The Internet, which is being recognized as 'most powerful communication medium on the planet' provides opportunities for data dissemination techniques. However, it is extremely important that this data dissemination to users is 'live'. A live quality control data can be referred to 'Intelligent Photographs' i.e., photographs in an electronic medium with associated attributes. The potential of the web as a medium for distributing images has been widely recognized. The images do not replace desktop demonstrations but rather enrich them by providing a visual means to extrapolate the small-scale demonstration.

K.3 ORGANIZATIONAL SET-UP

The focal point of the quality team organization is the advisory committee responsible for establish the team structure and developing the policies and procedures for the implementation process and team formation. Once the teams are established, the advisory committee continues to provide direction for maintaining the TQM process. In a zone or region a team leader, so-called project manager, will chair the team. A proposed organizational set up is given in Table K.1.

Table K.1: Responsibility of the various members in Quality Management Team

Project Title	Work Arena	Specific project responsibility
Project Manager	Internal, Regional	Total project coordination, directing and documenting.
Senior Experts (Construction, Structural)	Internal, Regional	Review and evaluation of digital images.
Junior Experts (Construction, Structural)	Internal/External, Focal	Review and evaluation of digital images.
Image Processors	Internal, Focal	Image processing
Photographers	External, Local	Site visit, shooting photo-graphs
Web page developers	Internal, Regional	Developing and maintenance of web site, compilation of the comments & images.

For the flow of information caused the project manager to interact with virtually all members of the team, giving him a central–as opposed to an epical–role. The following is a description of the complete evaluation process (Fig. K.1):

1. Project Manager (PM) instructs authorized Photographers (PG) to cover the concerned project at different stages of construction, e.g., foundation, plinth, etc. A web site is created and PG is provided a password.

2. PG covers the site, shoots photographs, scans, and uploads the photographs to the web site and informs the authorized image processor (IP).

3. Trained IP downloads, processes the images, and uploads the processed images. He then informs the PM about the availability of the processed photographs/ images at web site.

4. PM deputes the Junior Expert (JE) and informs him regarding the availability of the image for his comments.

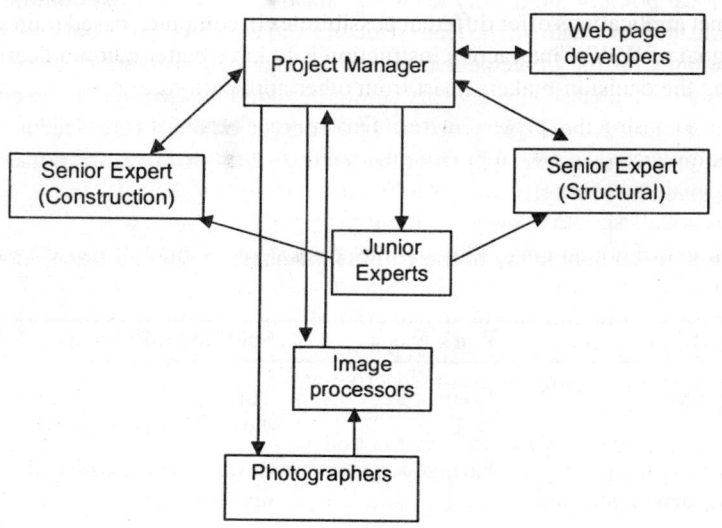

Fig. K.1 Simplified Organizational chart

5. JE evaluates the processed images and puts his comments. He informs the Senior Expert (SE) regarding the updating of web site.

6. SE reevaluates the images and modifies the comments of JEs, if required, and attaches his comments. He also informs to PM.

7. PM takes the necessary action, i.e., he informs the concerned client/ authority about the quality of work.

8. PM instructs the Web page developers (WPD) to compile the report.

Each authorized person, that is, PM, SE, JE, IP, and PG is given a login identity and a password. It is required from the security point of view and each user is authenticated at this stage. There is one website for each project and provision are made so that a junior expert (lower in hierarchy) cannot view the page of a senior expert (higher in hierarchy). However, reverse is not true. This provision is made for all the members of the evaluating team. Only PM can view all the photographs and the corresponding comments. The making of a single website results in saving the memory storage, easy-maintenance, and proper data integration. An evaluation of the performance of team members also can be done.

K.4 TRAINING

The computing technology has been undergoing transformation from its number crunching abilities to the recent Internet capabilities. Interactive Internet applications offer different possibilities in computer-based training. The goal of WWW interactive instruction is to have better communication among the decision-makers apart from other applications.

In a construction project, there are many types of professionals involved, including technical and non-technical, skilled and unskilled professionals. In order to have a thorough control over the quality of construction all the professionals should undergo training in regards to quality control through 'Intelligent Photographs'. The various professionals like project experts, regional or zonal experts, planners or architects, designers, quantity surveyors, including senior managerial staff, photographers, and image processing experts, should be given training in order to achieve control on the quality of construction in respective areas. The execution of training program can vary from project to project, but main objective of training lies in achieving quality control of construction process without being in physical contact of it. The basic aim of training programme lies in providing an incremental advancement in the knowledge and skill in harmony with the changing technological, economic, social and political scenario.

The senior and junior experts can be given training in context of monitoring of construction activities at regular intervals through the WWW using the image processing packages. They can be trained in way of checking, from digital image processing view, various aspects of construction like soil characteristics, site conditions, foundation work, masonry and concrete work, detailing of reinforcements, fixing of doors, windows, etc., alignment of building components, workmanship, safety of labors, etc.

An image-processing expert can be trained in various ways of enhancing image quality. It is required that he should possess knowledge of certain image processing software packages like Photoshop, Paint, Imaging, Photo Editor, Imaging, etc. These packages have sufficient facilities (tools) for better exposure of images. These packages provide various tools for altering texture of images into grains, mosaic tiles, patchwork or strained glass. The images can be viewed in different canvases or can be rotated as and when desired. The packages provide the facilities for adding various different colors, rendering/ lighting effects for exposing details of image, etc. However, the expert also should be capable of developing particular processing techniques, if required.

A photographer while taking photographs at construction site should take care of certain things. It is desired that the photographs be taken throughout the life span of the construction project, showing each and every detail at regular interval of construction activities. A proper care is needed in regards to the depth of the photograph, existing lighting facilities, selection of proper location in order to cover the maximum details.

K.5 CASE-STUDY

The present case study shows how the digital image processing can be used to highlights the defects in building components during construction activities. In any construction project, regarding construction quality management, the following aspects are important:

- Site selection.
- Earthwork:
    1. Properties of soil.
    2. Types: sandy, silt, clay.
- Masonry:
    1. Type: brick, stone, etc.
    2. Bonds.
    3. Joints.
    4. Curing.
- Reinforced Cement Concrete (RCC):
    1. Shape, size, and slope.
    2. Joints: numbers, continuity, development length, No. of bars.
    3. Reinforcement: Type, placing.
    4. Clear cover.
    5. Curing.
- Finishes:
    1. Joints, Corners.
    2. Grinding, polishing.
    3. Number of coats: Whitewash, colorwash, paint, etc.
- Sanitary:
    1. Leakage.
    2. Choking.

- Material:
  1. Quality.
  2. Quantity.
  3. Stacking.
  4. Procurement.
- General:
  1. Site cleanliness
  2. Quality of work.
  3. Availability of facilities: water, electricity, etc.
  4. Tools and Plants: condition, reliability, and optimum quantity.
  5. Workmanship.
  6. Dimension and the tolerances.
  7. Work progress.
  8. Safety.

Apart from these common aspects, the project may have some other special features.

A case study of local construction has been taken. A set of more than 100 photographs were shot at site of local construction, scanned and converted into digital images. Out of 100 photographs, some of them are selected for further investigations. The section describes the observations and results.

Figures K.2…K.25 shows the main components of construction process. The critical observations indicate the followings:

- The initial image of Fig. K.2 was not clear; hence image-processing technique was applied. It is clear from Fig. K.2 that there is some supply/sanitary pipeline underneath. This indicates that the site chosen for construction was not well planned and survey was not conducted properly. The image also provides the kind of soil at site. In case of doubt the experts can demand the value of soil bearing capacity.

- Fig. K.3 shows the foundation trench after the completion of PCC (plain cement concrete) work. It can be observed that loose soil spoiled the plain surface of PCC.

- Fig. K.4 is the digital photograph at the plinth level. Apart from indicating the poor site conditions, it also indicates the poor consolidation and compaction below plinth level. The raise of plinth from the ground also can be questioned and this can be compared with the minimum required level.

- Fig. K.5 indicates the type of workmanship, site condition, and the type of foundation provided. It also provides a rough estimation of reinforcement details at foundation level, which can be estimated quite accurately by zooming the concerned portion. The completed portion of the foundation clearly shows the honeycombing in cement concrete.

- Fig. K.6 shows again the type of foundation, depth of foundation, along with the poor formwork. In case the type and/or depth of foundation (depth to width ratio can be obtained by simple image processing techniques) is not found suitable for the structure, the work can be stopped. A zoom view shows that reinforcement (diameter of bars and spacing) is same in both directions. It also shows that hooks (bending) provided at the end of bars are not uniform. However, a lesson also can be learnt from the photograph that extra width of foundation trench is needed to execute the work.

- Fig. K.7 shows the partial filling of foundation column. It can be seen that the concrete layer is more or less covered by the soil. It is a very common site and the normal practice is that the layer of soil is not properly cleaned before further concreting of the foundation column.

- Fig. K.8 shows a number of construction shortcomings, like, (i) bricks quality is poor; (ii) bricks are not properly soaked in water; (iii) bonds and joints are not properly constructed; and (iv) water cement ratio is not proper in mortar. Joints are also not filled properly with mortar, which is also indicated in Fig. K.9.

- Fig. K.10 is to show the manual stone dressing due to unavailability of the proper plants and tools. It not only results in more time to execute the work but also quality of the work suffers. Safety of the labor can also be questioned.

- Fig. K.11 indicates the improper verticality of the column. It is necessary to avoid jack ling of bars. The figure is used to compare the reinforcement details of the columns with the actual design with further zooming. A poor connection of column with stone masonry cannot be overlooked in the figure, which is also indicated in Fig. K.12, along with the poor formwork.

- Fig. K.13 is a good example of efflorescence in the brick masonry. A poor joint connection at plinth level also can be viewed in the photograph.

- Fig. K.14 shows the details of the scaffolding that is normally used in developing countries. These scaffolding are made of wood, which make them environmental unfriendly and from safety point of view also, they are not appreciated.

- Fig. K.15 shows material stacking at site. An improper stacking of material always raise the cost of the project and will results in project delay, since it may result in wastage of material, need security against theft, and bringing them in proper shape before use.

- Fig. K.16 is a good example of fixing the reinforcement at the joints of beams and columns. The photograph also shows very clearly the quality of workmanship and material used. Fixing, spacing, and providing development length of the reinforcement bars at these joints are critical and proper care has not been taken.

- Fig. K.17 shows the poor joint of beam, column, and slab after finish. The photographer is not experienced enough to shoot the photographs showing such technical details. However, here image processing play an important role.

- Fig. K.18 is shot for showing the details of concealed electrical fitting. The photograph shows the fitting of a fan box. Along with this, it also shows the clear cover provided and the quality of reinforcement used in the work.

- Fig. K.19 shows that the concreting at steep slope that needs a technical hand to avoid the honeycombing. Figure also indicates the poor workmanship at different points, which is also shown in Fig. K.20, especially efflorescence after plastering.

- One of the main features of this study is to identify certain mixes. Here, an example is taken to identify the mix ratio of cement and fine aggregates in mortars. Fig. K.21, K.22, and K.23 shows the plaster mix of ratio 1:6, 1:4, and 1:3 respectively. It is quite clear from the photographs that Mix 1:6 is the brightest, and as the ratio of cement increases the brightness decreases. To further investigate the results, photographs were zoomed to 1600% till we obtained the image in form of pixels. Colour of pixels confirmed the results. Same procedure was followed to identify the different types of cement concrete mixtures and soils.

- Fig. K.24 shows the slope of the floor after completion. A pool of water can easily be identified at floor level, which is stationed due to poor slope of the floor.

- Fig. K.25 shows the vertical alignment of the building components. In this case the column's (Fig. K.25 (a)) alignment is checked by drawing a rectangular box (Fig. K.25 (b)). The check box does not coincide with the outlines of the column.

Web site was developed using Active Server Pages, Hyper Text Markup Language, and Database files and is placed at Birla Institute of Technology

(BITS) Web – Server (ASURA). It can be accessed through Local Computers Network. Active Data Objects technology is used to add database (Microsoft Access). Using this Web site, Project Manager could monitor the local construction sites with the help of Experts, and a local photographer. The various image-processing techniques were used either available in various soft wares or developed indigenously. Image processing techniques such as drawing horizontal, vertical, inclined (in any direction) lines with specified spacing were used to observe the photographs critically and hence, to reach to certain conclusions.

The advantages of the methodology are:

1. It is fast and necessary preventive action can be taken to improve the overall quality of the construction work, construction site, labor work condition, and their safety.

2. It will reduce the interface between the employees and customers, will increase efficiency and minimize corruption.

3. Sluggishness from any personal can be easily identified because all the actions are stored date and time wise.

4. It can be used to train the professionals concerned with construction industry in classrooms.

It is important to note the ethical issues involved in presenting manipulated photographs. Digital techniques can create convincing photographic images that are completely false. Apart from ethical issues, certain limitations cannot be overcome easily, like accurate calculations of different measurements. A Geographical Information system's approach and its tools may be useful for such purposes. A detailed cost analysis is not carried out, however, a rough estimate shows an overall cost reduction in terms of less no. of site visits, quality management by experts sitting at far off place. Moreover, cost gain in terms of quality, and safety would be enormous which cannot be quantified.

Fig.K.2. Pipe line at foundation level

Fig.K.3. PCC work at foundation level

Fig.K.4.Plinth level

Fig.K.5. Honey Combing at foundation

Fig.K.6. Formwork for trapezoidal foundation

Fig.K.7. Mixing of loose soil with concrete work

Fig. K.8. Masonry work

Fig. K.9. Masonry joints

Fig. K.10. Manual Stone dressing

Fig. K.11. Stone masonry and column tilting

Fig. K.12. Column - Stone masonry joint

Fig. K.12. Column - Beam joint and formwork

joint

formwork

Fig. K.14. Details of Scaffolding

Fig. K.15. Material Stacking

Fig. K.16. Beam - Column joint reinforcement detailing

Fig. K.17. Beam - Circular column - Slab joint

Fig. K.18. Concealed Electrical Fittings

Fig. K.19. Honey-combing at steep roof

Fig. K.20. Efflorescence in
cement plastering

Fig. K.21. Cement plaster (1:6)

Fig. K.22. Cement plaster (1:4)

Fig. K.23. Cement plaster (1:3)

Fig. K.24. Floor slope

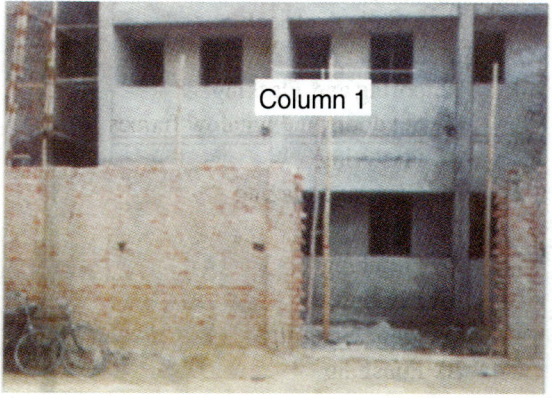

Column 1

Fig. K.25(a). Alignment of building components

Column 1

Fig. K.25 (b). Alignment of Column1 shown in Fig. 25(a)

Problem 1

Develop a network for the project with following activities and calculate the critical path

Duration	(days)
1. Contract bid	15
2. Contract award	5
3. Site survey	7
4. Foundation excavation	10
5. Site drainage	5
6. Footings and brickwork in foundation	15
7. Flooing	10
8. Fire and damp proofing	5
9. Rough plumbing	3
10. Rough electrical	3
11. Masonry	10
12. oofing and Flashing	15
13. Set doors and window frames	5
14. Plastering	4
15. Glass and glazing	3
16. Tiles	8
17. Plumbing	3
18. Painting	4
19. Floor covering	5
20. Finishing	7

2. Draw a network for the following activities which are part of construction of television tower and a building adjacent to the tower for transmission purpose. The site is at the top of a hill. Between tower and building there is a crushed gravel service road and an underground cable. A fuel tank also has to be installed on the building.

Activity	Duration (days)
1. Contract award	10
2. Survey site	15
3. Excavate for basement	20
4. Grade tower site	50
5. Procure steel, and guys	150
6. Procurement of electrical equipment & cable	180
7. Footings	60
8. Erection of tower and ele equipment	50
9. Install connecting cable	10
10. Install drain tile and storm drain	40
11. Backfill	12
12. Building Footing	30
13. Basement slab & fuel tank slab	50
14. Construction of walls of basement	10
15. Construction of floor beams	15
16. Construction of floor slab	20
17. Roof slab	22
18. Interior framing & utilities	55
19. Building interier etc	25
20. Main cable between tower and building	40
21. Install fuel tank	7
22. Septic tank	15
23. Drain tile and storm drain	20
24. Backfill, grade and surface	12
25. Connecting road	25
26. Clear up site	15
27. Handing over	6

Calculate floats for each activity, & find out the critical path.

If the duration of the activities given, are taken as most likely time& optimistic time & passimistic time for each activity are taken as given below, find out the critical path, and the probability to complete the project within a year.

Optimistic time: If duration is less than 10 days decrease it by 2 days. If duration is less than 30 days and more than 10 days decrease it by 4 days. If duration is less than 50 more than 30, decrease it by 7 days and for more ther 50 days decrease the duration by 10 days.

Pessimistics time Duration is less than 10 increase it by 2 days.

10 < duration < = 40, increase it by 7 days

40 < duration < = 75, increase it by 10 days

more than 75, increase it by 15 days.

3. 12 labourers and 12 plumbers are available for a water supply project. Prepare
   (a) CPM network
   (b) Resource allocation (series and parallel method)
   (c) Smooth the resources allocation curve.

Activity	Duration (Days)	Plumbers	Labourers
1 - 2	8	3	7
1 - 3	4	4	4
1 - 4	4	3	5
2 - 4	8	9	9
2 - 5	7	0	8
3 - 6	3	6	4
4 - 7	10	4	10
5 - 6	6	0	6
6 - 7	9	6	9

Problem 4

For the given Cascade Bar chart, draw the equivalent A-O-A and A-O-N network.

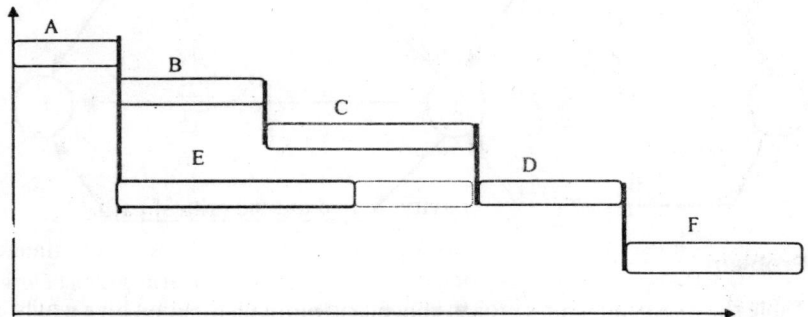

Problem 5

Duration and resources for each activity is given in table. Draw the limited resource allocation chart and smoothen it.

Activity	Duration (Week)	Resources
A	4	6
B	6	4
C	8	6
D	5	4
E	7	4
F	7	3

Problem 6

Prepare a LoB for a small project of 15 houses based on a rate of build of 3 houses/week assuming 8 hours per day and 6 days per week. A minimum buffer of 4 days should be assumed. Table below shows the operations together with the optimum man-hours and of no. of man for each operations.

Activity	Man-hours	Optimum no. of man per operations
A	120	6
B	320	4
C	200	4
D	60	2
E	40	2

The activities are related as given below :

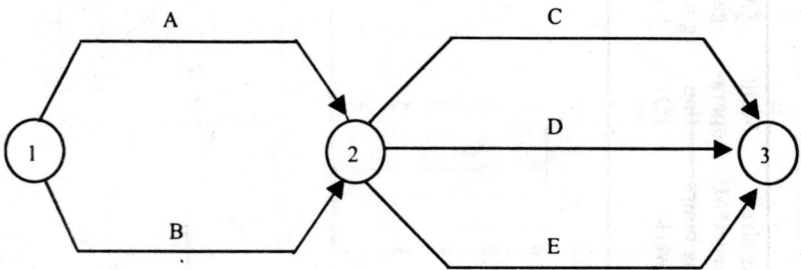

Problem 7

Table shows a contractor's project budget and profit distribution for a newly awarded project. The conditions of contract allow interim measurements to be made monthly and payment of the amount certified less 10% retention to be paid to the contractor. Annual interest rate is 12% on locked-up capital. If the contractor is given two options (given below), which one he will choose. Justify your answer.

Solution

Operation	Man hours (M)	Gang size (G = target*M/ working hours per week)	Men/opera- tion (Q)	Actual gang size (g = multiple of Q)	Actual rate/ week (U = (g/G)*target)	Time/ operation (T = M/ hours per day*Q)	Round- off	Time between first and last operation (s = n-1*working days per week/U)	Round off
A	120	12	3	12	4	5	5	36.25	37
B	290	29	6	30	4.137931	6.041667	6	35.04167	36
C	250	25	4	24	3.84	7.8125	8	37.76042	38
D	40	4	3	3	3	1.666667	2	48.33333	49
E	30	3	2	2	2.666667	1.875	2	54.375	55
F	220	22	5	20	3.636364	5.5	6	39.875	40

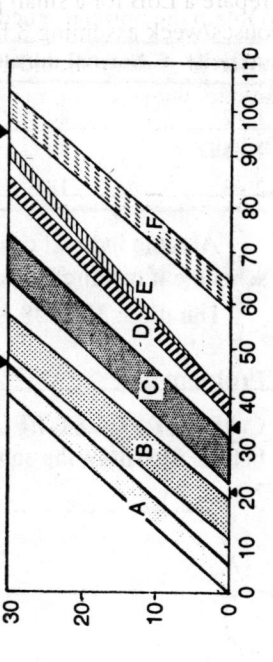

1. Payment after retention is made after one month and retention money is released 4 months after completion of work.
2. Payment after retention is made after two months and retention money is released 1 month after completion of work

Month no.	1	2	3	4
Value of work each month (00000)	2	3	4	8
Profit (% of value)	6	6	6	6

Problem 8

Obtain the optimal schedule for the network given below.

Activity	Normal duration	Crash duration	Normal cost	Crash cost
1 - 2	20	10	1300,000	1000,000
1 - 3	32	20	2220,000	1500,000
2 - 3	18	14	2200,000	2000,000

Assume indirect cost/unit duration is Rs. 100,000. Obtain the optimal schedule if the direct cost of activity 1-3 is given as follows:

Duration: 32 to 28, slope is 60,000 and from 28 to 24, it is 30,000.

Problem 9

Calculate EST, LST, EFT, LFT, FF, and TF for the precedence diagram given below. Also draw the equivalent CPM and BAR chart for the given diagram.

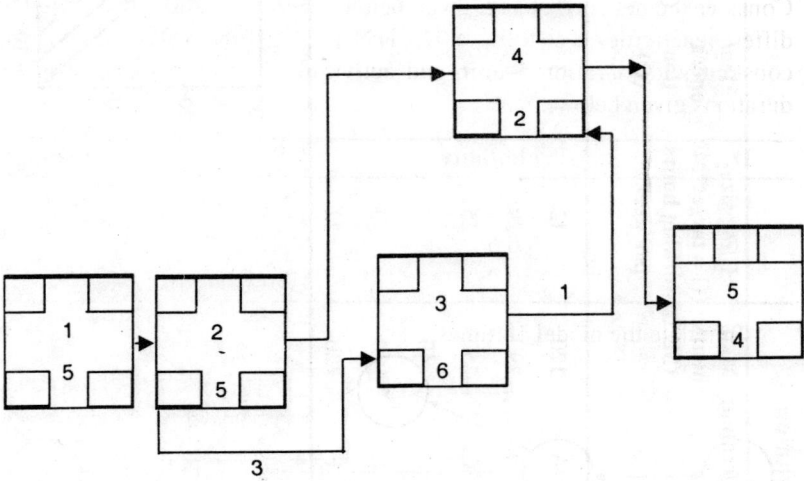

Problem 10

Differentiate among budgeted cost, actual cost, and valued cost with an example:

Problem 11

Given the following data :

Activity	Present duration	Crash duration	Slope (Rs./day)
(1,2)	3	1	300
(1,5)	6	5	700
(2,3)	5	3	700
(2,4)	4	2	600
(2,5)	3	1	100
(3,6)	7	4	500
(4,7)	3	2	600
(6,8)	4	2	600
(5,7)	8	4	300
(7,8)	5	3	400

Obtain the most economical schedule. The normal duration of activity (2,5) is 5 days, and total indirect cost is Rs. 3000/day. Draw an equivalent A-O-N network.

Problem 12

Consider the network of Fig. given below. The distribution of duration of different activities is as follows. D_{12} is N(12, 4), D_{34} is N(5.5, 0.25), D_{45} is constant with duration 5 unit, and activities 2-3 and 2-4 have discrete durations given below :

D_{23}	Probability	D_{24}	Probability
5	0.3	7	0.4
6	0.5	12	0.6
7	0.2		

Simulate the model 14 times.

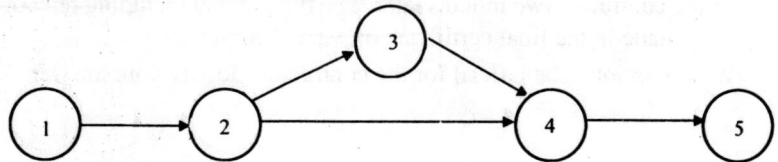

Problem 13

Table shows the activity, available options, and their time and direct costs. Assuming indirect cost to be Rs. 10000/ unit duration, obtain the optimum duration for the project.

Activity	Activity description	Option	Duration	Cost
1-2	Site preparation	Crew 1 + Equip. 1	14	230000
	Site preparation	Crew 2 + Equip. 2	20	180000
	Site preparation	Crew 3 + Equip. 3	24	120000
2-3	Forms and rebar	Method 1	15	30000
	Forms and rebar	Method 2	18	24000
	Forms and rebar	Method 3	20	18000
2-4	Precast concrete girder	Method 1	12	450000
	Precast concrete girder	Method 2	16	350000
4-6	Pour foundation & piers	Method 1	22	200000
	Pour foundation & piers	Method 2	28	150000
	Pour foundation & piers	Method 3	30	100000
6-7	Erect girders	Crane 1 + Crew 1	9	300000
	Erect girders	Crane 2 + Crew 2	18	220000

Problem 14

A contractor's budget of a project and profit distribution is presented in the table below :

Month No.	1	2	3	4	5	6	7
Value of work each month (00000)	2	3	4	8	9	8	5
Profit (% of value)	7	8	5	7	6	5	4

The interim measurements to be made monthly. The contractor has two options:

1. Payment of the amount certified less 10% retention is to be paid to the contractor one month later. Half the retention is made included in the final certificate on virtual completion and the other half is released six months later.

2. Payment of the amount certified less 10% retention is to be paid to the contractor two months later. The full amount including retention is made in the final certificate on virtual completion.

Which option is beneficial for the contractor? Justify your answer.

Problem 15

A contract consists of 100 bored piles. The activities and associated information for each pile are presented in the table below:

Activity	Man-days	Optimum No. of men per gang
1-2 (A)	20	10
1-2 (B)	12	4
2-3 (C)	15	8
2-3 (D)	10	6
2-4 (E)	12	6

The time limit provided in the contract is four weeks. Draw a line of balance diagram to schedule the work. Give a buffer of 1 day between two activities. Assume data as required.

Problem 16

A firm of civil engineering contractors undertakes to carry out as a subcontract the excavation work on a road construction project. The work involves forming cuttings and embankments described in the bill of quantities as follows:

1. Excavate materials on site and form embankments.

2. Excavate materials on site and cart away.

3. Excavate materials in borrow pit and form embankment on site.

The quantities for the above items are calculated from longitudinal section and cross sections, and are shown in table. The subcontractor plans to use tractors and scrappers, and estimates the cost of doing the work from the following rates:

- Dig and form bank upto 300m haul, Rs. $3/m^3$

- Dig and form bank with a haul over 300m and up to 600m, Rs. $8/m^3$

- Dig and form bank with a haul over 600m and up to 900m, Rs. $16/m^3$

- Dig and form bank with a haul over 900m and up to 1200m, Rs. $28/m^3$

- Dig in cutting and cart to tip, Rs. $6/m^3$

- Dig in borrow pit, bring to site and form embankments, Rs. $7/m^3$

Length along road 100m	Cut (m³)	Fill (m³)
0	0	0
3	40,000	
6	70,000	
9	30,000	
12		11,000
15		30,000
18		44,000
21		20,000
24	20,000	
27	10,000	
30		40,000

Chapter III

CIVIL ENGINEERING SYSTEMS

Civil Engineering is a creative profession. The role of civil engineer is essentially synthesis, planning and designing, moulding and shaping the domestic and industrial environment. In order to create and synthesise, civil engineers must be fully aware of the materials and their behaviour under working conditions. The education of a civil engineer is consequently much dominated by learning how things behave and how that behaviour may be determined by analysis. Knowledge of disciplines such as structural mechanics, hydromechanics, soil mechanics and their associated analysis is important only as an adjunct of the process of synthesis.

Faced with a completely designed, civil engineering project, most civil engineering graduates should be able to analyse how it will behave under working conditions. Usually this is done by establishing a mathematical model of the project which embodies the known mechanical laws (equilibrium, compatibility, material properties, conservation of energy etc.). This mathematical model is then manipulated and solved to yield values for the required behavior parameters (stresses, displacements, flows etc.). Usually the analysis yields a unique set of results which can be checked against ranges of acceptable values from codes of practice and other sources.

Faced with the problem of designing the same project instead of analysing it, things are very different and much more difficult. It would be very convenient if the process of analysis were completely reversible, but of course it is not. It is not possible to start with a set of acceptability criteria such as are specified by codes of practice, and to work backwards to a unique structure or project which satisfies those criteria, without at many stages making design decisions about the shape and dimensions of the structure. In design, a set of acceptability criteria does not define a unique solution except in the most trivial of examples. Generally there are many widely different designs that will satisfy the acceptability criteria and a single design will only be obtained by making decisions which eventually eliminate all alternatives but one. Design is, therefore, a decision-making process, unlike analysis which allows no scope for choice or decision-making.

It is this decision-making aspect of design which makes it so daunting to civil engineering students. How can beginners in the profession make the right decisions? Clearly an injudicious decision at any stage might ultimately lead to a design that is unnecessarily difficult or expensive to build or even to one that fails to meet the acceptability criteria. How can an ability to make good decisions be learned.

More precisely civil engineering systems is concerned with quantitative decision-making techniques of use in the planning, design, construction and operation of civil engineering projects. It is essentially concerned with the synthesis rather than with the analysis aspects of civil engineering. It relies on the same mathematical modeling approach as analysis. Instead of constructing a mathematical model for analysis and manipulating it to yield behavior parameters, a different mathematical model is constructed for synthesis purposes so that the manipulation and solution yields design decisions (numbers, sizes, configurations etc.). Civil engineering systems is concerned with mathematical models of this synthesis type.

Civil engineering systems is concerned with examining the mathematical structures of decision-making methods for all kinds of civil engineering problems, identifying similar mathematical forms and classfying decision-making methods according to these forms. An analogy can be drawn with structural mechanics. There is an infinite variety of structural forms-bridges, beams, columns, plates, shells, etc. - yet the analysis of all these different structures under applied load may be performed using only a few techniques. Examination of the behavior of these structures has shown that although all the structures are different they behave in similar ways. Thus a knowledge of linear elasticity, elasto-plastic theory, rigid-plastic theory and buckling is sufficient to enable an analysis to be performed on many structures. Analysis is, therfore, classificatory. Civil engineering systems attempts to do the same sort of classification with decision-making problems. It attempts to establish a basic framework encompassing all decision-making methods so that any decision-making problem can be examined, classified and, with the aid of a few appropriate methods, solved.

3.1.2 Decision Making

Many contemporary problems of design and operation in engineering, architecture, construction, and urban and regional planning are of such magnitude and complexity as to require the most systematic and rational approach possible.

Generally, these problems consist of a very large number of interacting variables many of which defy quantification. Therefore, it is necessary first to classify the variables appearing in real life problems into tangible or quantifiable and intangible or unquantifiable. The purpose of the systems approach is to develop methods, mathematical or otherwise, to deal systematically and rationally with the quantifiable parameters through statistical observation, testing, development of measuring techniques; and to provide a clear understanding of the situation at hand as an aid to the decision maker for subjectively evaluating the intangibles which are present in most real problems.

Physically, a system is composed of a large number of interacting components, each of which may or may not serve a different function but all of which contribute to a common purpose. Because the components usually involve many areas of knowledge, the only way to implement the systems approach to decision making is through team action; teams involving a spectrum of specialists in seemingly unrelated fields. System theory is intended to provide a common basis of understanding between disciplines and, as a result, the systems approach is permeating most fields of knowledge. The physical sciences and engineering have already developed a strong vehicle for interdisciplinary teamwork through the systems approach and recently, the life sciences as well as the humanities and the social science are beginning to apply to their problems the vast methodology that exists in systems analysis and design.

A complex problem can be handled with success when the systems approach is effectively applied. Cost effectiveness, which is a usual gauging device for measuring success, strives to produce a system with the lowest possible cost for a set level of effectiveness or, vice versa, the highest level of effectiveness for a set cost. Combinations of these goals are usual and trade-offs occur when such an approach is taken.

An example of the application of the systems approach to decision making, where no mathematics is needed but where the approach to problem solving recognizes the fundamental behavior of people in developing a realistic policy that can be successfully implemented, is the case of a congested urban area where two local authorities conflict in their plans for city operation and management. One of the authorities, the Planning Commission, attempts to dissuade suburbanites from bringing their personal automobiles into the Central Business District by promoting the use of the Mass Transit System, with the expectation of relieving traffic congestion and reducing the frequency and duration of traffic snarls. On the other hand, the second governmental group, the Port Authority, is charged with the planning, financing, construction, operation, and maintenance of all toll bridges and tunnels leading to the Central Business District. The Port Authority promotes the use of these facilities by offering to the public books of bridge and tunnel toll coupons which can be purchased by the frequent or routine driver at a savings of several Rupees per month. A conflict immediately develops because, while one authority tries to discourage the use of personal automobiles in the Central Business District, the other encourages it by offering savings in toll charges. A behavioural solution to this sticky situation was proposed by R.L. Ackoff, Operation Researcher Professor at the University of Fennsylyania: Implement a graduated scale of toll charges as a function of the number of empty seats in a private automobile.

The policy could be developed as follows: Assume a six-passenger car arrives at the toll booth:

1. If all six places are occupied, the car passes free of charge.
2. If one seat is empty - that is, only five passengers are in the car - the charge is, say, Rs. 5.
3. If two seats are empty, the charge is Rs. 10.
4. If three seats are unoccupied, the cost now climbs to Rs. 20, and so on until.
5. The driver is the only passenger in the car, in which case the toll is Rs. 25.

The figures quoted may or may not be realistic but the point is well made. The day after this type of policy implemented, a marked reduction in the number of cars visiting the Central Business District would most surely be observed. People would probably drive to the outskirts of the downtown area and then form car pools, or they would use the Mass Transit System. A suspected side effect might be an increase in sales of sports cars; with only two seats they would make an inexpensive mode of transportation. This, of course, represents at least a reduction in automobile size and can therefore be considered a desirable side effect. On the other hand, in the interest of fair play, a special tariff could be developed for nonstandard vehicles, and so on. These types of applications are primarily based on common sense, and many such decision rules can be developed without the use of mathematical techniques. However, in some of the most important applications of the systems approach, a great many complex techniques exist. This chapter concentrates on the development of some of the most important and useful of those techniques.

3.2 The Structure of Systems Analysis and Design

Systems analysis is the process of separating or breaking up a whole system into its fundamental elements or component parts. It involves a detailed examination of the system in order to understand its nature and to determine its essential features. Systems design, on the other hand, is the process of selecting the components and of contriving the elements, steps, and procedures for producing a system that will optimally satisfy the stated goals. In the context of this book, systems design is used as a basis for anticipating problems and for solving them at their planning, engineering, architectural, and construction stages. Systems synthesis is akin to design because it is the process of putting together composing, or combining parts or elements to form a whole system, completely blended to achieve its finest level of performance. Through systems synthesis, often, varied and diverse ideas, forces, or factors are combined into one coherent, consistent structure. In examining complex problems, one must recognize the following classification of elements occurring repeatedly in their solution through the systems approach:

1. **A set of decision and state variables:** The decision variables are those over which the analyst has complete control and which he can manipulate at will. The state variables are those which are dependent on the decision variables and which, consequently, cannot be directly controlled by the decision maker. Often, the classification of variables into decisions and states is an arbitrary one. However, once they have been so stratified, their behavior follows the stated pattern. This element is basically in the analysis phase of the problem solution and the significance of each variable that is, how sensitive the problem is to its settings - as well as whether or not the variable is quantifiable must be ascertained in this stage.

2. **An optimization model:** This solution element is necessary for understanding the problem at hand. It involves both analysis and synthesis and consists of the development of a conceptual model which is sufficiently analogous to the real problem but which, on the other hand, is simple enough to be amendable to quantitative analysis.

3. **A measure of effectiveness:** Called the objective function, this measure is formulated as a means for evaluating the degree of success or failure attained in fulfilling the problem goals. It relates various decision and state variables for the expressed purpose of ranking the outcome of the different decision sets.

4. **Generation of alternatives and optimal solution:** After the problem has been formulated quantitatively, the sets of decisions arrived at following a rational, systematic plan are evaluated by means of the objective function, and the one producing the most desirable results is selected. The different sets of decisions are the alternative plans of action and the selection of the most desirable outcome constitutes the optimization phase of the problem solution; the decision policy produc-

ing the best results is the optimal policy. Frequently, a system cannot be completely optimized. Near optimal results are often extremely valuable especially when the objective function is not too sensitive to changes in the values of the decision and state variables near the optimum. This phase of problem solution is primarily a design phase.

5. **Policy implementation:** This step involves the carrying out of the optimal policy into the real physical situation. It constitutes, in fact, the realization of the objective and the only reason for having gone through the previous four steps. Usually, because of additional knowledge gained or because conditions change, the analyst finds it necessary to recycle the process by returning to one of the previous steps. This recycling is required in adaptive or learning processes where newly acquired data permit the system to refine itself and to adapt to a changing environment.

3.3 Classification of Decision Systems

In engineering, architecture, construction, and planning, decision systems can be classified according to size and according to predictability of behaviour as follows:

I. **According to size:**
1. Simple systems are those which involve only a relatively small number of quantifiable decision and state variables.
2. Complex systems involve a large number of decision and state variables. However, the variables are, by and large, of the quantifiable type.
3. Exceedingly complex systems: These consist of a large number of decision and state variables, most of which are of a non-quantifiable nature.

II. **According to behavioural predictability:**
1. Deterministic systems are those for which every input produces a predictable response. Furthermore, the system's responses to identical inputs are themselves identical.
2. Stochastic systems involve randomly determined sequences of observations, each of which is considered as a sample from a probability distribution. Stochastic variation implies system randomness in passing from one state to an adjacent state. A stochastic system's response to a specified input is not reproducible at will - that is, one cannot expect exactly the same behaviour when the system is subjected to identical inputs.

3.4 Mathematical Decision Making Models

A typical system analysis model will consist of a single objective function and a set of constraint equations. The objective function consists of a set of design, decision, or control variables. The objective function may be expressed as a mathematical relationship.

$$y = f(X_1, X_2, ..., X_n)$$

where X_1, X_2, ..., X_n are designated as a set of n control variables. The variables, which may represent the assignment of the no. of workers, an amount of money and a volume of material, are all non-negative values. They are introduced as

$$X_1 > = 0; X_2 > = 0; X_n > = 0$$

The financial, institutional and physical limitations are represented by a set of m constraint equations.

$$G_1 (X_1, X_2, ..., X_n) (< = >) B_1$$

$$G_2 (X_1, X_2, ..., X_n) (< = >) B_2$$

....

....

$$G_m (X_1, X_2, ..., X_n) (< = >) B_m$$

or in a vector form $\hat{G}_i (x) (< = >) \hat{B}_i$

Now the model becomes

$$y = f(X)$$

$$G(X) (< = >) B$$

$$X > = 0$$

The function f (X) and the set of functions represented by G (X) may be either linear or non-linear set of functions of X.

Any combination of control variables that satisfies the set of constraint conditions is called a feasible solution and solution that does not satisfy all constraint equations is called infeasible solution. An optimum solution is a feasible solution that satisfies the goal of the objective function as well.

In multiobjective problems we have a number of objectives instead of one. The different solution techniques, students have already learnt in "Optimization Techniques" and that's why they are not discussed in detail here, but the problems are solved using those techniques.

3.5 Linear Mathematical Models

The linear mathematical model can be represented as follows:

$$y = f(x)$$
$$y = A_1x_1 + A_2x_2 + \ldots\ldots + A_n x_n$$
$$G(X) \, (> = <) \, B$$
$$B_1 \, (> = <) \, A_{11} X_1 + A_{12} X_2 + \ldots.. + A_{1n} X_n$$
$$B_2 \, (> = <) \, A_{21} X_1 + A_{22} X_2 + \ldots.. + A_{2n} X_n$$
$$\ldots\ldots\ldots\ldots\ldots\ldots\ldots\ldots\ldots\ldots\ldots\ldots\ldots\ldots\ldots\ldots\ldots\ldots$$
$$\ldots\ldots\ldots\ldots\ldots\ldots\ldots\ldots\ldots\ldots\ldots\ldots\ldots\ldots\ldots\ldots\ldots\ldots$$
$$B_m \, (> = <) \, A_{m1} X_1 + A_{m2} X_2 + \ldots.. + A_{mn} X_n$$

where A's and B\s are constants
$$X > = 0$$
$$X_1 > = 0; \ X_2 > = 0; \ \ldots..; \ X_n > = 0$$

Example 1 - Earth moving operations: It is necessary to use a fleet of large earth moving vehicle to level a large and uneven site prior to the start of Construction Operation. The objective is to move the earth between cut and fill locations in such quantities that the site is levelled as cheaply as possible.

There are three areas on the site, 1, 2 and 3 & there are three locations, A, B and C where fill material is required. There is also a dump D where excess cut material can be taken. Tables show the cut material produced by the site 1, 2 and 3, and the fill material is required at three sites A, B and C.

Site	Cut material produced
1	600
2	8000
3.	9000 m³

Location	Fill material required
A	3000 m³
B	6000 m³
C	9000 m³

Distances (km) between cut and fill locations

		Fill locations			
		A	B	C	D
Cut	1	0.8	0.4	0.4	0.8
Sites	2	0.7	0.5	0.7	0.6
	3	0.4	0.3	0.6	0.5

1. Consider first the cut material taken away from the sites.

From site 1 $\quad X_{1A} + X_{1B} + X_{1C} + X_{1D} = 6000$ $\qquad$ .... (1)

From site 2 $\quad X_{2A} + X_{2B} + X_{2C} + X_{2D} = 7000$ $\qquad$ .... (2)

From site 3 $X_{1A} + X_{3B} + X_{3C} + X_{3D} = 9000$ .... (3)

2. Consider now the fill material required at different locations.

at location A $X_{1A} + X_{2A} + X_{3A} = 3000$ .... (4)

at location B $X_{1B} + X_{2B} + X_{3B} = 6000$ .... (5)

at location C $X_{1C} + X_{2C} + X_{3C} = 9000$ .... (6)

Equations 4 to 6 ensure that all material requirements at fill locations are met. Consequently any set of values of the 12 variables which satisfies all the equations 1 to 6 will represent a possible transportation schedule which will level the site. These six equations are, therefore, a complete mathematical specification of the restrictions on an acceptable transportation schedule.

Since 12 variables may satisfy the six equations 1 to 6 in an infinite number of ways, there is an infinite number of possible transportation schedules for this problem. Just one of them, derived in a casual fashion is shown in Table 3.1.

Table 3.1: Quantities carried (m) between cut and fill locations

		Fill locations			
		A	B	C	D
	1	1000	1000	2000	2000
Cut	2	1000	2000	3000	2000
locations	3	1000	3000	4000	1000

Now the total cost, C is given by,

$$C = 0.8\,X_{1A} + 0.4\,X_{1B} + 0.4\,X_{1C} + 0.8\,X_{1D}$$
$$+ 0.7\,X_{2A} + 0.5\,X_{2B} + 0.7\,X_{2C} + 0.6\,X_{2D}$$
$$+ 0.4\,X_{3A} + 0.3\,X_{3B} + 0.6\,X_{3C} + 0.5\,X_{\#D} \qquad \text{.... (7)}$$

Since it is easy to cost out different possible schedules it is logical to try to find the very best possible schedule, the one with an absolute lowest cost. This, the fourth step of the systematic approach, requires that the best possible decisions are made. In this problem the best decisions would be that set of values of the 12 variables which satisfy all the equations 1 to 6 and at the same time give a minimum value of the total cost function. Expressed more formally, if x is the vector of variables whose fifteen elements are the Xijs, the problems is to

Minimize equation 7 over variables X

To calculate the total profit, calculate the profit element wise.

	Profit		
	A	**B**	**C**
1	1402.5	1202.5	—
2	1105	905	1305
3	997	—	881

Net profit $1402.5\, X_{A1} + 1202.5\, X_{B1} + 1105\, X_{A2} + 997\, X_{A3}$
$+ 905\, X_{B2} + 1305 X_{C2} + 881\, X_{C3} = P$

It is a complete mathematical model which can be solved by the different optimization techniques, discussed later on in this chapter.

3.6 Optimum Structural Design

Introduction

The design process may be devided into the following four stages.

1. Formulation of functional requirements
2. The conceptual design stage.
3. Optimization
4. Detailing

There may be many possible designs that satisfy the functional requirements so optimization in the present context is an automated design procedure giving the optimal or best values of desired design quantities, considering suitable criteria and constraints. Computers are effectively used for this part of the design, since the optimization methods are numerical.

3.6.1 General Formulation of the Problem

A structural system can be described by a set of quantities, some of which are variables . Those quantities that are fixed during the automatic design are called preassigned perameters and quantities that are not preassigned are called design variables. The preassigned paramatar represent those parameters which are fixed by the design procedure because the experience shows that they produce good results . The design variables represent the following.

1. The mechanical or physical properties of the material.
2. The topology of the structure.
3. The configuration or geometric layout of the structure.
4. The cross sectional dimensions

If a design meets all the requirements placed on it, it will be called as a feasible design. The restrictions that must be satisfied in order to produce a feasible design are called constraints and behavior constraints. There are two types, namely design constraints and behaviors constraints. Design constraints are imposed on design variables. Behavior requirements impose behaviors constraints. Both these may be expressed as a set of inequalities.

$$I_i(\{ x \}) \leq 0 \hspace{6cm} j = 1 , \ldots, m.$$

m = no of inequality constraints

{ x } = vector of variables, which may or may not be equalities such as

$$h_i (\{ x \}) = 0 \hspace{6cm} j = 1, \ldots, k.$$

In order to find the best feasible design it is necessary to form a function of the variables to use for comparision of alternatives. The objective function is the function whose best value is sought in an optimization procedure. It is function of x, (F{ x }) also

$$\max F(\{ x \}) = -\min (-(\{ x \})).$$

The objective function should reflect the goal of the design problem. Weight and cost of the structures are commonly used objective functions: In general, the objective function is a non linear function of the design variables.

3.6.2 Linear Programming Problem : If the objective function F({ x }) and the constraints $I_i (\{ x \}) = 0$, i=1,....m are linear function, than it is called a LPP. For solving a LPP, simplex method can be used.

1. Simplex method: The simplex method is an iterative method of solving. Before we start the solution, we must have

1. The standardised problem.

Example:

Mix $x_0 = -12x_1 - 15x_2$.

Subject to $4x_1 + 3x_2 < 12$

$2x_1 + 5x_2 < 10$,

$x_1, x_2 > 0$.

Converting this to the standard form,

Mix $x_0 = -12x_1 - 15x_2$

Subject to $4x_1 + 3x_2 + S_1 = 12$

$2x_1 + 5x_2 + S_2 = 10$.

$x_1, x_2, S_1, S_2 > 0$.

S_1, S_2 are slack variables which are used to convert inequalities into equalities.

2. Matrix A should have an identity matrix as a submatrix.

Example: Continuing the previous example, the data are entered in a table as given below.

Basic	x_0	x_1	x_2	S_1	S_2	Solution
x_0	1	12	15	0	0	0
S_1	0	4	3	1	0	12
S_2	0	2	5	0	1	10

[A]

x_0-row is the objective function written after transferring all terms to the left. The last column shows the right hand side of constraint after the introduction of slack and surplus variables. First column shows basic variables i.e. slack, surplus.

3. The objective function must be expressed in terms of non basic variables.

Then, the solution for maximization problem is as follows :

1. The variable (or column) with most negative coefficient in x_0-row enters as basic variable. This is to ensure a largest increase in objective function.

2. The leaving variable is determined by

$$\text{Min} \left(\frac{\text{right hand side}}{\text{Entering column with} > 0} \right)$$

This is for feasibility (>=0).

3. The optimal solution is reached when all the entries below variables in x_0-row are greater than or equal to zero.

For a minimization problem.

1. The variable (of column) with most positive coefficient in x_0-row enters as basic variable.

2. The leaving variable is determined by the same variable as in a minimization problem.

3. The optimal solution is reached when all the entries below variables in x_0-row are <= 0.

Example : Continuing the previous example, the problem has been converted to a standard format.

It is clear from the rules that x_2 is a leaving variable since it is most positive entry in x_0 row.

Leaving variable is determined by Min (12/3, 10/5) = 10/5 which indicates S_2 leaves. Now, the pivotal element of x_2 column and S_2 row is 5. So, divide pivot row by 5, then with this row, make x_2 column $(0; 0; 1)^T$. This is the first iteration. The second iteration is done on the same lines to give the results.

Basic	x_0	x_1	x_2	S_1	S_2	Solution
x_0	1	6	0	0	-3	-30
S_1	0	1	4/5	0	1 -3/5	6
S_2	0	2/5	1	0	1/5	2
x_0	1	0	0	-15/7	-12/7	-300/7
x_1	0	1	0	5/14	-3/14	15/7
x_2	0	0	1	-1/7	2/7	8/7

Now, all entries below variables in x_0 row are $<= 0$. Thus the optimal solution is

$x_1 = 15/7$, $x_2 = 8/7$, and Min $x_0 = -300/7$.

2. Artificial Variables Technique

If the LPP has constraints with $<$ or of a mixed type with $>$, $<$ and $=$, then the simplex method cannot be applied directly since the standard from would not have an identity matrix for those cases. We will have to introduce new variables which are called as artificial variables.

Example :

Min $x_0 = 3x_1 + 2x_2$

Subject to $x_1 + x_2 \geq 2$.

$x_1 + 3x_2 \leq 3$.   (1)

$x_1 - x_2 = 1$.

$x_1, x_2 > 0$.

Transforming this to standard form,

Min $x_0 = 3x_1 + 2x_2$

Subject to $x_1 + x_2 - S_1 = 2$.

$x_1 + 3x_2 + S_2 = 3$.   (2)

$x_1 - x_2 = 1$.

$x_1, x_2, S_1, S_2 \geq 0$.

This doesnot have an identity matrix as submatrix. So, we creat one by artificially adding variables R_i to the system resulting in

Min $x_0 = 3x_1 + 2x_2$

Subject to $x_1 + x_2 - S_1 + R_1 = 2$.

$x_1 + 3x_2 + S_2 = 3$. $\hspace{4cm}$ (3)

$x_1 - x_2 + R_3 = 1$.

$x_1, x_2, S_1, S_2, R_1, R_3 > 0$.

But the solution of (3) would also be a solution of (1) & (2) only when the artificial variables are all equal to zero. There are two methods to achieve this.

3. Big-M method

The objective function is modified by adding to it M times or -M times sum of all artificial variables where M > 0 is a big numbers in a minimization and maximization problems respectively.

So, the modified problem of (3) is

Min $x_0 = 3x_1 + 2x_2 + MR_1 + MR_3$.

Subject to $x_1 + x_2 - S_1 + R_1 = 2$.

$x_1 + 3x_2 + S_2 = 3$. $\hspace{4cm}$ (4)

$x_1 - x_2 + R_3 = 1$.

all variables > 0.

The simplex method is applied to (4)

The table is made, now

Basic	x_0	x_1	x_2	S_1	R_1	S_2	R_3	R.H.S.
x_0	1	-3	-2	0	$-M_1$	0	-M	0
R_1	0	1	1	-1	1	0	0	2
S_2	0	1	3	0	0	1	0	3
R_3	0	1	1	0	0	0	1	1

But this table should be modified before applying the simplex method since the objective function is not expressed in terms of non basic variables. Adding M times R_3 row to x_0 row, the starting table is obtained. Then, the normal simplex method can be applied.

Basic	x_0	x_1	x_2	S_1	R_1	S_2	R_3	R.H.S.
x_0	1	-3+2M	-2	M_1	0	0	0	3M
R_1	0	1	1	-1	1	0	0	2
S_2	0	1	3	0	0	1	0	3
R_3	0	1	-1	0	0	0	1	1

Two Phase Method

In the first phase, the objective function is to minimize the artificial variables with respect to the original constraints and a best feasible solution is arrived by simplex method. Using this as the initial BFS for the second phase, the original objective function is entered and after some row operations. The solution is arrived.

Duality in LPP

A LPP in canonical form :

$$\text{Min } x_0 = f(x) = c^T x,$$

Subject to $A x \geq b$

$x \geq 0.$

$$c = (c_1, c_2, \ldots \ldots c_n)^T, \, x = (x_1, \ldots \ldots x_n)^T,$$

$$A = (a_{ij})_{mxn}, \, b = (b_i \ldots \ldots b_m)^T.$$

The dual problem of above LPP is

$$\text{Max } Y_0 = \phi(Y) = b^T y$$

Subject to $A^T Y \leq C.$

$Y \geq 0.$

$$Y = (Y_1, \ldots \ldots Y_m)^T$$

Dual simplex method: When a particular basic solution is not feasible i.e. one or more basic variables are negative, but satisfies optimatity criterion, we use dual simplex method to get the solution.

The procedure to be followed is

1. Express the problem in following format
    (a) Min or Max $f(x) = c^T x,$
        Subject to $Ax = b, \, x \geq 0.$
        one or more b are negative and A has identity matrix as sub matrix.
    (b) $f(x)$ in term of non basic variables.
2. The bearing variable x_r is given by
    $x_r = 1 \leq i \leq m \{ x_r, x_i < 0 \}.$
3. The entering variable x_k is selected by
    $$x_k = \min_j \left(\mid \frac{z_j - c_j}{\alpha_r^j} \mid; \alpha_r^j < 0 \right)$$

4. The next table is obtained as in regular simplex method.

6. Simple methods

If there are equality constraints in NLPP, they should be utilised to eliminate variables if it is possible. Some constraints have bounds upon the values of the variables. These can be simplified by the following methods giving unbound variables.

Constraint	Substitution
1. $x_i >= 0$	$X_i \longleftrightarrow Y_i^2$
2. $a >= x_i >= 0$	$x_1 \longleftrightarrow a \cos^2 Y_i$
3. $a >= x_i = -a$	$x_i \longleftrightarrow a \sin Y_i$
4. $a >= x_i >= b$	$x_i \longleftrightarrow (c + b)/2] + [(a - b)/2] \sin Y_i$

(b) **Deletion of constraints :** Engineering knowledge and experience in design can be used to discard the constraints which will not be active at the optimum. This can be used as trail and error method if one is not sure whether the constraint is active or slack at the optimum. We may assume it to be slack and once the solution is found, we can determine whether our assumption is correct or not. If not, the constraint is an equality one end can be used to eliminate a variable

Example Min $f = (x_1 + 2x_2 + 3x_3 - 4)^2 + (2x_1 + x_3)^2 + 4x_2^2$

subject to $x_1 + x_2 + x_3 < 1.5$

$-x_1 - 2x_2 + 3x_3 < 2$

Solution

Assume both constraints are slack.

Equating each of the first derivatives of f to zero and solving the three equations, we get minima of f at (-0.8, 0, 1.6) as zero. Substituting this values in constraints.

$-0.8 + 0 + 1.6 = 0.8 < 1.5$ Slack

$0.8 - 0 + 4.8 = 5.8 < 2.$ Active

So, the new constraints is

$-x_1 - 2x_2 + 3x_3 = 2.$

so $x_1 = -2 -2x_2 + 3x_3.$

The problem is Min $f = (6x_3 - 6)^2 + (-4x_2 + 7x_3 - 4)^2 + 4x_2^2.$

Soving this, we get f = 1.415 at x_2 = 0.472, x_3 = 0.908,

x_1 = -0.218.

Both the constraints are slack at this point.

3.6.3 Non linear Programming

1. Quadratic Programming : The following is quadratic programming problem :

Min $f(x) = P^T x + x^T cx$.

Subject to $Ax \le b$, $x \ge 0$.

Where $c = (c_{ij})$ is n x n symmetric positive definite or positive semi definite matrix.

$P = (P_1, ...P_n)^T$, $A = (a_{ij})_{mxn}$ n

$X = (x_1, x_2.........x_n)^T$, $b = (b_1, b_2........b_m)^T$.

A quadratic form $x^T cx$ is positive definite if $x^T cx > 0$ for all $x \ne 0$.

If $x^T cx$ is such that $x^T cx \ge 0$ for all $x \ne 0$ and there is

atleast one non-zero x such that $x^T cx = 0$ than $x^T cx$ is said to

be positive semi definite.

2. Kuhn-Tucker Conditions

For a CNLPP, with the form,

Min $f(x)$,

subject to, $g_i(x) \le 0$, i =1 to m,

$x \ge 0$,

Than the necessary and sufficient conditions for x to be the solution of above problem are

$$\frac{\delta L(x, \lambda)}{\delta x_j} \ge 0; \ x_j \frac{\delta L(x, \lambda)}{\delta x_j} = 0$$

$$\frac{\delta L(x, \lambda)}{\delta \lambda_i} = g_i(x) \le 0; \ \lambda_i \frac{\delta L(x, \lambda)}{\delta \lambda_i} = 0$$

$x_j \ge 0$, j=1 to n, $\lambda_i \ge 0$, i = 1 to m.

The lagrangian function for the above problem is

$L(x,\lambda) = f(x) + \lambda^T (Ax - b).$

$= p^Tx + x^Tcx + \lambda^T(Ax-b)$

$\lambda = (\lambda_1, \lambda_2, \ldots \ldots \lambda_m)^T$ are Lagrangian multipliers

Then , by aplying Kuhn-Tucker conditions, we get the system of equations which, after introduction of block or surplus variables needed can be solved by simplex method

Example:

min $f(x) = 2x_1^2 + 2x_2^2 - 4x_1 - 4x_2.$

Subject to $2x_1 + 3x_2 < 6.$

$x_1, x_2 > 0.$

Here $P = (-4,-4)^T$, $C = \begin{bmatrix} 2 & 0 \\ 0 & 2 \end{bmatrix}$

Here C is positive definite.

Solution : Lagrangian $L(x,\lambda) = 2x_1^2 + 2x_2^2 - 4x_1 - 4x_2$

$+ \lambda_1 (2x_1 + 3x_2 - 6).$

Applying Kuhn-Tucker conditions,

$\dfrac{\delta L}{\delta x_1} = 4x_1 - 4 + 2\lambda_1 \geq 0; \; x_1(4x_1 - 4 + 2\lambda_1) = 0$

$\dfrac{\delta L}{\delta x_2} = 4x_2 - 4 + 3\lambda_1 \geq 0, \; x_2(4x_2 - 4 + 3\lambda_1) = 0,$

$\dfrac{\delta L}{\delta \lambda_1} = 2x_1 + 3x_2 - 6 \leq 0, \; \lambda_1(2x_1 + 3x_2 - 6) = 0$

$x_1, x_2, \lambda_1 \geq 0.$

Now $4x_1 + 2\lambda_1 - \mu_1 = 4$ $x_1 \mu_1 = 0,$

$4x_2 + 3\lambda_1 - \mu_2 = 4,$ $x_2 \mu_2 = 0,$

$2x_1 + 3x_2 + s_1 = 6,$ $\lambda_1 S_1 = 0.$

$x_1, x_2, \lambda_1, \mu_1, \mu_2, s_1 > 0.$

A solution of the latter is a solution of the former. We solve the latter by artificial variable technique by introducing artificial variables. We get, phase-I of solution.

Min $r_0 = R_1 + R_2$,

Subject to $4x_1 + 2\lambda_1 - \mu_1 + R_1 = 4$,

$4x_2 + 2\lambda_1 - \mu_2 + R_2 = 4$,

$2x_1 + 3x_2 + S_1 = 6$.

All variables > 0.

To satisfy, (1) both x_1, μ_1 or both x_2, μ_2 (λ_1 & S_1) should not be in basic, except in degeneracy of one variable. So, when choosing, entering & bearing veriables, this should be kept in mind.

Table

Basic	r_0	x_1	x_2	γ	μ_1	μ_2	R_1	R_2	S_1	Soln
r_0	1	4	4	5	-1	-1	0	0	0	8
R_1	0	4	0	2	1	0	1	0	0	4
R_2	0	0	4	3	0	-1	0	1	0	4
S_1	0	2	3	0	0	0	0	0	1	6

Table is the starting table of phase - I. By rule, λ_1 enter the basis,

but, $\lambda s_1 = 0$ restricts its entry since S_1 is not leaving. So, next promising entry x_1 enters and R_1 leaves. **We continue the solution and the next table gives the solution of phase - I. By Kuhn-Tucker theory**, it is also the solution of the problem.

3.6.7 Non-Linear Systems

In Civil Engineering by far the largest source of non-linear decision making problems is the design phase of a project. Since almost all design problems are non-linear, it is therefore the examples chosen to introduce and illustrate the systems are design examples.

The design phase of a project may be viewed as having two parts: firstly macro-design followed by micro-design. Engineering design is always a process of "optimisation". Good designers always strive to produce the best possible design.

The first step in the systematic approach is to determine the decision making and assigning a mathematical variable to each decision. To synthesise a design the assigned mathematical variables are the physical quantities of the design: the numbers, sizes, dimensions and geometry of the components of the design.

The second step is to impose the restrictions upon the variables by the problems and construct a mathematical model of these relationships, which is often a difficult task. The restrictions can originate from a variety of sources as:

    (a) The properties of the materials.
    (b) Limits imposed by codes of practice.
    (c) The mechanical behaviour.
    (d) Limits imposed by the availability of materials.
    (e) Limits imposed by the boundaries of the design.
    (f) Limits imposed by the construction methods.
    (g) Limits imposed by aesthetic considerations.

But in most design synthesis problem it is not possible to express all the restrictions in mathematical terms. The objective technological restrictions can be modelled mathematically but the subjective restrictions cannot be easily modelled.

The third step requires that some criterion representing the excellence of the design must be chosen and a mathematical objective function written to express this criterion as a function of all the design variables. The criterion depends upon the particular problem. The fourth step is to solve the problem to yield optimum values for all the variables. The resulting optimum values of the design variables then represent the synthesised design - an initial design to be analysed and modified until it becomes a final design.

Example: Beam design: Design a simply supported elastic beam of rectangular cross-section and known span L to carry a line uniformly distributed load of w KN/m. The material of the beam weighs p KN/cum.

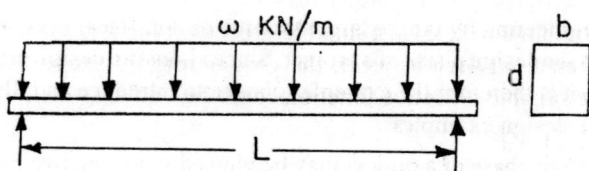

Solution

The design must select values for the breadth b and depth d, so they are the design variables.

Requirements:

    (a) It must not deflect excessively.
    (b) Maximum bending stress and shearing stress should exceed the limits.

Design considerations:

(a) deflection = L/360.

(b) max. deflection occurs at the center of beam = $(5 w L^4)/(384 E I)$

(c) for a rectangular cross-section $I = (b d^3)/12$.

(i) Deflection requirement $(5 w L^4)/(32 E b d^3) = < D$ .... (1)

(ii) Maximum permissible bending stresses.

Maximum bending moment on the beam (Live load + Dead load) is $M = (w + b d) L^2/8$

the constraint is

$[(w + b d) L^2/8] (d/2) (12/bd^2) = < f_b$

$(2 w L^2) + (3 L^3) < f_b$

(iii) If average permissible shearing stress = τ

$[(w L/2)/(b d) + (L/2)] = < \tau_{av}$ .... (3)

and if the maximum permissible shearing stress is τ_{max}. then the constraint

and if the maximum permissible shearing stress is max. then the constraint

$[(3 w L/4)/(b d) + (3 L)/4] = < \tau_{max}$

Apart from these, there may also be limits placed upon the over all dimensions or upon the relative properties of the design i.e.

$B_{min} = < B = < B_{max}$

$D_{min} = < D = < D_{max}$

$(B/D)_{min} = < (B/D) = < (B/D)_{max}$

Many possible feasible designs exist and in order to choose the best, some criterion of efficiency or economy may be used. A often used criterion is "Cost". To minimize the cost, a suitable objective function has to be chosen. If, a beam material such as steel, the cost is roughly proportional to the volume, the suitable objective function could be

$C = L b d$

Then the problem objective is to find the values of b and d which minimize the equation $C = L b d$ and do not violate any of the constraint inequalities.

Non-Linear Unconstrained Optimisation Methods

3.7.1 The Classical Differential Method

Consider the problem

Minimise $f(X_1, X_2, ..., X_n)$

over variables $X_j, j = 1, 2, ...n$

Necessary and sufficient conditions for a minimum, are

(a) $\delta f \delta x_j = 0$ $j = 1, 2, ... n$

When a function has n variables a stationary point must be stationary with respect to all variables simultaneously.

(b) To ensure that the stationary point is minimum:

for one variable $\delta^2 f/\delta x^2 > 0$

for n variables

$$H \text{ (Hessian matrix)} = \begin{bmatrix} \dfrac{\delta^2 f}{\delta x_1^{\,2}} & \dfrac{\delta^2 f}{\delta x_1 x_2} & \dfrac{\delta^2 f}{\delta x_1 \delta x_n} \\[2mm] \dfrac{\delta^2 f}{\delta x_n x_1} & \dfrac{\delta^2 f}{\delta x_n x_2} & \dfrac{\delta^2 f}{\delta x_n^{\,2}} \end{bmatrix} \text{ is + ve definite}$$

For a maximum, H must be negative definite. The matrix H is positive (negative) definite if all the determinants $d_1, d_2, d_3 ...., d_n$ are positive negative.

Disadvantages

(a) Function $f(X)$ must be analytically differentiable.

(b) Both $f(X)$ and its first derivatives must be continuous functions. This rules out problems involving the optimisation over number or sizes of pipes, rolled steel sections etc.

(c) Solving a set of n simultaneous non-linear equations.

3.7.2 Zeroth - Order Methods

It is the only method applicable to solving the problems with discontinuities. The specific requirement of a zeroth order method is that it must be possible to evaluate the value of $f(X)$ for the given numerical values for the components $x_j, j = 1, 2, ... N$ or X. Each numerical evaluation of $f(X)$ at a given X is called a trial and the objective of it is to locate the point X* at which $f(X)$ achieves a minimum (maximum) value using minimum no. of trials.

Single variable problems: It is convenient to start by studying problems involving a function $f(X)$ of only one variable X.

3.7.3 Grid Search and Random Search

First choose a wide range of X and divide it equally in a finite no. of intervals. Evaluate the value of function $f(X)$ and find a point X*, where the value of function is lower in value then at any other point. Logically a minimum of f (X) must lie somewhere between the two values of X immediately adjacent to X*. Divide the range again in a finite no. of interval and repeat the method till minimum value is received.

In random search, generate a random set of values of X and evaluate f (X) at each random point. Both are crude and insufficient methods.

3.7.4 Sequential Search (Bracketting the Minimum)

Using a subscript notation to denote sequential values of the variable X, a starting value of X is chosen, X_0, and a step length 's' (constant or vari-able). The function f (X) is evaluated at $X_1 = X_0 + s$, $X_2 = X_0 + 2s$ ..... until a function increases over the previous trial is obtained. The last three trials then bracket the minimum. The choice of step length s is important. It should not be too small (No. of trial increases) nor it should be too large (bracket length will be large). A variable step length can be used to overcome the problem.

3.7.5 Fibonacci Search Method

The method is superficially a complicated one but actually it is easy to apply. First find out the no. of trials by examining the ratio of the initial interval to the desired final interval. The first two trials are then placed at spacings proportional to the two Fibonacci numbers (1, 1, 2, 3, 5, 8, 13, 21 ...). The remainder of the search is then automatic. After each elimination the next trial is always placed symmetrically to the existing trial in the reduced interval unless it is the last trial in which case it is placed as close as possible to the existing trial in the interval.

Example: Find a minimum of the function

$$f(X) = X^4 - 4X^3 - 6X^2 - 16 X + 4$$

within an interval of length 0.1 starting from point X = 0.

Solution

First bracket the minimum. A basic step length of s = 0.1 is selected.

Trial	Step length (*0.1)	Position (X)	f (X)
1.	-	$X_1 = 0$	4
2.	1	$X_2 = X_1 + s = 0.1$	2.3361
3.	1	$X_3 = X_2 + s = 0.2$	0.5296
4.	2	$X_4 = X_3 + s = 0.4$	- 3.5904
5.	3	$X_5 = X_4 + s = 0.7$	- 11.2719
6.	5	$X_6 = X_5 + s = 1.2$	- 28.6874
7.	8	$X_7 = X_6 = s = 2.0$	-68
8.	13	$X_8 = X_7 + s = 3.3$	- 139
9.	21	$X_9 = X_8 + s = 5.4$	- 36.9104

Table shows the minimum of f (X) to be bracketed by trials 7 and 9. The length of the interval = (13 + 21) x 0.1 = 3.4. Now we shall reduce the interval of length 3.4 to one of length 0.1 within which the minimum lies. Now only (9-2) trials will be needed to bracket the minimum.

	LH trial	Internal trial	RH trial	New trial
Trial no.	7	8	9	10
spacing	-	13 x 0.1 from LH	-	13 x 0.1 - RH
X	2.0	3.3	5.4	5.4 - 1.3 = 4.1
f (X)	- 68.0	- 139.2959	- 36. 9104	- 155.5679

Next trial (11) will be 8, 10, and 9 (LH, Internal and RH respectively) which gives f (X) = - 138.1584. 138 < 155, so next trial no. 12 will be made from 8, 10, and 11 which gives f (X) = - 154.4144. If we continue like this trial no. 15 gives f (X) = - 155.5919 and the interval will be 0.1.

3.7.6 Curve Fitting Methods

Suppose a function f (X) of a single variable has been evaluated at three points X_1, X_2 and X_3 and values of function f (X) at X_1, X_2 and X_3 are known. A general quadratic curve F (X) = aX + bX + c may be fitted so

$$f (X_1) = aX_1 + bX_1 + c$$

$$f (X_2) = aX_2 + bX_2 + c$$

From the three equations the coefficients a, b and c can be found. Instead of trying to search numerically for the minimum of f (X) a minimum of the fitted function F (X) can be found very easily. If F (X) is a reasonable approximation to the shape of f (X), the minimum value calculated from F (X) will be close to the true value of X* which minimises f (X) then

$$d f/dx = 2 aX + b = 0 \qquad \text{(for a quadratic function)}$$

$$\frac{d^2f}{dx^2} = 2a \text{ (positive)}$$

The value of X* minimises f(X), the approximating quadratic function and the fourth trial is made at $X_4 = - b/2a$ and the function f (X) is calculated. Now according to the results new three values can be taken (for example X_2, X_3 and X_4) and the same process can be repeated till the time the sequence of new trial points coverages on a small region. If it is used wisely, it is very efficient and quick method but if the function chosen is not a good representative of the actual function it often fails to converge and it may diverage. A higher order polynomial such as a cubic equation is likely to approximate to a general function better than a simple quadratic. It is quite possible to use a cubic - fitting method although there are some added difficulties.

The general cubic function is

$$f(X) = aX + bX + cX + d$$

$$X^* = [-b + sqrt(b \times b - 3ac)]/(3a)$$

provided b x b > 3ac

Problems with more than one variable

Sequential line minimisation methods: A starting point is chosen consisting of values for the variables X_i, i = 1, 2, ... N which comprise X. N-1 of the variables are kept constant while the value of one variable X_i, is allowed to vary. A minimum of f (X) with respect to X_i is calculated by any one variable minimisation methods. Once it is completed, the procedure is repeated for another variable till the time all variables are covered.

Simplex method: The name is misleading because the method has absolutely nothing to do with the simple method. A simplex is defined as a pattern of N + 1 points which donot all lie in the same N-dimensional hyperplane. For a two-variable problem a simplex has three points arranged in a triangle, for 3-variable problem for four points define a tetrahedron.

First order methods: Earlier we have discussed to minimize the function using the values of the function. To know more about the function first partial derivatives can be calculated. The method is applied assuming that the exact values of all first derivatives of the function can be calculated. It is possible to evaluate the value of a function of one variable by making two trial functions evaluation very close together at X and x + Δx. Then an approximate derivative value is given by

$$df/dx = (f(X + \Delta X) - f(X))/\Delta X$$

For the numerical derivative to be accurate X must be very small.

3.8 Non linear constained optimization methods

Constrained non linear optimization problems are always more difficult than unconstrained ones. There are meny methods available which can be used effectively by the careful choice of method according to the problem. The similar problems can be made easier by elimination of constraints and normal of variables. Lagrangian methods convert a constrained optimization problem into an uncronstrained by solving a concerging sequence of unconstrained problems. Sequential linear proramming approximates all the non-linear functions in the problem by linear functions. Geometric programming. is very useful in engineering design problems and it is concentually different from other methods.

PANALTY FUNCTION METHODS

Like lagrangion methods, the constraints are added to the objective function by means of penalty coefficients and an unconstrained minimum of this new objective function is determined. This process is repeated several time. Each time using different values of coefficients is selected so as to make the unconstrained solution converge to real solution.

EQUALITY CONSTRAINED PROBLEMS :

The problem is

$$\text{Min } f(x_1,........x_n)$$

Subject to $c_i(x_1,............,x_n) = 0$ $i = 1, 2......, I$

This is converted into

$$\text{Min } P(x,R) = f(x_1,.......,x_n) + R \sum_{i=1}^{I} [Ci(x_1,...x_n)]^2$$

P(x,R) is the penalty function and it is to be minimised over variable x. The penalty coefficient R is a constant. Suppose R to be zero, then, the values of $c_i(x_1,......x_n)$ donot affect the solution and the penalty function is zero. But the solution we get might not satisfy the constraints. If we keep the value of 'R' sufficiently high, then to get minima, the values of $c_i(x_1,......x_n)$ should be zero, Thus , the solution will satisfy the constraints.

Example : $\text{Min } f = 1/2\,(x_1^2 + x_2^2 + x_3^2)$

Subject to $g_1 = x_1 + 2x_2 + 3x_3 - 1 = 0$

$g_2 = 3x_1 + 2x_2 + x_3 - 2 = 0$

Solution

The problem is converted to

$$\text{Min } P(x,R) = 1/2\,(x_1^2 + x_2^2 + x_3^2) + R[(x_1 + 2x_2 + 3x_3 - 1)^2 + (3x_1 + 2x_2 + x_3 - 2)^2]$$

Table

R	x_1	x_2	x_3	S_1	S_2	f	P
0	0	0	0	-1	-2	0	0
10^{-2}	0.9959	0.00547	0.00004	-0.88935	-1.69025	0.00497	0.04145
1	0.46712	0.24476	0.03663	0.06653	-0.07249	0.13973	0.14941
10^2	0.49964	0.24995	0.00026	0.00032	-0.00092	0.15606	0.15615
10^4	0.49999	0.24999	0.0000	0.0000	-0.00001	0.15625	0.15625
∞	0.5	0.25	0	0	0	0.15625	0.15625

Inequality constrained problems :

The problem is,

$$\text{Min } f(x_1,.....,x_n)$$

$$\text{Subject to } c_i(x_1,.......,x_n) \le 0 \qquad\qquad i = 0, 1, ...I \qquad (1)$$

The panalty function form is

$$\text{Min } P(x,R) = f(x_1,......x_n) - R\,[c_i(x_1,......x_n)]^{-1} \qquad (2)$$

To keep $P(x,R)$ larger than f, a requirement for minimization problems, it is necessary to subtract the penalty since the constraints can have negative values. The penalty function will be very large near the constraint boundaries. So, essentially the inequality constraints keep the solution away from the boundaries. From a large value of R, the problem is solved several times with consectively lower values of R allowing the solution to approach the constraints which are right (=0) at the optimum from the negative side. Those which are slack would remain slack with large negative values of c_i bacause thier reciprocals will be small and will be negligible when multiplied by a small R.

The solution of problem (1) is determined by a sequence of solutions of (2) is unconstrained, so the unconstrained optimization methods can be used to solve it.

However, there are some difficulties in the numerical solution of penalty functions. They are listed below.

1. An feasible point which satisfies the constraints is needed to start the unconstrained minimization. For, engineering problems, it can be taken as the initial feasible design. But, fortunately, only one feasible starting points is needed for the whole sequence of minimizations.

2. The initial value of R and the reduction scheme for R has to be chosen carefully since the pace of concregence to the solution depends very much on this choice. There are some specific techniques available but they are beyond the scope of this text.

Normalisation of constraints helps in improving the general efficiency and speed of convergence of penalty function solutions. Normalisation consists of altering all constraints so that the constant term is unity.

$$C_1(x) = \delta(x) - \delta_{max} < 0$$
$$C_2(x) = \sigma(x) - \sigma_{max} < 0.$$

Normalised constraints :

$$c_1(x) = \frac{\delta(x)}{\delta_{max}} - 1 \leq 0$$

$$c_2(x) = \frac{\sigma(x)}{\sigma_{max}} - 1 \leq 0$$

The penalty function described above is called interior penalty function since it establishes a barrier and the minima have to be in the feasible region. But, there are exterior penalty function methods which operate in the infeasible region and converge to a feasible solution. A typical exterior penalty function is

$$\text{Min } P(x,R) = f(x_1,.....x_n) + R \sum_{i=1}^{I} [f_i(c_i)]^2$$

The main advantage is that, this method doesn't require a starting feasible point. But, the sequence of minimization can't be terminated until the feasible solution is arrived since all the intermediate solutions are infeasible unlike the interior penalty function methods.

Linearisation methods :

These consist of solving the NLPP with the help of linear simplex method by replacing NLPP with a series of LPP and solving them by simplex method to get the solution of NLPP.

Sequential linear programming :

The general inequality constrained non-linear problem is

$$\text{Min } f(x_1,.......x_n)$$

$$\text{Subject to } c_i(x_1,.........x_n) <= 0 \ i = 1,.....I.$$

Suppose $x^1 = (x_1^1,........x_n)$ is a feasible point. Using Taylor series expension truncated after linear terms, the functions f and c_i, i=1....I can be linearised. Now, the problem is

$$\text{Min } f(x^1) + \sum_{j=1}^{n} (\delta f/\delta x_j) \ x \ (x_j - x_j^1) \leq 0$$

$$\text{Subject to } \quad C_i(x^1) + \sum_{j=1}^{n} (\frac{\delta C_i}{\delta x_j})_x (x_j - x_j^1) \leq 0$$

Rearranging, we get

$$\text{Min } \ f(x^1) - \sum_{j=1}^{n} x_j (\frac{\delta f}{\delta x_j \ x^1}) + \sum_{j=1}^{N} (\frac{\delta f}{\delta x_j \ x_j}) x^1$$

subject to
$$S_i(x) - \sum_{j=1}^{n} x_j^1 \left(\frac{\delta C_i}{dx_{j-x_j}} \right) + \sum_{j=1}^{n} \left(\frac{\delta C_i}{dx_{j-x_j}} \right) xj \leq 0 \, x^1$$

The term within { } in objective function is a constant so it can be neglected. In each of constaints, the { } term can be transfered to the right hand side and made positive, so we get

$$\text{Min} \sum_{j=1}^{n} a_j x_j$$

Subject to
$$\sum_{j=1}^{n} b_{ij} x_j > C_i \quad i = 1 \ldots.. I,$$

$$\sum_{j=1}^{n} b_{ij} x_j > C_i \quad I_1 = 1 \ldots.. I,$$

This problem is identical to the LPP and can be solved by simplx method, subsequently. The resulting solution should be checked for any violations of constraints and if not, it should be used to drive the next LPP and solving them again for a new optimum. This method is not without disadvantages. A few are listed below.

1. A feasible starting solution must be found for the nonlinear constraint set. But, this may not lead to a feasible solution after the first linear approximation. While trying to restore feasibility, we may increase the value of objective function excessively.

2. The solution of non linear problems may lie on a constraint vertex or constraint boundaries. LPP always have solutions at constraint vertices. So, the sequence of LPP must converge upon a vertax of nonlinear constraints. Thus an optimum on a constraint boundary cannot be found by this method. This can be remedied by adding new sets of more limits constraints to enable the optima sequence ventura away from the constraint vertices and in theory, the optima inside the unconstrained region can be found.

3.8.2 Newton's Raspon Method

It requires that the function f be twice differentiable. It begins with a point x_1 that is initial estimate to the stationary point of the equation $f'(x) = 0$. A linear approximation of the function $f'(x)$ at the point x_1 is taken, and the point at which the linear approximation vanishes is considered as the next approximation. Given the point x_n to be the current approximation, the linear approximation of the function $f'(x)$ at x_n is given by

$$f'(x : x_n) = f'(x_n) + f''(x_n)(x - x_n)$$

setting it to be zero now, we get the next approximation point as

$$x_{n+1} = x_1 - \frac{f'(x_n)}{f''(x_n)}$$

Figure explains the general method of Newton method. Sometimes depending upon the starting point and the nature of the function, it diverges rather than converges.

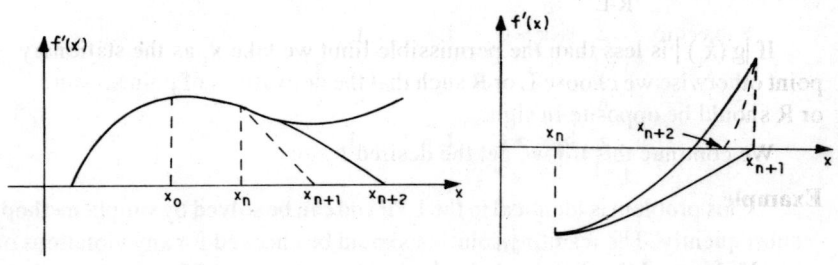

Newton Raspon method (convergence   Newton Raspon method (divergence)

Example: Minimize $f(x) = 2x^2 + \dfrac{16}{x} + 4$

starting at the point $x_1 = 1$.

$$f'(x) = 4x - \frac{16}{x^2}$$

$$f''(x) = 4 + \frac{32}{x^3}$$

at $x = 1$, $f'(x_1) = -12i$ $f''(x_1) = 36$

$$x_2 = 1 - \frac{-12}{36} = 1.33$$

at $x_2 = 1.33$, $f'(x_2) = -3.73$, $f''(x_2) = 17.6$

$$x_3 = 1.33 - \frac{-3.73}{17.6} = 1.54$$

It should be continued till we get the desired result (with tolerance).

3.8.3 Secant Method

The method combines, Newton's method with the Bisection method. (Two point are chosen such that their first derivatives are of opposite sign & the stationary point is tried at the mid point of the region defined by these two points.) To find the stationary point of a function $g(x)$ if we have two points L and R in region (m,n) so the derivaties of the points are of oposite sign then the next approximation is given by

$$x_1 = R - \frac{g'(R)}{g'(R) - \dfrac{g'(L)}{R-L}}$$

If $|g'(x_1)|$ is less than the permissible limit we take x_1 as the stationary point otherwise we choose L or R such that the derivatives of points x_1 and L or R should be opposite in sign.

We continue this till we get the desired result.

Example :

Minimize $g(x) = 2x^2 + \dfrac{16}{x} + 4$

Interval 1 to 8

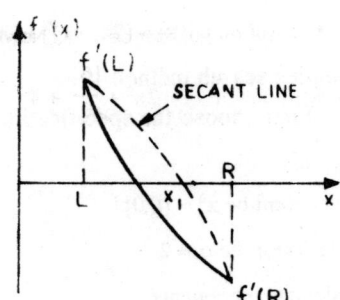

Solution :

$$g'(x) = 4x - \frac{16}{x^2}$$

Step 1: $R = 8 \Rightarrow g'(R) = 32 - \dfrac{16}{64} = 32 - \dfrac{1}{4} > 0$

$$= 127/4$$

$$L = 1 \Rightarrow g'(L) = 4 - \frac{16}{1} = -12 < 0$$

Step 2: $x_1 = 8 - \dfrac{127/4}{\dfrac{127}{4} - \dfrac{(-12)}{7}}$

$$= 8 - \frac{31.75}{31.75 + 1.71} = 8 - .948$$

$$= 7.05$$

Step 3: $g(x_1) = 4 \ (7.05) - \dfrac{16}{(7.05)^2}$

$$= 27.878 > 0$$

Second iteration : Region is limited to (1 to 7.05)

Step 1: $x_2 = 7.05 - \dfrac{27.878}{27.878 - \dfrac{(-12)}{6.05}}$

$$= 6.116.$$

Step 2: $g'(x_2) = 4 \ 6.116 - \dfrac{10}{(6.116)^2}$

$$= 24.03 > 0$$

We continue this till we get the desired result.

Example → Minimize $f(x) = (2x_1 - 1)^2 + (3x_2 - 2)^2$

by S^2 simplex search method

Solution: First choose the specification of an initial point and a scale factor.

Let initial point be $x^0 = \{0,0\}^T$

and scale factor be $\alpha = 2$

Then calculate increments

ΔL_1 and ΔL_2

$$\Delta L_1 = \left[\frac{\sqrt{(N+1)} + (N-1)}{(N \ 2)} \right] \alpha$$

$$\Delta L_2 = \left[\frac{\sqrt{(N+1)} - 1}{(N\sqrt{2})} \right] \alpha$$

N = problem dimension. In this case it is 2.

$$\Delta_1 = \frac{\sqrt{3} + 2}{2\sqrt{2}} \ X \ 2 = 1.93$$

$$\Delta_2 = \frac{\sqrt{3} - 1}{2\sqrt{2}} 2 = 0.5176$$

With these two increaments other two vertices are calculated

$$x^1 = [\,0 + 0.5176, 0 + 1.9318\,]^T$$

$$= (0.5176, 1.9318)^T$$

$$x^2 = [\,0 + 1.9318, 0 + 0.5176\,]^T$$

$$= (1.9318, 0.5176)^T$$

function values $f(x^0) = 5$, $f(x^1) = 14.4063$

$$f(x^2) = 8.400$$

Calculate the centroid of the points & then calculate the replacement $x^{(3)}$

$$x_c = \frac{1}{N} \sum_{\substack{i=0 \\ i \neq j}}^{N} x^i$$

$$= \frac{1}{2} \, (x^{(1)} + x^{(2)})$$

$$x^{(3)} = x^{(1)} + x^{(2)} - x^{(0)}$$

$$x_{new}^{(j)} = x^{(j)} + \lambda \, (x_c - x^{(j)})$$

for $\lambda = 2$

$$x^{(3)} = (2.4494, 2.4494)^T$$

At new point $f(x^3) = 43.80$

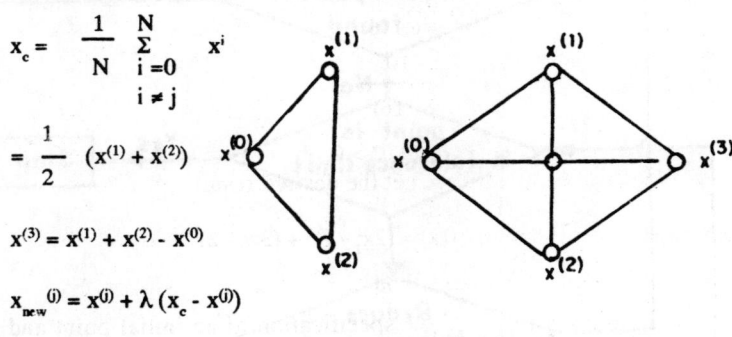

Rules

1. If the selected "worse" vertex was generated.
2. If a given vertex remains unchanged for more than m iterations, reduce the size of the simplex by some factor.

Find the minimum of

$$g(y) = 8y_1^2 + 4y_1y_2 + 5y_2^2$$

from $y^0 = [-2, -2]^T$

Solution

Step 1: $\Delta y = [1,1]^T =$ step increments

The Hooke-Jeeves Pattern Search Method

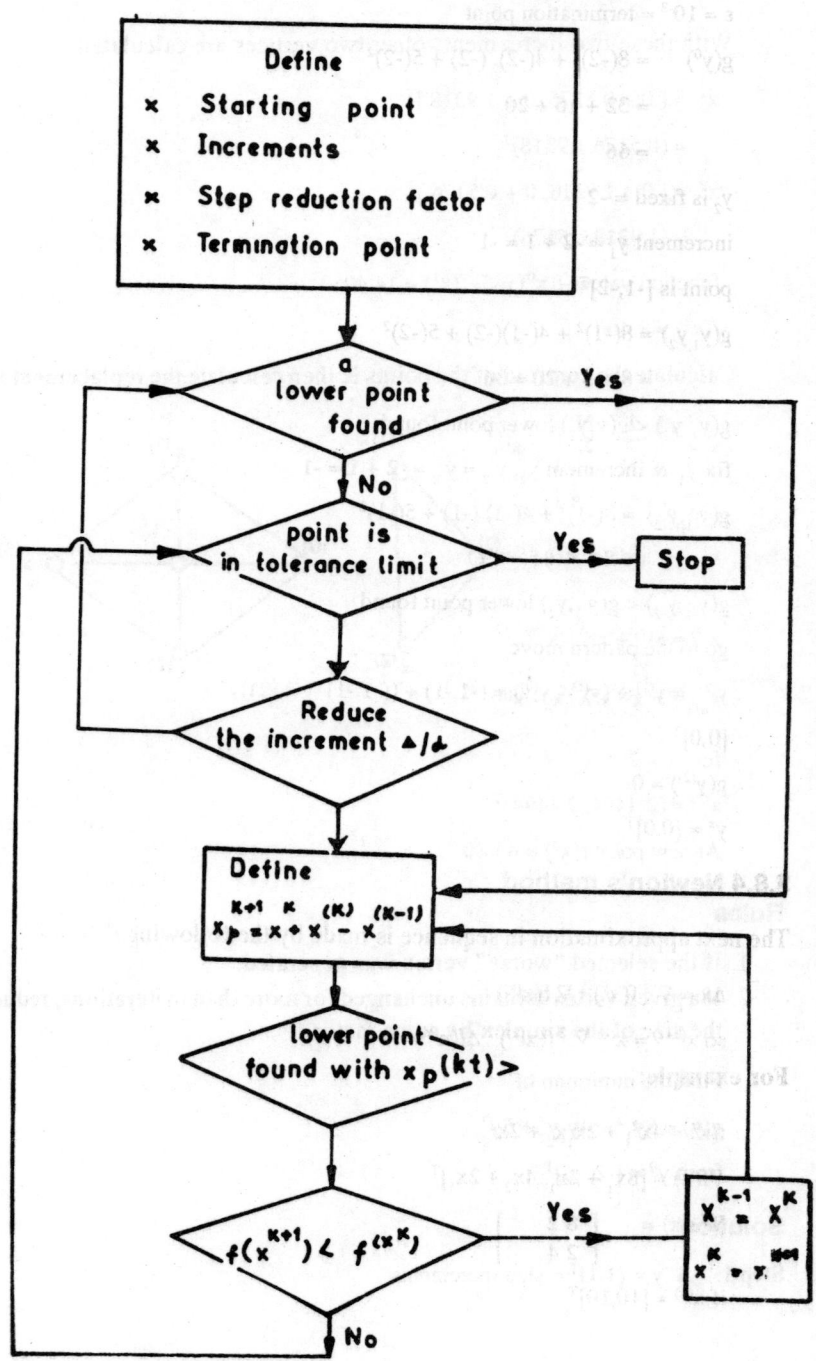

α = step reduction factor = 2

$\varepsilon = 10^{-3}$ = termination point

$$g(y^0) = 8(-2)^2 + 4(-2) \cdot (-2) + 5(-2)^2$$

$$= 32 + 16 + 20$$

$$= 66$$

y_2 is fixed = -2

increment $y_1 = -2 + 1 = -1$

point is $[-1,-2]^T$

$$g(y_1,y_2) = 8(-1)^2 + 4(-1)(-2) + 5(-2)^2$$

$$= 8 + 8 + 20 = 36$$

$g(y_1',y_2) < g(y_1,y_2)$ lower point found

fix y_1' & increment y_2, $y_2' = y_2 = -2 + 1 = -1$

$$g(y_1',y_2') = 8(-1)^2 + 4(-1)(-1) + 5(-1)^2$$

$$= 8 + 4 + 5 = 17$$

$g(y_1',y_2') < g(y_1',y_2)$ lower point found

go to the pattern move

$$y^{(2)}_p = y^{(1)} + (y^{(1)} - y^{(0)}) = (-1,-1) + ((-1,-1) - (-2,-2))y^2 =$$

$$[0,0]^T$$

$$g(y^{(2)}) = 0$$

$$y^x = [0,0]^T$$

3.8.4 Newton's method

The next approximation in sequence is made by the following relation

$$\Delta x = \nabla^2 f(x^k)^{-1} \nabla f(x^{(k)})$$

$$\text{so } x^{(k+1)} = x^{(k)} \nabla^2 f(x^{(k)})^{-1} \Delta f(n^{(k)})$$

For example:

$$f(x) = 4x_1^2 + 2x_1x_2 + 2x_2^2$$

$$\nabla f(x) = [8x_1 + 2x_1, 4x_2 + 2x_1]^T$$

$$\nabla^2 f(x) = \begin{bmatrix} 8 & 2 \\ 2 & 4 \end{bmatrix}$$

if $x^{(k)} = [10,10]^T$

$$x^{k+1} = [10,10]^T - \frac{1}{28} \begin{bmatrix} 4 & -2 \\ -2 & 8 \end{bmatrix} [100,60]$$

$$= [10,10]^T - \frac{1}{28} [280,280]$$

$$= [10,10]^T - [10,10]^T = [0,0]^T$$

3.8.5 Modified Newton's Method

If the chosen point is not close to the optimum point the method will be a long process & chances of divergence exist. To reduce the steps, in Newton's method a reduction facros α is introduced which is called modified Newton's method. According to this method

$$x^{(k+1)} = x^{(k)} - \alpha^{(k)} \, \nabla^2 \, f(x^{(k)})^{-1} \, \nabla \, f(x^{(k)})$$

Example: The sample mean

To ensure safety of a concrete structure three samples of compressive strength at 28 days are taken. The average or mean value of this sample must exceed the design strength of the concrete. Since the mean is a weighted sum of numbers, this procedure allows for one or two samples to be less than the design strength and still pass the strength test.

The following samples of 28 day strength were obtained : 18.8,

19.1, 21.2 N/mm^2

(a) The sum of squared deviations can be calculated about any value. Show that when it is calculated about the mean, the sum of the square deviations is a minimum.

(b) Determine if this sample meets design specifications for 20 N/mm^2 concrete.

Solution:

(a) Formulation : Let x = unknown average or mean value of the three samples. Let $x_1 = 18.8$, $x_2 = 19.1$ and $x_3 = 21.2$, be the sample 28 days strengths.

The deviation about the mean e is defined to be the difference between the observed and the mean value. Thus.

$$e_1 = x_1 - x = 18.8 - x$$

$$e_2 = x_2 - x = 19.1 - x$$

$$e_3 = x_3 - x = 21.2 - x$$

The sum of the square deviations about the mean σ is

$$= e^2_1 + e^2_2 + e^2_3 = (18.8 - x)^2 + (19.1 - x)^2 + (21.2 - x)^2$$

The objective is to minimize σ where x is the control variable.

The necessary and sufficient conditions for a minimum are

$$\frac{d\sigma}{dx} = 2(18.8 - x)(-1) + 2(19.1 - x(-1) + (21.2 - x)2(-1) = 0$$

$$\frac{d^2\sigma}{dx^2} = 2 + 2 + 2 > 0$$

Since $d^2\sigma/dx^2 > 0$ the conditions for a minimum are satisfied.

Simplifying $d\sigma/dx = 0$ gives :

$$x = \frac{(18.8 + 19.1 + 21.2)}{3}$$

$$= 19.7 \text{ N/m}^2$$

b) Adequacy Text: If the average strength of the sample exceeds 20 N/m^2 we may assume that the entire batch exceeds the minimum strength requirement. Since $x < 20$ the design requirement is not met.

Example Pressure Piping

The cost of piping is directly proportional to the amount of material used. Determine the minimum cost, or equivalently, the minimum-weight. Pipe to sustain a pressure of 15 N/mm^2. To accommodate the flow, the cross-sectional area of the pipe must be 5cm^2. The allowable stress of the steel piping material is 120 N/mm^2. Formulate a mathematical model and solve for the minimum weight per unit length of pipe.

Solution:

The total weight of pipe will be equal to

$$W = \rho.Al ; A = 2\pi rt$$

Where ρ is the unit weight of steel. r is the pipe radius; t is the pipe wall thickness; and l is the length of pipe. Thus, the objective is to minimize $w = W/l$. Since the cross section is 5cm^2 the radius of the pipe may be determined.

$$\pi r^2 = 5$$

$$r = \sqrt{\frac{5}{\pi}} = 1.26 \text{ cm}$$

Thus, the weight per unit length is

$$w = \rho A = \rho(2\pi rt)$$

$w = 7 \times 2\pi \times 1.26\ t$

$w = 55.41t$

Utilizing thin-wall pipe theory, the stress σ in the pipe is

$$\sigma = \frac{Pr}{t}$$

where $P = 15\,N/mm^2$ = internal pressure

$$\sigma = \frac{15 \times 1.26}{t} = \frac{18.9}{t}$$

The mathematical model is

Minimize $w = 55.41t$

$$\sigma = \frac{18.9}{t} \leq 120$$

$t > 0$

This is a nonlinear model because $18.9/t$ is a nonlinear function. The constraint equation may be rearranged as

$$\frac{18.9}{120} < t\ ;\ 0.1575 < t$$

The model becomes

Minimize $w = 55.41t$

$0.1575 \leq t$

$t \geq 0$

This is a linear model. Since w increases with t, the minimum value of t will occur at an extreme point, or

$t = 0.1575$ cm.

The minimum weight per unit length is

$W = 55.41 \times t$

$W = 8.72$ gm/cm.

Example: A Statically Determinate Minimum-Weight Truss

Consider the truss shown in Figure. Formulate a mathematical model to design a simple truss of minimum weight. The critical buckling and maximum allowable tensile stresses of compression and tension members are 1 KN/cm^2 and 2 KN/cm^2, respectively. The truss is to be constructed of steel. All compression and tension members are assumed to have the same cross-sectional area.

Solution:

Control Variables : The control variables are defined as

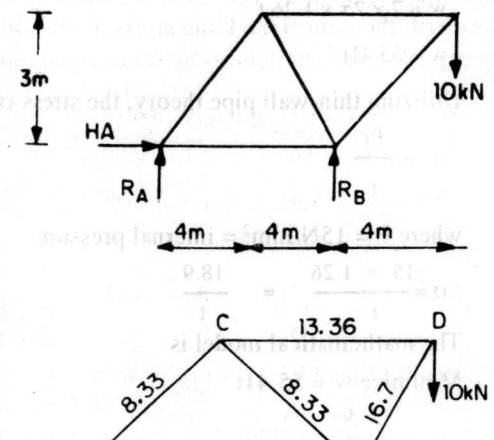

$x_1 = A_1$ = cross-sectional area of a compression member (cm^2)

$x_2 = A_2$ = Cross-sectional area of a tension member (cm^2)

The control vector x is $x^1 = [A_1 \ A_2]$.

Member Forces and Stresses: Use the method of joints to determine forces in each member.

$\Sigma F_x = 0$: -CD + 4/5 BD = 0

$\Sigma F_y = 0$: 3/5 BD - 10 = 0

The member forces are BD = 16.7 KN (compression) and CD =13.36 KN (tension), The stress in the members will be equal to the member force divided by the cross-sectional area of the member. In this case, the tension member CD will have a stress σ_{CD} equal to

$$\sigma_{CD} = \frac{13.36}{A_2}$$

The compression member CD will have a stress equal to

$$\sigma_{BD} = \frac{16.7}{A_1}$$

The member forces and reactions are shown in Figure. The stress in remaining members AB, AC, and BC will be equal to

$$\sigma_{AC} = \frac{83.3}{A_2} \quad : \sigma_{AB} = \frac{66.7}{A_1} \quad : \sigma_{BC} = \frac{83.3}{A_1}$$

respectively

Constraint Equations : The stress in each member may be equal to, but must not exceed, the critical buckling stress or allowable tensile stress. Thus, we may express these restrictions by set of equations :

Member AC : $\dfrac{8.33}{A_2} \le 2$ or $A_2 > 4.17$

Member AB : $\dfrac{6.67}{A_1} \le 1$ or $A_1 > 6.67$

Member BC : $\dfrac{8.33}{A_1} \le 1$ or $A_1 > 8.33$

Member CD : $\dfrac{13.36}{A_2} \le 2$ or $A_2 > 6.68$

Member BD : $\dfrac{16.7}{A_1} \le 1$ or $A_1 > 16.7$

Objective function: The weight of each member is equal to the density of steel times the volume of each member. The density of steel is approximately 12gm/cc. The equation for weight of the truss is the sum of the weight of each individual member; thus,

$$W = [V_{ab} + V_{ac} + V_{bc} + V_{bd} + V_{cd}] \times 12$$

Where V = volume of each member.

The objective function is

$$W = [500A_2 + 800A_1 + 500A_1 + 800A_2 + 500A_1] \times 12$$

Mathematical Model: The problem formulation is complete, Summarizing the equations results in the following mathematical model :

Minimize $W = 21600A_1 + 15600A_2$

Subject to the constraints

$A_2 \ge 4.17$

$A_1 \ge 6.67$

$A_1 \ge 8.33$

$A_2 \ge 6.25$

$A_1 \ge 16.7$

→ min $W = 458220$ gms.

Example: Maximum Fluid Flow Through a Pipe

The manning equation is an empirically derived formula used to determine the flow in open channels, sewer lines, flumes and so forth.

$$Q = \frac{1}{n} \, A \, R^{2/3} \, S^{1/2}$$

where Q is the flow in meter cube/sec: n is the channel roughness coefficient; A is the cross-sectional area of flow ; R is the hydraulic radius ; and S is the slope of the channel. Consider a 1m-diameter concrete pipe (n = 0.013) on a slope of 0.0001.

(a) Formulate an optimization model to determine the maximum flow in the pipe.

(b) Use the Newton method to determine the optimum solution.

Solution:

(a) Our objective is to determine the maximum flow.

$$\text{Maximum } Q = \frac{1}{n} A R^{2/3} S^{1/2}$$

This equation may be rewritten in terms of the cross-sectional area A and wetted perimeter P.

Thus

$$\text{Maximize } Q = \frac{(1)}{(n)} \frac{A^{5/3} \, S^{1/2}}{P^{2/3}}$$

where A and P will be written in terms of the control variable x, which is depicted in the diagram of the cross section of the pipe.

The wetted perimeter P is

$$P = \pi r + 2rx = (\pi + 2x).\,r$$

The area A is

$$A = \frac{\pi r^2}{2} + \int dA$$

where dA is the incremental area .

or

$$dA = x\,dy$$

$$A = \frac{\pi r^2}{2} + 2 \int (r^2 - r^2 \sin^2 \alpha)^{1/2} r \cos\alpha \; x\,dx$$

Evaluating the integral gives

$$A = \frac{\pi r^2}{2} + r^2\left(\alpha + \frac{\sin 2\alpha}{2}\right)$$

The mathematical model in terms of the control variable x is

$$\text{Maximize } Q = \left(\frac{1}{n}\right) S^{\frac{1}{2}} \left(\frac{\pi r^2}{2} + r^2 \left(\alpha + \frac{\sin 2\alpha}{2}\right)\right)^{5/3} (\pi r + 2r\alpha)^{-2/3}$$

$$0 \leq \alpha \leq \pi/2$$

(b) The optimum solution for maximum flow must satisfy the following condition :

$$\frac{dQ}{dx} = 0$$

or

$$\frac{dQ}{d\alpha} = \left(\frac{1}{n}\right) S^{\frac{1}{2}} \left(\frac{5}{3}\right)\left(\frac{\pi r^2}{4} + r^2 \left(\frac{\sin 2\alpha}{2}\right)\right)^{2/3} (r^2(1 + \cos 2\alpha)$$

$$(\pi r + 2r\alpha)^{-2/3} + \frac{1}{n} \quad S^{\frac{1}{2}}\frac{-2}{3} (2r) \left(\frac{\pi r^2}{4} + r^2 \left(\frac{\sin 2\alpha}{2}\right)\right)^{5/3}$$

$$(\pi r + 2rx)^{-5/3} = 0$$

After substituting all known values of S, n, and r and simplifying the expression reduces to

$$\frac{dQ}{dx} = (\pi + 2x)(3 + 5 \cos 2x) - 2 \sin 2x = 0$$

The iteration equation for the Newton method is

$$\alpha^{k+1} = \alpha^k - \frac{f(\alpha^k)}{f'(\alpha^k)}$$

where

$$f(\alpha) = \frac{dQ}{d\alpha} = (\pi + 2\alpha)(3 + 5 \cos 2\alpha) - 2 \sin \alpha$$

$$= \frac{1}{n} S^{1/2} \left(\frac{\pi r2}{2} + r^2\left(\alpha + \frac{2 \sin 2\alpha}{2}\right)\right)^{5/3} (\pi r + 2r\alpha)^{-2/3}$$

$$= \frac{1}{0.013} (0.0001)^{\frac{1}{2}} \left(\frac{\pi k}{8} + \frac{1}{4}\left(1 - \alpha + \frac{\sin(2 \times 63.025)}{2}\right)\right)^{5/3}$$

$$\left(\frac{1.1}{2}\right) + 2)^{-2/3}$$

$$= \frac{1}{0.013} \quad (0.0001)^{\frac{1}{2}} \frac{(0.645)}{(1.927)}$$

For $\alpha = 1.1$ Radians $Q = 0.257$ m^3

next approximation - $\alpha^{k+1} = 1.1 - \frac{f(\alpha^k)}{f'(\alpha^k)}$

$$= 1.1 - \frac{(\pi + 2(1.1))(3+5\cos 2\alpha) - 2\sin\alpha}{(6 - 10\pi \sin 2\alpha - 10\alpha \sin 2\alpha + 10 \cos 2\alpha - 2\cos\alpha)}$$

$$= 1.06$$

$$Q = \frac{1}{0.013} \left(0.0001 \right)^{\frac{1}{2}} \left(\frac{\pi}{8} + \frac{1}{4} \left(1.06 + \frac{\sin(60.73 \times 2)}{2} \right) \right)^{5/3}$$

$$\left(\frac{\P}{2} + 1.06 \right)$$

$$= \frac{1}{0.013 \, 2.} \; (0.01) \; \frac{(0.764)}{63} = 0.223 \text{ m}^3$$

Example : Minimum cost aggregate mix

A contractor has to supply the material for which he has two alternatives.

1. He takes out the material from first pit and supplies, for which he has to bear Rs 100/m^3.

2. Material from pit 2, for which he has to spend Rs 150/m^3 for loading, delivering and unloading.

The minimum quantity to be delivered is 1000 m^3. The mix that he has to deliver should have

1. not more than 75% gravel

2. minimum 60% and

3. Not more than 5% silt.

The pit 1 material consists of 70% gravel and 30% sand. The pit 2 material consists of 30% gravel, 60% sand and 10% silt.

* Formulate a minimum cost model.

* Determine the proportion of sand gravel and silt in the optimum solution.

Solution:

(a) **Formulation:** We define the control variables to be

x_1 = amount of material taken from pit 1

x_2 = amount of material taken from pit 2

The cost function is

minimize $C = 100 \, x_1 + 150 \, x_2$

Let $x_1 + x_2$ equal the total amount of standard mix delivered to the project site. The delivery constraint is

$x_1 + x_2 \geq 1000$

The mixture must contain at least 60% sand.

$0.3x_1 + 0.6x_2 \geq 0.6 \, (x_1 + x_2)$

The products $0.3x_1$ and $0.6x_2$ are the amounts of sand taken from pits 1 and 2, respectively. The term $0.6(x_1 + x_2)$ is the amount of sand in the mix. Similarly,

$0.7x_1 + 0.3x_2 \leq 0.75 \, (x_1 + x_2)$

the constraint equation for silt is

$0.1x_2 \leq 0.05 \, (x_1 + x_2)$

The minimum cost model may be written as

Minimize $C = 100x_1 + 150x_2$

$x_1 + x_2 \geq 1000$	(delivery)
$0.3x_1 + 0.6x_2 \geq 0.6(x_1 + x_2)$	(sand)
$0.7x_1 + 0.3x_2 \geq 0.75(x_1 + x_2)$	(gravel)
$0.1x_2 \geq 0.05(x_1 + x_2)$	(silt)
$x_1 \geq 0$	
$x_2 \geq 0$	

A Minimum Weight Truss

Determine the angle α and the cross-sectional areas of the members 1,2 and 3 of a minimum weight truss. Assume that the compression members, 1 and 2, are of equal cross-sectional area and of equal length. The stress in members must not exceed compressive or tensile limits of 10 N/m^2

(a) Formulate a mathematical model.

(b) Use the newton method to search for the optimum solution.

The objective function to minimize the total weight, or equivalently, the total volume of the truss is

$V = A_1 l_1 + A_2 l_2 + A_3 l_3$

where $A_1, A_2,$ and A_3 are cross-sectional areas of the members 1,2 and 3 and $l_1, l_2,$ and l_3 are the member lengths, respectively. The length of member

3 is known, $l3 = 6m$. The lengths of members 1 and 2 are variables with $l1 = l2$. From trigonometry,

$$\cos a = \frac{3}{1_1} = \frac{3}{1_2}$$

or

$$1_1 = 1_2 = \frac{3}{\cos a}$$

With this relationship and the assumption that $A_1 = A_2$, the objective function becomes

$$\text{Minimize } V = 2A \left(\frac{3}{\cos \alpha} \right) + 6A_3 = {}^* 6 \left(\frac{A_1}{\cos \alpha} + A_3 \right)$$

From statics, it may be shown that the vertical reactions at A and B are equal $R_A = R_B = 2.5N$.

The method of joints will be used to determine the member force in the members. At joint C, the free-body diagram is as shown in figure. The sum of the vertical and horizontal forces is

$$5 = F_1 \sin \alpha - F_1 \sin \alpha = 0$$

$$F_1 \cos \alpha - F_2 \cos = 0 \tag{1}$$

$$F_1 = F_2 \tag{2}$$

The member force is determined

$$F_1 = F_2 = \frac{2.5}{\sin \alpha} \quad ; F_3 = 2.5 \tan \alpha \tag{3}$$

Substitute Eq. (2) into this expression and simplify :

$$F_3 = 5 \frac{\cos \alpha}{\sin \alpha}$$

The force F_3 is a tensile force.

The stresses in members 1,2 and 3 are limited to an allowable stress of 10 N/m^3. Thus, the constraint set will consist of the following equations :

$$\frac{F_1}{A_1} = \frac{F_2}{A_2} < 10$$

$$\frac{F3}{A3} < 10$$

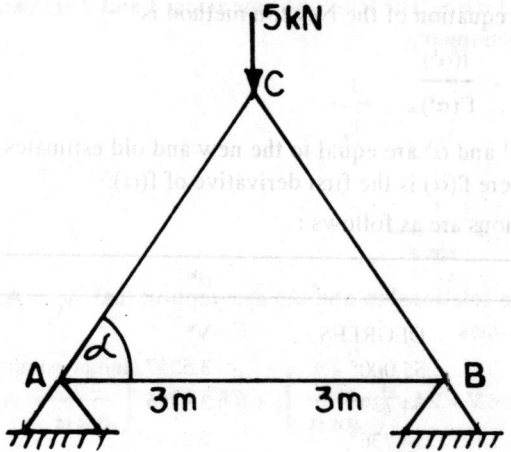

or

$$\frac{2.5}{A_1 \sin\alpha} < 10 \text{ or } A_1 < \frac{1}{4\sin\alpha}$$

$$\frac{2.5 \cos\alpha}{A_3 \sin\alpha} < 10 \text{ or } A_3 \geq \frac{1\cos\alpha}{4\sin\alpha}$$

The mathematical model becomes

Minimize V = 10 $\left(\dfrac{A_1}{\cos\alpha} + A_3 \right)$

$$A_1 \geq \frac{1}{4\sin\alpha}$$

$$A_3 \geq \frac{1\cos\alpha}{4\sin\alpha}$$

$$A_1 \geq \frac{1\cos\alpha}{4\sin\alpha}$$

A_1 and A_3 from the constraint set may be substituted into the objective function :

Minimize V = 10 $\left(\dfrac{1}{4\cos\alpha \sin\alpha} + \dfrac{1\cos\alpha}{4\sin\alpha} \right)$

or

Minimize V = $\dfrac{1.25}{\sin\alpha}$ $\left(\dfrac{1}{\cos\alpha} + \cos\alpha \right)$

(c) The optimum solution will occur at the point where $f(\alpha) = 0$, The iteration equation of the Newton method is

$$\alpha^{k+1} = \alpha^k - \frac{f(\alpha^k)}{f'(\alpha^k)}$$

where α^{k+1} and α^k are equal to the new and old estimates of α, respectively, and where $f'(\alpha)$ is the first derivative of $f(\alpha)$.

The iterations are as follows :

		α^k	
k	DEGREES	V^k	
0	55.000°	3.5357 (Initial estimate)	
1	54.739°	3.5355	
2	54.736°	3.5355	

This table illustrates that the initial estimate is a good one, and the Newton method had little difficulty converging to a solution in a minimum number of iterations. The solution is a critical point.

$$\alpha^* = 54.736° \quad V^* = 3.604$$

From angle α, Areas, & forces can be calculated

$$A_1 = 0.306 \text{ cm}^2, \quad A_3 = 0.176 \text{cm}^2$$

Example: The estimated duration to complete the different activities for pipeline construction is given. Find the optimum total time to complete the project

	Activity	Duration (Weeks)
L	Survey & layout pipelines	2
M	Obtain pipes	5
N	Obtain values & connections	4
O	Dig trench	3
P	Construct value chambers	3
Q	Lay pipes	4
R	Fit valves and connection	2
S	Concreting	2
T	Backfill	1
U	Finish value	1
V	Chambers & test	0
W	Clear & leave site	1

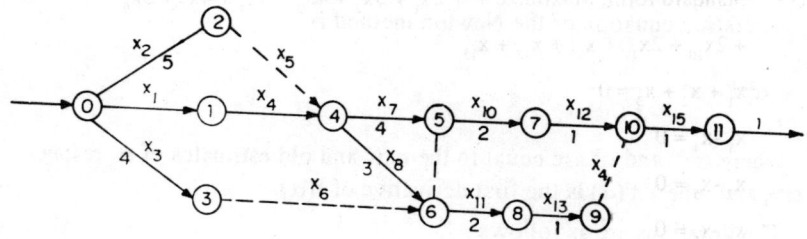

Solution:

Unit flows are placed on the start and finish nodes 1 & 12. x_i is used to indicate whether a job lies on the critical path or not.

The control variables are

$$x_i - \begin{array}{l} 0 \text{ Job i is on critical path} \\ \\ 1 \text{ Job i is not on critical path} \end{array}$$

The earliest time to complete the project is equal to the longest time path connecting the start and finish, our objective function is

Maximize $D = 2x_1 + 5x_2 + 4x_3 + 3x_4 + 4x_7 + 3x_8 + 2x_{10} + 2x_{11} + x_{12} + x_{13} + x_{15}$

At each node apply the conservation of flow condition

i.e. flow in = flow out.

Node 0 $\rightarrow$ $x_1 + x_2 + x_3 = 0$

Node 1 $\rightarrow$ $x_1 = x_4$

Node 2 $\rightarrow$ $x_2 = x_5$

Node 3 $\rightarrow$ $x_3 = x_6$

Node 4 $\rightarrow$ $x_4 + x_5 = x_7 + x_8$

Node 5 $\rightarrow$ $x_7 = x_{10} + x_9$

Node 6 $\rightarrow$ $x_8 + x_6 = x_9 + x_{11}$

Node 7 $\rightarrow$ $x_{10} = x_{12}$

Node 8 $\rightarrow$ $x_{11} = x_{13}$

Node 9 $\rightarrow$ $x_{13} = x_{14}$

Node 10 $\rightarrow$ $x_{12} + x_{14} = x_{15}$

Node 11 $\rightarrow$ $x_{15} = 1$

Standard form, Maximize $\longrightarrow 2x_1 + 5x_2 + 4x_3 + 3x_4 + 4x_7 + 3x_8$

$+ 2x_{10} + 2x_{11} + x_{12} + x_{13} + x_{15}$

$x_1 + x_2 + x_3 = 0$

$x_1 - x_4 = 0$

$x_2 - x_5 = 0$

$x_3 - x_6 = 0$

$x_4 + x_5 - x_7 - x_8 = 0$

$x_7 - x_9 - x_{10} = 0$

$x_6 + x_8 - x_9 - x_{11} = 0$

$x_{10} - x_{12} = 0$

$x_{11} - x_{13} = 0$

$x_{13} - x_{14} = 0$

$x_{12} + x_{14} - x_{15} = 0$

$x_{15} = 0$

A transportation problem : A contractor has three sites where different projects are going on. He has to assign workers to the sites each day. The travel time from dispatch locations to work site is given by the network diagrams. Traveling time is considered as productive work hours. To maximize the productive work hours formulate a mathematical model.

The dispatched workers are 60 & 50 respectively & worker required at three sites 3,4 & 5 are 40,40 & 30 The traveling time from dispatched locations to the site is also shown.

Solutions $\rightarrow$ Let us assume the control variables.

$x_1 , x_2 \longrightarrow x_6$. Then minimize traveling time

$D = 20x_1 + 30x_2 + 40x_3 + 60x_4 + 40x_5 + 30x_6$

At Each node, apply the conservation of flow

for node 1, $60 = x_1 + x_2 + x_3$

for node 2, $50 = x_4 + x_5 + x_6$

for node 3, $x_1 + x_4 = 40$ (demand)

for node 4, $x_2 + x_5 = 40$ (demand)

for node 5, $x_3 + x_6 = 30$ (demand)

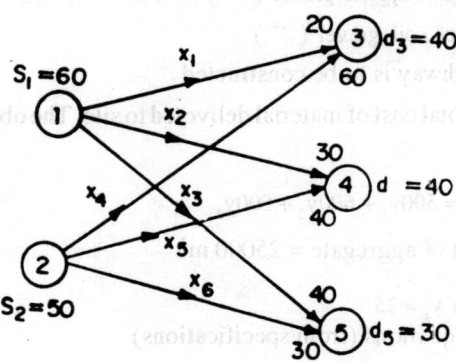

In case supply is less than the demand, inequality sign be used.

The network analysis can be used to many other types of problems also like to assign the equipments at different sites, to find out the maximum capacity of a water distribution system etc.

Example: From the grain size distribution it has been found out that the roadway aggregate should have the following specifications :

Aggregate specification

Percent volume		Gravel	Sand	Silt
	Lower limit	10	70	0
	Upper limit	20	90	20

Three sources of material are identified for the road construction from Pilani to Chirawa (14km)

Sources	% gravel	% sand	%silt	cost/m³
Pilani	20	70	10	500
Chirawa	30	65	5	600
Rajgarh	—	100	—	800

The soil on which the construction has to be done contains 5% gravel, 25% sand and 70% silt supply 25000m³ of aggregate

   (a) Formulate a minimum cost aggregate mix model.
   (b) Find out the minimum cost.
   (c) Determine the highest price you charge from Pilani.

y_1 = amount of Pilani gravel (1000m³)

y_2 = amount of Chirawa gravel ('')

y_3 = amount of Rajgarh gravel (")

y_4 = amount of soil gravel (")

on which highway is to be constructed

Let C be the total cost of material delivered to site. The objective function

is

Minimize $C = 500y_1 + 600y_2 + 800y_3$ (1)

Total amount of aggregate = 25000 m³

$y_1 + y_2 + y_3 + y_4 = 25$ (2)

Constraints equations (from specifications)

For gravel

$0.2y_1 + 0.3y_2 + 0.05y_4 \leq 0.20 (25) = 5$

$0.2y_1 + 0.3y_2 + 0.05y_1 \geq 0.10 (25) = 2.5$

For sand

$0.70y_1 + 0.65y_2 + y_3 + 0.25y_4 \geq 0.90(25) = 22.5$

$0.70y_1 + 0.65y_2 + y_3 + 0.25y_4 \leq 0.70(25) = 17.5$

For silt

$0.1y_1 + 0.05y_2 + 0.70y_3 \leq 0.2(25) = 5$

$0.1y_1 + 0.05y_2 + 0.70y_3 \leq 0 = 0$

The mathematical model is

Min $C = 500y_1 + 600y_2 + 800y_3 = 5y_1 + 6y_2 + 8y_3$

$y_1 + y_2 + y_3 + y_4 = 25$

$0.2y_1 + 0.3y_2 + 0.05y_4 \geq = 5$

$0.2y_1 + 0.3y_2 + 0.05y_4 \geq = 2.5$

$0.70y_1 + 0.65y_2 + y_3 + 0.25y_4 \leq = 22.5$

$0.70y_1 + 0.65y_2 + y_3 + 0.25y_4 \leq = 17.5$

$0.1y_1 + 0.05y_2 + 0.70y_3 \leq = 5$

$0.1y_1 + 0.05y_2 + 0.70y_3 \leq = 0$

For $\geq$ to constraints introduce slack variables y_5, y_6 and y_7

For $\leq$ to constraints introduce surplus variables y_8 & y_9

For equality constraint, artificial variables y_{10}, y_{11} and y_{12} are introduced.

The artificial objective function is

$$a = y_{10} + y_{11} + y_{12}$$

Example : A beam has to be welded with a steel column. The beam has to support 100kN Select the dimensions of the beam so that the cost of the whole system is minimum .

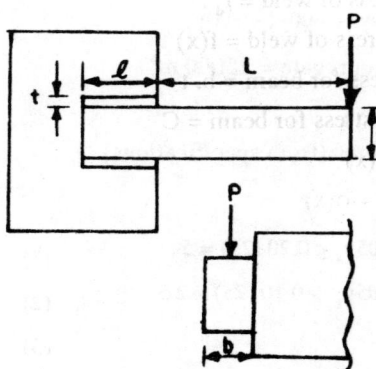

Solution: Consider the variables shown in fig

$$x = [\, x_1, x_2, x_3, x_4 \,]^T = [\, h, l, t, b]^T$$

The cost of the system

(a) material

(b) labour

(c) set up

Welding labour cost - assume machine cost is = Rs 500/hr.

If 15 cm³ is laid in 10 minutes

$$\text{cost } C_1 = \left(500 \;\frac{Rs}{hr.} \right) \left(\frac{1 \; hr}{60 \; min} \right) \left(10 \; \frac{min}{Cm_3} \right) V$$

$$= \left(\frac{250}{3} \right) V$$

V_0 = volume of weld.

material cost $C_2 = C_3 V_0 + C_0 V_1$

C_3 = Rs/volume of weld material

C_0 = Rs/volume of bar material

V_1 = Volume of beam

$$V_0 = 2 \left(\frac{1}{2} t^2 l \right) = t^2 l$$

$$V_1 = hb(h + l)$$

Cost function $= C(x) = t^2 + C_3 t^2 l + C_4 hb(l + L)$

$$= x_3^2 + C_3 x_3^2 x_2 + C_4 x_1 x_4 (x_2 + L)$$

If design shear stress of weld $= f_d$.

maximum shear stress of weld $= f(x)$

Design normal stress for beam $= b. t$

Maximum normal stress for beam $= C$

buckling load $= P_c(x)$

bar end deflection $= \sigma(x)$

$fd - f \geq 0$	(1)
$\sigma b - \sigma \geq 0$	(2)
$x_4 - x_3 \geq 0$	(3)
$x_2 \geq 0$	(4)
$x_3 \geq 0$	(5)
$P_c - 0 \geq 0$	(6)
$x_3 - 0.25 \geq 0$	(7)
$0.6 - \delta \geq 0$	(8)

equation 3 indicates that weld thickness is less than beam thickness

equation 4 indicates that length of weld control be zero or less

equation 5 indicates that thickness of weld control be zero or less

equation 6 indicates that buckling load is not encoded

equation 7 indicates that a min. thickness is required for weld.

Weld stress consists of

(a) primary stress - $f' = P/\sqrt{2} \, x_2 \, x_3$

(b) secondary torsional stress $f'' = \dfrac{MR}{J}$

   M = moment of P about the c.g.

   J = polar moment of inertia

$$M = P \left(L + \frac{l}{2} \right) = P \left(L + \frac{x_2}{2} \right)$$

$$R = \left\{ \frac{x_2^2}{4} + \left(\frac{x_1 + x_3}{2} \right)^2 \right\}^{\frac{1}{2}}$$

$$J = 2\left[1/\sqrt{3}\, x_2 \left(\frac{x_2^2}{12} \right) + \left(\frac{x_1 + x_3}{2} \right)^2 \right]$$

total weld stress $= ((f^1)^2 + 2f'f'' \cos\theta + f'')^2)^{\frac{1}{2}}$

$\cos\theta = (x_2)/2R$

bar bending stress $= \dfrac{6PL}{x_4 x_1}$

Beam Buckling load $(P_c) = \dfrac{4\ \sqrt{EI\alpha}}{L^2} \left(1 - \dfrac{X_1}{2\,L} \sqrt{\dfrac{EI}{\alpha}} \right)$

E = Young modulus $= 200 \times 10^6$ KN/m^2

$I = \dfrac{1}{12}\ x_1 x_4^3$

$\alpha = \dfrac{1}{3}\ g x_1 x_4^3$

g = Shearing modulus.

bar deflection $= \delta = \dfrac{4PL^3}{Ex_1^3 x_4}$ cantilever beam.

Putting all these values the mathematical model can be written at the joint.

Minimize $C = \alpha_1\, x_3^2 x_2 + \alpha_2 x_1 x_4\, (L + x_2)$

subject to $\dfrac{f}{f_d} - \left(\dfrac{1}{2x_3^2 x_2^2} + \dfrac{(2L + x_2)}{x_2^2 x_2\, [x_2^2 + 3x_3 + x_4)^2]} \right.$

$\left. \dfrac{\alpha_4 (2L + x_2)^2\, [\, x_2^2 + (x_1 + x_3)^2\,]^{\frac{1}{2}}}{x_3^2 x_2^2 (x_2^2 + \alpha_3\, (x_1 + x_3)^2\,)} \right) \geq 0$

$x_1^2 x_4 - \alpha_5 \geq 0$

$x_4 - x_3 \geq 0$

$x_1 x_4^3\, (1 - \alpha_6 x_3) - \alpha_7 \geq 0$

$x_1 - 0.25 \geq 0$

$x_1^3 x_4 - \alpha_8 \geq 0$

$x_2\ \&\ x_3 \geq 0$

Where α_1, α_2 ——α_8 are constants and their values depend upon the permissible stresses and other conditions.

Example : For a circular sendimentation tank where bottom is of comical hopper shape, determine the proportions to hold a volume of 7480m³ for minimum area of the bottom and sides.

Solution : From fig : $d = \dfrac{2}{3}\ r$ and $s = \dfrac{\sqrt{13}}{3}\ r$

total volume $= \pi r^2 h + \dfrac{1}{3}\ d\pi r^2 = \pi r^2 h + \dfrac{2}{9}\ \pi r^3$

total area $A = 2\pi rh + \pi rs = 2\pi rh + \dfrac{\sqrt{13}}{3}\ \pi r^2$

Model: Minimize $A = 2\pi rh + \dfrac{\sqrt{13}}{3}\ \pi r^2$

subject to volume $V = \pi r^2 h + \dfrac{2}{9}\ \pi r^3$

from the Lagrangian - $L = 2\pi rh + \dfrac{(\sqrt{13})}{3}\ \pi r^2 + \lambda\ (\pi r^2 h + \dfrac{2}{9}\ \pi r^3 - V)$

conditions: $\dfrac{\delta L}{\delta h} = \dfrac{\delta L}{\delta r} = \dfrac{\delta L}{\delta \lambda} = 0$

$\dfrac{\delta L}{dh} = 2\lambda\pi r + \lambda\pi r^2 = 0$

$\dfrac{\delta L}{\delta r} = 2\pi h + \dfrac{2}{3}\ \sqrt{13}\ \pi r + 2\ \pi rh + \dfrac{2}{3}\ \alpha\ \pi r^2\ \lambda = 0$

$\dfrac{\delta L}{\delta \lambda} = \pi r^2 h + \dfrac{2}{g}\ \pi r^3 - V = 0$

from the sbobr equations

$\lambda = -2/r$

$h = \dfrac{1.6}{3}\ r$

$V = (34/45)\pi r^3 = \dfrac{34}{45}\ \times\ \dfrac{22}{7}\ r^3$

for $V = 7480m^3 \Rightarrow r^3 = \dfrac{7480 \times 45 \times 7}{34 \times 22}$

$= 10 \times 45 \times 7 = 3150$

r = 14-15m

$$h = \frac{2}{3} \ (14\text{-}15) \ m$$

Example : The cost of the sedimentation process is determined in terms of volumentric flow rate of industrial waste as follows :

1. The initial cost 4Q
2. The plant operational cost $0.9Q^2t^3$
3. The extra cost depending upon the quality of waste $10Qe^{-t}$.

   Find out the value of t (detention time) for the minimum cost if the value of Q = 1m³/s.

Solution : Minimize C $\quad = 4Q + 0.9Q^2t^3 + 10Qe^{-t}$

$$= 4 + 0.9t^3 + 10e^{-t} \ (Q = 1)$$

subject to t > 0

(i) Solution by Newton's Raspon Method

$f(x) = 4 + 0.9x^3 + 10e^{-x} \ (t = x)$

$f'(x) = 2.7x^2 - 10e^{-x}$

$f''(x) = 5.4x + 10e^{-x}$

$$X_{k+1} = x_k - \frac{f'(x)}{f''(x)}$$

	X_k	$f'(x)$	$f''(x)$	X_{k+1}
1.	1.0	-.978	9.078	1.1077
2.	1.1077	0.0098	9.284	1.1066
3.	1.1066	4.9X10	-4	

t = 1.1066 & cost C $\quad = 4 + 0.9 \ (1.1066)^3 + 10e^{-1.1066}$

$$= 5.55$$

Example: Rigid plastic design of frameworks

The portal frame shown in figure must be designed in steel on a rigid-plastic basis to have a factor of safety of 3.0 against total collapse under the loading shown. The two columns are to be of identical section while the beam may have different section.

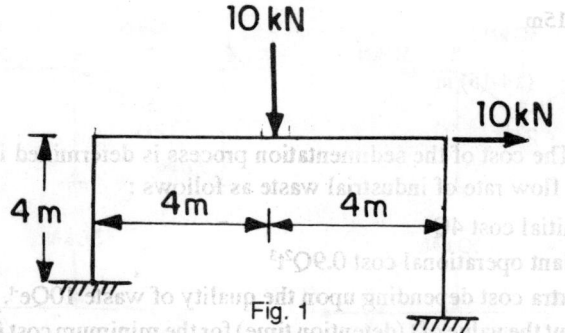

Fig. 1

Since the member lengths are known, the design process consists of selecting appropriate cross-sectional sizes. Restrictions on the values that the two variables may take come from the requirements that the structure must have a factor of safety of 3.0 against collapse under the given loading. Figure 2 shows the six possible collapsed mechanisms of the frame and the energy-balance requirement associated with each kinematic mechanism. There are three general failure mechanisms possible:

(a) beam mechanism;

(b) sway mechanism, and

(c) combined mechanism in which both the beam and sway failure occur simultaneously.

The hinges at joints B and C of the frame will always occur in the weaker member at the joint since less energy will be needed to produce a plastic hinge in that member. For example, consider the mechanism shown in figure 2. The work done by the loads on that mechanism is all done by 30 kN load (10 kN actual load X 3.0 safety factor). The work done by the 30 kN load is:

$$\text{work done} = 30 \times 4m \times \theta$$

$$= 120\,\theta \text{ kN m}$$

This assumes that deformations are small so that $\sin\theta \sim \theta$. The energy absorbed by the two hinges at the tops of the columns is equal to their fully plastic moments multiplied by the angular rotations at the hinges. Thus

$$\text{energy absorbed (column hinges)} = 2 \times m \times \theta$$

$$= 2M\,\theta \text{ kNm}$$

Similarly for the hinge in the middle of the beam

$$\text{energy absorbed (beam hinge)} = M_1 \times 2\theta$$

$$= 2M_1\,\theta \text{ kN m}$$

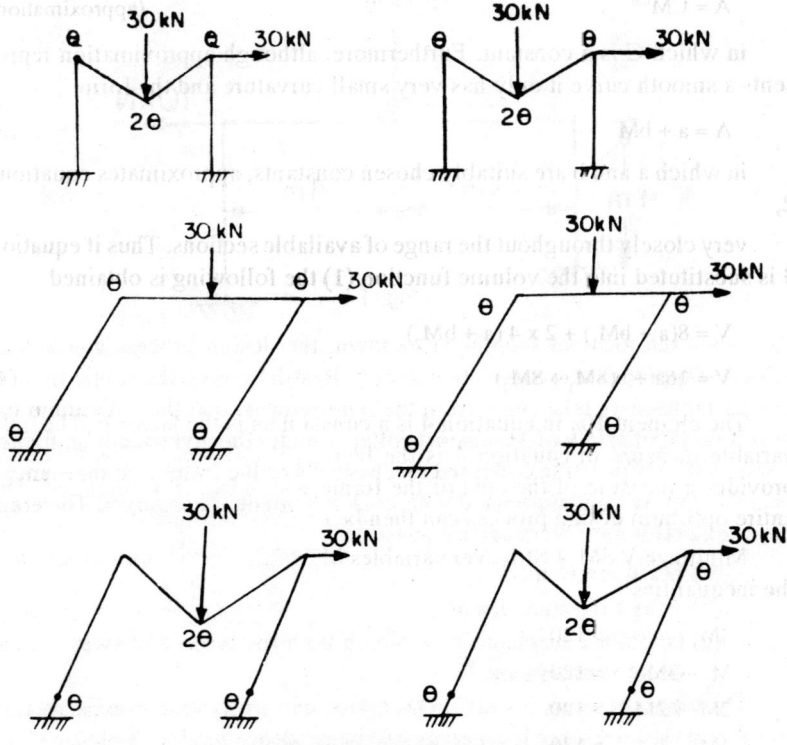

Fig. 2. Possible collapse mechanism.

In order that the frame shall either be safe or just collapse in this mechanism it is required that the total energy absorbed in all the hinges should be at least equal to the work done by the factored applied loads. Therefore

$$2M\theta + 2M\theta > 120\theta = \theta$$

and, cancelling the θs, this gives the requirement that

$$2M_1 + 2M_2 > 120$$

Similar relationships can be obtained from all the other possible collapse mechanisms. These six relationships form a set of restrictions on the values that the designed can choose for M_1 and M_2 and they are given at the end.

The cost of steel is proportional to the weight of steel and weight is proportional to volume. Volume of steel used provides a useful measure of the cost of the frame. If A_1, A_2 are the cross-sectional areas of the beam and columns respectively, the volume of the frame is

$$V = 8A_1 + 2 \times 4A_2 m^3 \tag{1}$$

$$A = CM^{0.6} \qquad \text{(approximation)}$$

in which C is a constant. Furthermore, although approximation represents a smooth curve it only has very small curvature and the form

$$A = a + bM \tag{3}$$

in which a and b are suitably chosen constants, approximates equations 2,

very closely throughout the range of available sections. Thus if equation 3 is substituted into the volume function (1) the following is obtained

$$V = 8(a + bM_1) + 2 \times 4 (a + bM_2)$$

$$V = 16a + b(8M_1 + 8M_2) \tag{4}$$

The element 16a in equation 4 is a constant as is the factor b. The only variable measure in equation 4 is the factor $(6M_1 + 6M_2)$. This therefore provides a measure of the cost of the frame which can be minimized. The entire optimum design process can then be expressed formally as

Minimize V $8M_1 + 8M_2$ over variables M_1, M_2 subject to non-violation of the inequalities

$$4M_1 \qquad\quad > 120$$
$$M_1 + 3M_2 \quad > 120$$
$$2M_1 + 2M_2 \quad > 120$$
$$4M_2 \qquad\quad > 120$$
$$4M_1 + 2M_2 \quad > 240$$
$$2M_1 + 4M_2 \quad > 240$$
$$M_1, M_2 \qquad > 0$$

Example : Allocating a Tower Crane

A contractor has work on 4 separate sites A, B, C and D and is considering hiring a large tower crane for 4 months to help in the construction work, because of the limited mobility of the crane it can only be moved from site to site at the end of each month. The crane is initial at the hiring yard, Y, and must be returned there after 4 months. The contractor has estimated the benefits to him in cash terms of using the crane on each of the 4 sites in each of the 4 months.

The lender has quoted a cost of 8000 to lend of the crane for 4 months, (the contractor to pay all moving costs). How should the contactor allocate the crane among his sites to maximize his total returns over the 4 months?

Table 1: Cost (in 100 units) of moving the crane between locations

	Y	A	B	C	D
Y	0	4	3	5	6
A	4	0	6	4	3
B	3	6	0	8	5
C	5	4	8	0	6
D	6	4	3	6	0

Table 2: Monthly benefits from use of crane on each site (in 1000 units)

		Month			
		1	2	3	4
Site	A	5	6	9	8
	B	11	10	5	4
	C	10	6	8	9
	D	9	10	10	4

This problem can be represented in serial form but with five stages. Each stage of moving the crane from its present location to a new location and, except for the last stage when it must return to the yard. The stage variable is the location of the crane, A, B, C, D or Y. The initial stage and final state must be Y, possible state values at other stages are A, B, C or D.

Stage I

$S_0 = Y$ $S_1 = A, B, C, D$
$S_1 = A =$ $5 - 4 = 1$
$S_1 = B =$ $11 - 3 = 8$
$S_1 = C =$ $10 - 5 = 5$
$S_1 = D =$ $9 - 6 = 3$

Stage II

$S_1 = A, B, C, D,$ $S_2 = A, B, C, D$

$$S_2 = A, \text{MAX} \begin{pmatrix} 1 + 6 - 0 \\ 8 + 6 - 6 \\ 5 + 6 - 4 \\ 3 + 6 - 3 \end{pmatrix} = \text{MAX} \begin{pmatrix} 7 \\ 8 \\ 7 \\ 6 \end{pmatrix} = 8$$

$$S_2 = B, \text{MAX} \begin{pmatrix} 1 + 10 - 6 \\ 8 + 10 - 0 \\ 5 + 10 - 8 \\ 3 + 10 - 4 \end{pmatrix} = \text{MAX} \begin{pmatrix} 5 \\ 18 \\ 7 \\ 9 \end{pmatrix} = 18$$

$$S_2 = C, MAX \left\{ \begin{array}{c} 1 + 6 - 0 \\ 8 + 6 - 8 \\ 5 + 6 - 0 \\ 3 + 6 - 6 \end{array} \right\} = MAX \left\{ \begin{array}{c} 3 \\ 6 \\ 11 \\ 3 \end{array} \right\} = 11$$

$$S_2 = D, MAX \left\{ \begin{array}{c} 1 + 10 - 3 \\ 8 + 10 - 4 \\ 5 + 10 - 6 \\ 3 + 10 - 0 \end{array} \right\} = MAX \left\{ \begin{array}{c} 8 \\ 14 \\ 9 \\ 13 \end{array} \right\} = 14$$

Stage III

$$S_2 = A, B, C, D \qquad S_3 = A, B, C, D$$

$$S_3 = A, MAX \left\{ \begin{array}{c} 8 + 9 - 0 \\ 18 + 9 - 6 \\ 11 + 9 - 4 \\ 14 + 9 - 3 \end{array} \right\} = MAX \left\{ \begin{array}{c} 17 \\ 21 \\ 16 \\ 20 \end{array} \right\} = 21$$

$$S_3 = B, MAX \left\{ \begin{array}{c} 8 + 5 - 6 \\ 18 + 5 - 0 \\ 11 + 5 - 8 \\ 14 + 5 - 4 \end{array} \right\} = MAX \left\{ \begin{array}{c} 7 \\ 23 \\ 8 \\ 15 \end{array} \right\} = 23$$

$$S_3 = C, MAX \left\{ \begin{array}{c} 8 + 8 - 4 \\ 18 + 8 - 8 \\ 11 + 8 - 0 \\ 14 + 8 - 6 \end{array} \right\} = MAX \left\{ \begin{array}{c} 12 \\ 18 \\ 19 \\ 16 \end{array} \right\} = 19$$

$$S_3 = D, MAX \left\{ \begin{array}{c} 8 + 10 - 3 \\ 18 + 10 - 4 \\ 11 + 10 - 6 \\ 14 + 10 - 0 \end{array} \right\} = MAX \left\{ \begin{array}{c} 15 \\ 24 \\ 15 \\ 24 \end{array} \right\} = 24$$

Stage IV

$$S_2 = A, B, C, D \qquad S_3 = A, B, C, D$$

$$S_4 = A, MAX \left\{ \begin{array}{c} 21 + 8 - 0 \\ 23 + 8 - 6 \\ 19 + 8 - 4 \\ 24 + 8 - 3 \end{array} \right\} = MAX \left\{ \begin{array}{c} 29 \\ 25 \\ 23 \\ 29 \end{array} \right\} = 29$$

$$S_4 = B, MAX \left\{ \begin{array}{c} 21 + 4 - 6 \\ 23 + 4 - 0 \\ 19 + 4 - 8 \\ 24 + 4 - 4 \end{array} \right\} = MAX \left\{ \begin{array}{c} 19 \\ 27 \\ 15 \\ 24 \end{array} \right\} = 24$$

$$S_4 = C, MAX \left\{ \begin{array}{c} 21 + 9 - 4 \\ 23 + 9 - 8 \\ 19 + 9 - 0 \\ 24 + 9 - 6 \end{array} \right\} = MAX \left\{ \begin{array}{c} 26 \\ 24 \\ 28 \\ 27 \end{array} \right\} = 28$$

$$S_4 = D, MAX \left\{ \begin{array}{c} 21 + 4 - 3 \\ 23 + 4 - 4 \\ 19 + 4 - 6 \\ 24 + 4 - 0 \end{array} \right\} = MAX \left\{ \begin{array}{c} 22 \\ 23 \\ 17 \\ 28 \end{array} \right\} = 28$$

Stage V

$$S_4 = A, B, C, D \qquad S_5 = Y$$

$$S_5 = Y, \text{MAX} \left(\begin{array}{c} 29 - 4 \\ 24 - 3 \\ 28 - 5 \\ 28 - 6 \end{array} \right) = \text{MAX} \left(\begin{array}{c} 25 \\ 21 \\ 23 \\ 22 \end{array} \right) = 25$$

Maximum total return = 25 x 1000 = Rs. 25000

hire cost = Rs. 8000

Total profit = 25000 - 8000 = 17000

Optimal allocation of crane:

$$Y \to B \to B \to D \to A \to Y$$

1st month - site B

2nd month - site B

3rd month - site D

4th month - site A

We can decide the other policy also by keeping the constraints in mind.

Example

A skip must be designed to carry concrete from a batching plant to the site of a large pour. The skip must have the form of a box, rectangular in plan, of length x_1, breadth x_2, and with sides of depth x_3. The top of the skip is open. The sides of the skip cost Rs. 500/m and the strengthened base costs Rs. 1500/m. The skip is to be used to transport a total of 1500m³ of concrete in several journeys, each round trip costing Rs. 500. Design the skip so that the total costs of the large pour are minimized.

The volume of the skip is $x_1 x_2 x_3$ and, since 1500m³ of concrete have to be transported, the number of trips which must be made is clearly $1500/x_1 x_2 x_3$. The total cost will be

$$500 (2x_1x_3 + 2x_2x_3) + 1500x_1x_2 + 75000/x_1x_2x_3$$

$$\text{Minimize} = 1000x_1x_3 + 1000x_2x_3 + 1500x_1x_2 + 75000/x_1x_2x_3$$

From this problem the dual constraints can be written immediately as

$$W_1 + W_2 + W_3 + W_4 = 1$$

for x_1, $W_1 + W_3 - W_4 = 0$

for x_2, $W_1 + W_2 - W_4 = 0$

for $x_3, + W_2 + W_3 - W_4 = 0$

These four equations in the for unknown weights can be easily solved to give

$W_1 = W_2 = W_3 = 1/5; W_4 = 2/5$

No examination is necessary to obtain these weights which must be optimal weights. Thus, from the objective function

$$D = \left(\frac{1000}{1/5} \right)^{1/5} \left(\frac{1000}{1/5} \right)^{1/5} \left(\frac{1000}{1/5} \right)^{1/5} \left(\frac{75000}{2/5} \right)^{2/5}$$

$$= 23122.6$$

$$\ln(x_1) + \ln(x_2) = \ln\left(\frac{23122.6}{5000} \right)$$

$$\ln(x_2) + \ln(x_3) = \ln\left(\frac{23122.6}{5000} \right)$$

$$\ln(x_1) + \ln(x_3) = \ln\left(\frac{23122.6}{7500} \right)$$

$$\ln(x_1) + \ln(x_2) - \ln(x_3) = \ln\left(\frac{23122.6}{1875000} \right)$$

from 1 to 4

$$\ln(x_3) = \ln\left(\frac{23122.6}{1875000} \right) + \ln\left(\frac{23122.6}{5000} \right)$$

$$= 2.09 + 1.53 = -0.558 \quad \Rightarrow x_3 = 1.74 \text{ m}$$

$$\ln(x_3) = \ln\left(\frac{23122.6}{1875000} \right) + \ln\left(\frac{23122.6}{5000} \right)$$

$$= -2.09 + 1.125 \quad \Rightarrow x_2 = 2.622$$

$$\ln(x_3) = \ln\left(\frac{23122.6}{1875000} \right) + \ln\left(\frac{23122.6}{5000} \right)$$

$$x_1 = 1.74 \text{ m}$$

so the width = 1.74 m

length = 1.74 m

and depth = 2.622 m.

total cost will be Rs. 23122.6

and the total trips = 15000/1.74 x 1.74 x 1.74 x 2.622 = 188.9

Example: A pumped pipeline

A new oil refinery is proposed at a site that is 15 km from a port at which tankers will unload. The oil is to be taken from port to refinery by means of a pipeline. Get the mathematical optimisation model.

The pressure head is dissipated along the length of the pipeline by friction between oil and pipe. For a pipe of length l and internal diameter d the frictional head loss h is approximately give by

$$h = \frac{\alpha l Q^2}{d^5}$$

Where Q is the volume of oil flowing in the pipe per second. It can be assumed that some desired volumetric flow rate Q is known the and pump must provide a total pressure head H which must be atleast as large as h to maintain flow. The pump power output, P, can be expressed as

P = WQH

Where W is constant, the cost of a pipe is proportional to its diameter and that the cost of a pumping station is proportional to the power it produces. The cost of a single pump and a length l of pipeline is

$C = C_1 l d + C_2 WQH$

Where C_1 and C_2 are unit cost coefficients for the pipeline and for the pump, respectively. The mathematical model:

Minimize $C = C_1 l d + C_2 WQH$

subject fo $H > \dfrac{\alpha l Q^2}{d^5}$

here diameter, and head are variables and other can be considered constants. If $d = x_1$ and $H = x_2$

Minimize $C = (C_1 l) x_1 + (C_2 WQ) x_2$

subject to $(\alpha l Q^2) x_2^{-1} x_1^{-5} < 1$

$$\sum_{t=1}^{T_0} W_0 t = 0; \qquad W_{01} + W_{02} = 1$$

for x_1 $\qquad W_{01} \rightarrow 5W_{11} = 0$

for x_2 $\qquad W_{02} \rightarrow 5W_{11} = 0$

These three equations can be solved for the three dual variables w to give

$W_{01} = 5/6$, $W_{02} = 1/6$ & $W_{11} = 1/6$

Thus, since $W_{01} = 5W_{02}$, the optimal design will be one in which the pipeline costs five times as much as the pumps. The total cost will be

$$C = \left(\frac{C_1 l}{5/6} \right)^{5/6} \left(\frac{C_2 WQ}{1/6} \right) (2lQ^2)^{1/6}$$

$$= 6LQ^{\frac{1}{2}} \left(\frac{C_1^5 C_2 WQ}{55} \right)^{1/6}$$

The values of the optimal primal variables are then found to be

$$D^* = Q^{\frac{1}{2}} \ 5 \left(\frac{C_2}{C_1} LW \right)^{1/6}$$

$$H^* = L/Q^{\frac{1}{2}}(C_1/5WC_2)^{5/6} \, L^{1/6}$$

Example

Consider the design of a steel beam. Suppose the length of the beam l is known and that the variables for which values must be found are as shown on figure . Assume also that the maximum live bending moment which the beam must carry is M and the maximum live shearing force is V.

Since the beam is made of steel, the cost of the steel plates will dominate the cost of the beam and will be approximately proportional to the volume of steel used. Minimize

$$A = x_1 x_2 + x_3 x_4 + x_1 x_5$$

The maximum bending moment to be carried is composed of the known live bending moment M plus the dead weight bending moment M_0. A maximum bending moment is then

$$M_{max} = M_1 + M_0 = M_1 + \frac{\rho l^2}{8} (x_1 x_2 + x_3 x_4 + x_1 x_5)$$

ρ is density of the steel of the beam.

The bending stresses in the extreme fibres of each flange must not exceed some known permissible values as specified by codes of practice. Firstly the position of the horizontal neutral axis of the section must be found. Assume that y is the depth of the neutral axis measured from the top surface of the top flange. Then

$$Y = \frac{\left(\dfrac{x_1 x_2^2}{2} \right) + x_3 x_4 (x_2 + x_4/2) + x_1 x_5 (x_2 + x_4 + x_5/2)]}{(x_1 x_2 + x_3 x_4 + x_1 x_5)}$$

Knowing y, the moment of interia of the cross-section can be expressed as

$$I = \frac{1}{12}(x_1 x_3^3 + x_3 x_4 + x_1 x_6^3) + x_1 x_2 \left(y - \frac{x_2}{2}\right)^2 + x_3 x_4$$

$$(x_2 + \left(\frac{x_4}{2} - y\right)^2 + x_1 x_5 \left(x_2 + x_4 + \frac{x_5}{2} - y\right)^2$$

The maximum stress in the extreme fibres of the top and bottom flanges

$$(M_{max} Y)/I \text{ and } \frac{M_{max}}{I}(x_2 + x_4 + x_5 - 4)$$

The top flange is in compression and the bottom flange is in tension. Two stress constraints can be written:

$$\frac{M_{max} Y}{I} < f_c \text{ and } \frac{M_{max} Y}{I} < f_t$$

Where f_c and f_t are maximum permissible stress values determined from code. Stresses are not given as constants but depend on section properties. Such as slenderness ratio, l/r, (length/radius of gyration) and d/t (over-all depth/compression flange thickness).

The other constraints could be

* min thickness of plates
* Overall depth $(x_2 + x_4 + x_5)$
* Width of flange (x_1)
* Depth/width ratio $(x_2 + x_4 + x_5)/x_1$
* Provide vertical and being stiffness to prevent buckling.

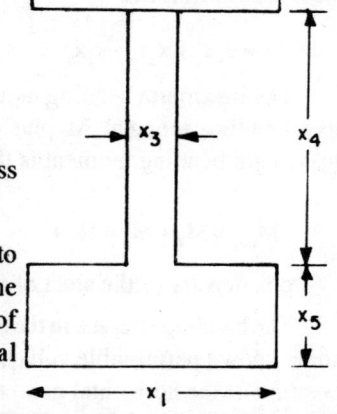

The example was just to show how to design a mathematical model for a beam. The concept can be extended to other types of beam, columns, slabs and other structural members.

Considering next the shearing stresses, codes often prescribe a maximum average shearing stress, and a peak shearing stress, max, neither of which may be exceeded in the beam.

$$\frac{V}{(x_1 x_2 + x_3 x_4 + x_1 x_5)} < \tau_{av}, \frac{fl}{2}$$

Peak shearing stress occurs at the neutral axis of the cross-section and leads to the constraint.

$$\frac{\left[V + \frac{\rho l}{2} \; (x_1 x_2 + x_3 x_4 + x_1 x_5) \; (x_1 x_2 \; (Y - \frac{x_2}{2}) + x_3)^2 \; \frac{(Y - x_2)^2}{2} \right]}{I x_3} \leq \tau_{max}$$

The maximum stress in the extreme fibres of the top and bottom flanges

EXERCISE

1. A contractor requires an aggregate mix of sand and gravel which contains not less then 10% not more than 48% gravel. The in situ soil contains 45% gravel and 55% sand. Pure sand may be purchased at Rs. 200/m³. A total mix of 1500 m³ is needed.

   Formulate the mathematical model.

2. Obtain the mathematical model to find the maximum volume of a sphere subject to an internal pressume of 5000 N/cm². The allowable stress of the material is 20000 N/cm².

3. Write the steps involved in system analysis approach.

4. In a project, 200 metre pipe is required. Three contractor bid for the supply of pipe. The rates and limitation are given for each of them. Formulate the mathematical model and get the optimum solution.

Limit	Cost	Limit
Contractor 1	Rs. 2000/metre	Maximum - 40 m
Contractor 2	Rs. 3000/metre	Unlimited
Contractor 3	Rs. 2500/metre	Maximum - 100 m

5. The plant produces three types of precast wall panels. The specifications are given below.

	Labour	Machinery	Insulation	Rate
Type 1	2 hours	3 hours	Not attached	Rs 1000/unit
Type 2	1.5 hours	2 hours	Attached	Rs 3000/unit
Type 3	2.5 hours	1.5 hours	Not attached	Rs 1200/unit

Labour rates are Rs 30/hour, machinery cost Rs. 100/hour and the insulation material cost Rs 1500/m³. For 12 units of panel maximise the profit if the plant has available 250 hours of machinary time, 175 hours of labour and 175 m³ of insulation material.

6. A contractor is supposed to complete a project in 40 weeks. It has been decided that if the project remain unfinished the contractor will pay Rs 50000 per week thereafter. In CPM chart it has been found out that Activities A, C, D, G, I are the critical with durations respectively 9, 6, 5, 7, 8 weeks. By employing more resources for each activities will an extra cost will incum as follows

Activity A

1 weeks extra cost 40000.

2 weeks extra cost 48000.

3 weeks extra cost 55000.

4 weeks extra cost 65000.

5 weeks extra cost 78000.

For activity $C = 20000 \left(2 + \dfrac{W}{2} \right)$

W = reduction of weeks.

For activity $D = (25000 + 1000W + 2000W^2)$

W = reduction of weeks

For activity $I = 45000 + 8000$ per week reduction

For I = 55000 for first week

65000 for second week

70000 for third week

73000 for fourth week

75000 for fifth week

Assume that always these five activities remain critical, minimize the extra cost for the contractor.

7. A moving load of value 20 KN is placed upon a statically indeterminate beam as shown in fig.

(a) Formulate the constrainted optimization problem to determine the maximum moment at the middle support.

(b) Find the location of force where the moment is maximum.

$$M = \frac{Fx (1 - x) (1 + x)}{4l^2}$$

8. A tank has to be built under ground. The different costs are given below (including the labour cost.)

(a) Excavation - Rs. 75/m³.

(b) 15 cm ground slab - Rs. 800/m².

(c) 25 cm walls (for work) - Rs. 125/m²

(d) Reinforcing steel (total) Rs. 2,00,000.

   (i) Obtain the mathematical model.

   (ii) Determine the maximum tank volume for the total cost = Rs. 5,00,000

9. Sikandrabad & Hyderabad produce 2×10^7 & 3×10^7 liters/day of waste water respectively. The BOD level of the waste water is 195 mg/l. For both the cities in all there are three treatment plants. The capacity of the plants are $(0.5, 2$ & 4×10^7 litre/day & the efficiency of removing BOD is 90, 80 & 65 % respectively.

The authority wants to operate the process at minimum cost & wants that BOD should not exceed 25 mg/l.

The unit treatment and all other costs are given below for different pipelines.

Unit annual treatment and other costs

(Rs. $25000/10^7$ litres/day - yr)

Pipelines	I	II	III	IV	V	VI
Cost (Rs.)	1000	1500	2100	1200	1400	2000

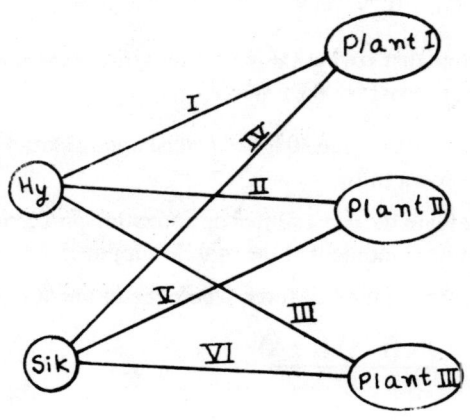

10. Design a steel beam made off of three steel plates which are connected by rivets. The rivets are connected in Diamond shape
    (a) Formulate a mathematical model to minimize the wt of the beam
    (b) Check the optimum solution by different methods.

Chapter IV

CONSTRUCTION MATERIALS, AND TECHNOLOGY

Section A

POZZOLANAS

A.1. Introduction

The need for establishment of a building materials industry in the developing world so as to cater to the emerging demand for housing for the burgeoning population has been the main impetus for the present interest in alternate or non-conventional building materials. Pozzolanas form one such type of materials. However, it would be erroneous to assume that the recognition of Pozzolana as a building material has been confined to our era.

Many centuries ago the ancient Greeks and Romans knew that certain volcanic deposits, if finely ground and mixed with lime and sand yielded a mortar which had superior strength and good resistance to water. As the best variety of this earth was obtained from the neighborhood of Pozzuoli (In Latin Puteoli) the material acquired the name Pozzolana.

In the modern context Pozzolanas are defined as natural or artificial materials which contain silica and or alumina, which though not cementitious themselves, but when finely ground and mixed with lime, the mixture will set and harden at ordinary temperatures in the presence of water, like cement.

Pozzolanas can replace 15 to 40 % of portland cement without significantly reducing the long term strength of the concrete and can be used as stabilizer in improving quality of materials such as soils.

A.2. Classification

The American society for testing and materials (ASTM) standard on cement 1971 designation 618-71 in a more precise way defines Pozzolanas as siliceous or siliceous aluminous materials, which in itself may posses little or no cementitious value but will, in finely divided form and in the presence of moisture, chemically react with Calcium Hydroxide at ordinary temperatures to form compounds possessing Cementitious properties.

Pozzolana activity, therefore, depends on the fixative property of the material in respect of Calcium Hydroxide and its ability to harden under water as a consequence of the changes the above reaction produces.

Such a definition permits us to include laterites and other red and yellow soils which have been known to possess pozzolanic properties ,in the class of pozzolanas.

Pozzolanas are grouped under two main categories:

(a) Natural pozzolanas

(b) Artificial pozzolanas.

Though the above materials have similar pozzolanic activity they differ greatly from one another in origin, chemical composition and mineralogical composition.

Natural pozzolanas may further be divided into two main groups

(a) those derived from volcanic rocks in which amorphous constituent is glass produced by fusion. These are, for example,volcanic ashes and tuffs etc.

(b) others are derived from rocks or earth for which the silica constituent contains opal, either from precipitation of silica from solution or from the remains of animals. Examples of these are diatomaceous earths, opaline silica etc.

The most important artificial pozzolanic materials are "Fly Ash" or pulverised fuel ash (PFA), burnt clays and shales burnt bauxite, rice husk ash and granulated blast furnace slag.

Fly ash is one of the most widely used artificial pozzolana and is a residue from the combustion of solid pulverised fuels in big thermal power plants.

A 2.1 Fly Ash

Fly ash is a residue from combustion and its essentially vitreous physical state and low alkali content makes it a useful pozzolanic material, provided it contains admissible amounts of unburnt carbon matter. Where coal is used as fuel for industrial purposes the ash can profitably be used as a pozzolana. Since naturally occuring pozzolanas are geographically confined, their universal application and study becomes severely restricted.

The finely ground coal in thermal power plants is injected with high speed with a stream of hot air into the furnace. The carbonaceous content is burnt instantaneously and the remaining matter comprising silica, alumina and iron oxide melts in suspension, forming white spherical particles. On rapid cooling while being carried away by the flue gases these spherical particles when collected constitute pulverised Fuel Ash or PFA. Depending on the type of coal PFA contains varying proportions of lime, low lime. PFA being pozzolanic and high lime PFA having cementitious properties by itself. Its of interest to note that the glassy hollow spherical particles of PFA have the same fineness of OPC hence no further grinding is necessary. Other advantages of using PFA in combination with OPC are :

1. With increasing age , higher strength than concrete are developed.
2. PFA does not adversely influence the structural performance of concrete members.
3. Compared to OPC concrete PFA concrete is less permeable and lighter and has a better surface finish.
4. PFA concrete is more resistant to Sulphate attack and Alkali-Silica reaction.
5. Concretes in which 35-50% by weight of OPC is replaced by PFA have shown satisfactory performances.
6. Aggregates derived from Fly Ash show excellent bonding in PFA concretes,contributing considerably to their performance and durability.

About 25 million tonnes of Fly Ash is produced per annum at 60 thermal power plants in India, at present. Hence, this Fly Ash can be utilised in a number of ways. The important building materials which have been produced from Fly Ash are;

1. **Portland Pozzolana Cement:** The Portland Pozzolana Cement conforming to IS:1489-1976 can be manufactured using Fly Ash either by intergrinding portland cement clinker, Fly Ash and gypsum or by intimately blending together portland cement and Fly Ash in measured proportions. Such cements are low heat giving and are Sulphate resistant and hence are suitable for marine and hydraulic structures and mass concrete constructions.

2. **Precast Fly Ash Concrete Building Units:** Fly Ash concrete with 20% less cement has been used in the manufacture of cement concrete building blocks, flooring and roofing units such as cored units, channel units and cellular units.

3. **Sintered Fly Ash Light Weight Aggregate (SFALA):** The SFALA is produced by pellitisation of Fly Ash or by sintering the Fly Ash pellets at 1200 degrees C in a vertical shaft kiln or in a moving grate sintering strand. This aggregate is suitable for use in the production of structural light weight concrete and precast building units.

4. **Lime Fly Ash cellular Concrete:** This concrete consists of fine grained silicate structure having small non communicating air cells. It has good fire resistance and like wood it can be sawn, chiselled, screwed and nailed. Its building units of bulk density of 700 kg/m³ are suitable for load bearing walls for 2 to 3 storeys in a multistoreyed building.

5. **Fly Ash Building Bricks:** Good quality high strength building bricks can be produced from Fly Ash using sodium silicate, cement or lime as binder. The mixture of Fly Ash binder and coarse fillers such as bottom ash, sand etc. in suitable proportions is moulded into bricks under pressure in suitable proportions. These are then burnt at 1100 degrees C. These bricks are usable in place of common bricks for all types of brick masonry.

A.2.2 Ground Granulated Blast Furnace Slag

Blast furnace slag is a molten material which settles above the Pig Iron at the bottom of the furnace. It is produced from the various inputs in the furnace when it reaches 1400 to 1600 degrees C. The slow cooling of the slag produces a crystalline material which is used as a aggregate. Rapid cooling with air or water under pressure forms glassy pellets and granules smaller than 4mm which posses hydraulic properties when finely ground.

The ground slag is blended with OPC to produce portland blast furnace cement (PBFC) whereby the slag content can reach 80%. Though early strength of PBFC concrete is lower than ordinary concrete the final strength is likely to be higher. The slow reactivity of PBFC develops less heat and can be advantageous in situations where thermal cracking is a problem, as in mass concreting.

A.2.3 Rice Husk Ash

India is producing about 20 million tonnes of Rice Husk annually. The combustion of such agricultural residues removes the organic matter and produces in most cases, a silica-rich ash. Rice Husk of all agricultural wastes, yields the largest quantity of ash around 20 % by weight- which also has the largest silica content-around 93% by weight. It is this high silica content that gives the ash its Pozzolanic Properties. Since only non-Crystalline silica posses these properties, the temperature and duration of combustion are of importance in producing Rice Husk Ash (RHA). Amorphous silica is obtained by burning the Rice husks at less than 100 Degree C. A incinerator for this purpose has been developed by the Cement Research Institute of India.

A Pozzolanic material from Rice-Husk and clay, on mixing with lime gives a very-good cementitious material (Lime-Pozzolana) cement and when blended with portland cement produces portland-pozzolana cement. The ash can replace up to 30% of cement in mortar or concrete. Alternatively it can be mixed with 30% to 50% of Hydrated lime to be used like cement in mortar, renderings and unreinforced concrete. Addition of ground Rice Husk 5 to 10% and rice Husk Ash 5 to 15% w/w has been found effective to check drying losses of bricks from plastic red & black soil.

In another process the ash obtained from heap burning is mixed with about 20 to 50% by weight of Hydrated lime. This is grounded for 6 or more hours is a ball mill to produce ASHMOH a hydraulic binder suitable for masonry, foundations and general unreinforced concreting work.

A.2.4 Manufacture of Rice Husk Ash

The Combustion of agricultural residues produces Rice Husk Ash. This combustion which is carried out under controlled conditions of temperature takes place in a special Incinerator.

A typical incinerator is made of bricks with many openings to allow good air flow through the Rice Husk mass. The inner surface is covered with a 16 gauge fine wire mesh. The Husks are filled in from the top and the ash removed from the bottom discharge door. A pyrometer monitors the temperature which can be controlled by shutting or opening the holes maintaining a temperature around 650°C for 2-3 hours. The reactive ash is dark grey to white, depending upon the residual carbon in it which has no negative effects if below 10%. To improve its reactivity the ash is grounded in a ball mill for about one hour or longer if it contains crystalline silica.

Among the numerous applications of Rice Husk Ash to low cost housing is the construction of a low cost House with RHA and Lime substituting cement completely in the production of hollow load-bearing block, mortar and plaster. 30% of the portland cement in the precast concrete lintels and roof beams were substituted by RHA.

A.3. EVALUATION OF POZZOLANIC ACTIVITY

Table A.1. Percentage composition of some pozzolanas

Pozzolana	SiO_2	Al_2O_3	Fe_2O_2	CaO	MgO	Na_2O	SO_3	Ignition loss
a) Burnt Clay	62.8	17.2	7.6	2.3	2.5	4.5	2.7	1.7
b) FlyAsh	44.8	18.4	11.2	11.6	1.1	3.1	2.0	8.0
c) Rice Husk Ash	85.6	2.5	0.3	1.0	1.0	2.5	1.5	

The most important chemical ingredients of pozzolanas are silica and alumina as shown in Table A.1. All pozzolanas react with lime in the presence of water to form compounds with cementitious properties. In assessing the ability of pozzolanic material to react with lime it is essential to consider its chemical bonds and its physical state.

Silica and alumina are, in effect, vulnerable to calcium hydroxide when their structural bonds are weak and unstable in the original material, as is the case with volcanic glass, in the zeolitic structure deriving from their alteration. The same applies to clay materials when, through total dehydration by heat treatment, the bonds between silica and alumina are relaxed or annihilated. If, instead, Si ,Al,and O are bound in the lattices of the individual minerals formed by crystallization of the magma, or in those of the silica-aluminous material which constitute natural clays, the calcium hydroxide acts far more slowly or not at all.

In case of cementitious action, the parts played by silica and alumina cannot be considered separately since there are silicious materials with very litle or no alumina which have high and rapid lime fixing properties but with poor hardening and strength properties when made into mortars or concrete. The presence of reactive alumina considerably enhances the strength properties of pozzolanic cements, particularly during brief curing. Ironoxides lower the calcining temperature of artificial pozzolanas.

Since by definition pozzolanic materials must have the ability to combine with lime to form cementitious compounds, to measure these properties, it is desirable for these actions to take place as rapidly and intensely as possible. Unfortunately there are as yet no satisfactory tests for predicting either the ability of the material to fix calcium Hydroxide or the ability to form cementitious binder. Many attempts are made to assess the value of pozzolanas by the solubility of their content in acid or alkali solution have all failed to give results that showed correlation between solubility and strength developed by pozzolana in lime or concrete. It means that in evaluating pozzolana both properties of the strength and lime fixation of the material should be considered separately. This requirement means determining under standard conditions, the mechanical strength of the mortars of concrete and the extent to which free Calcium hydroxide is reduced. These tests must be practical and quickly performed to give results for acceptance or rejection of the pozzolanic material. The evaluation of the ability of the pozzolana, however, to combine with calcium hydroxide in the paste, mortar or concrete after reasonably short period, is inaccurate whether this test is performed by dissolving extraction of chemical agents or by thermogravimetric analysis. This inaccuracy arises because when pozzolanic binders harden, a portion of the calcium hydroxide remains embodied and protected by the newly formed pseudo-gelatinous products which impede its passage into the solution. Attempts have, therefore been made to devise means whereby these problems could be avoided.

One solution incorporated a test designed to determine whether the pozzolanic material added to portland cement, whatever be its nature and mix ratio, will actually fix the calciumhydroxide freed by hydrolysis of the portland cement. The type of pozzolana mixed with the portland cement and the mix ratio are satisfactory if the point representing the analysed solution comes below a solubility isothem.

Another investigator proposed strength tests in which the pozzolana cement mortar cubes are cured at 50 degrees C and 18 degrees C. The difference between the strength developed at the two temperatures affords a measure of the value of the pozzolana present

Compression or tension tests of sand lime-pozzolana mortars have been used for evaluating pozzolana activity. Many investigators have used various ratio of lime to pozzolana and have cured the specimen under different types of curing conditions. Maximum strengths are obtained at the early stages with a lime-pozzolana ratio of about 1:4. For longer periods, however, high ratios of lime tend to give higher strength. Ratios of 1:2 and 1:3give maximum strength at one year.

Since the activity of a pozzolana is measured by its ability to react and form a cementitious material with lime, the ability of the lime will obviously affect the rate of strength development as it is with portland cement concrete. A litle rise in temperature however, shows a marked increase in strength development in lime-pozzolana mortars.

A.4. POZZOLANA MANUFACTURE AND ENERGY SAVINGS

Natural pozzolanas can be used directly without any processing. But they are generally more reactive when heated for a few hours at temperatures ranging between 300-700 degrees C and are more reactive when finely ground. The artificial ones like burnt clays and shales however require proloned heating at temperatures ranging between 500-1000 degrees C. The maximum reactivity depends on the type of the predominant clay material present in the sample. The temperatures for maximum activity for the three main groups of clay materials, montmorillonite, kaolinite, and illite are respectively 600-800; 700-800 and 900-1000 degrees C respectively. At these temperature ranges the chemically bond hydroxyls are lost resulting in the collapse of the structure. Consequently large amounts of free surfaces are released for reaction.

The heating can be conducted in either a rotary kiln, vertical shaft kiln or in a fluidised bed calciner. There are many other simple technologies available for calcining clays and shales. In all of these technologies the control of the calcining temperature is crucial. For converting rice husk to pozzolanas special burners or stoves where temperatures can be controlled are required. Much work has been carried in the designing of suitable stoves by the Cement Research Institute of India.

Energy requirements for manufacturing a kilo of pozzolana requirements are higher for the wet process than for the dry process. Favourable energy consumption conditions are obtained with modern rotary kilns which are equipped with special devices for achieving complete utilsation of all available sources of heat. Energy requirements for manufacturing a kilo of clinker can be as low as 840Kcal for such kilns. But for pozzolana production this type of technology is not required. Table A.2 shows the energy consumed during the manufacture of lime, pozzolana and cement. It is clear from the table that much more energy is required for producing clinker even if energy required for grinding is neglected. This means that pozzolana can be produced at a relatively lower cost. There is, therefore, some cost benefit to be derived in using pozzolana cements locally produced for low cost housing.

Note : Cement is produced by finely grinding the clinker

Table A.2. Energy Inputs of Materials

Material	Solid Fuel Kcal/Kg	Electrical Energy Kcal/Kg	Total Energy Kcal/Kg
Lime	722	3	725
Hydrated Lime	547	50	597
Fly Ash	0	0	0
Surkhi	0	50	50
Burnt Clay (pozzolana)	228	56	284
Rice Husk Ash	0	50	50
Portland Cement	1440	120	1560

Section B

FERROCEMENT

B.1 Introduction

Ferrocement is a highly versatile construction material which has established itself as an appropriate material for socially-relevant applications in developing countries. It is a composite material comprising cement mortar which is reinforced with welded and chicken mesh. The raw materials for ferrocement are available in most countries. Ferrocement can be fabricated into almost any shape to meet the needs of the user.

Ferrocement is principally the same as reinforced concrete (RCC), but has the following differences:

Its thickness rarely exceeds 25 mm, while RCC components are seldom less than 100 mm.

A rich portland cement mortar is used, without any coarse aggregate as in RCC.

Compared with RCC, ferrocement has a greater percentage of reinforcement, comprising closely spaced small diameter wires and wire mesh, distributed uniformly throughout the cross-section.

Its tensile-strength-to-weight ratio is higher than RCC, and its cracking behaviour superior.

The major advantages of ferrocement construction are :
- structures can be thin and light,
- they can be easily precast,
- they are amenable to repairs in case of local damage,
- Considerable saving in formwork particularly for complex shapes, and
- saving in cost.

The construction technique is simple and the workmen can be easily trained for construction of ferrocement structures.

A comparison of the basic properties of flexural members in steel, concrete, timber, etc. are given in Table B.1.

Table B.1

Physical properties of material

Material strength	Flexural in N/mm²	Modulus of elasticity in N/mm²	Density in kg/m³
Timber	8.5-18.0	0.060-0.12x10⁵	500-1000
Ferrocement	10.0-25.0	0.30-0.50x10⁵	2500
Concrete	2.7	0.22x10⁵	2300
Steel	150.0	2.00x10⁵	7900

B.2 What is Ferrocement ?

It is a kind of composite material in which cement and a filler material (usually brittle in future, eg. sand), called matrix, is reinforced with fibres (usually steel meshes) dispersed throughout the composite resulting in a better structural performance than that of its original. The composite material possesses certain unique properties such as higher tensile strength, better resistance to cracking, and improved durability.

B.2.1 Applications

Where this material can be useful ?

(a) (i) Water storage tanks
   (ii) Grain storage bins/silos
   (iii) Boats and small ships
   (iv) Pipes
   (v) Service core units
   (vi) Canal lining elements
   (vii) Elements for roofing houses
   (viii) As lost forms for beams
   (ix) Utility products : sinks/wash basins., planks etc.
(b) As substitute to timber
   (i) Rafters
   (ii) Trusses
   (iii) Cupboards, etc.

The first step is to prepare the skeletal framework onto which the wire mesh is fixed with a thin tie wire (or in some cases, by welding). A minimum of two layers of wire mesh is required, and depending on the design, up to 12 layers have been used (with a maximum of 5 layers per cm of thickness).

The sand, cement and additives are carefully proportioned by weighing, mixed dry and then with water.

After checking the stability of the framework and wire mesh reinforcement, the mortar is applied either by hand or with a trowel, and thoroughly worked into the mesh to close all voids.

Thicker structures can be done in two stages, that is, plastering to half thickness from one side, allowing it to cure for two weeks, after which the other surface is completed.

Compaction is achieved by beating the mortar with a trowel or flat piece of wood.

Care must be taken not to leave any reinforcement exposed on the surface, the minimum mortar cover is 1.5 mm.

Each stage of plastering should be done without interruption, preferably in dry weather or under cover, and protected from the sun and wind. As in concrete construction, ferrocement should be moist cured for at least 14 days.

B.3 Mortar Composition

The essential ingredients of the mortar which represents about 95% of ferrocement are portland cement, sand, water, and in cases an admixture.

Most locally available, standard cement types are suitable, but should be fresh, of uniform consistency and without lumps or foreign matter. Special cement types are needed for special uses, eg sulphate-resistant cement in structures exposed to sulphates (as in seawater).

Only clean, inert sand should be used. Particle sizes should not exceed 2 mm and uniform grading is desirable to obtain a high-density workable mix. Lightweight sands (eg volcanic ash, pumice, inert alkali-resistant plastics) can also be used, if high strengths are not required.

Fresh drinking water is the most suitable. Seawater should not be used.

Admixtures can be used for water reduction, thus increasing strength and reducing permeability (by adding so-called "superplasticizers"); for water-proofing; for increased durability (e.g. by adding up to 30% fly ash); or for reduced reaction between mortar and galvanized reinforcements (by adding chromium trioxide in quantities of about 300 parts per million by weight of mortar).

The recommended mix proportions are: sand/cement ratio of 1.5 to 2.5, and water/cement ratio of 0.35 to 0.5, all quantities determined by weight. For watertightness (as in water - or liquid-retaining structures) the water/cement ratio should not exceed 0.4. Great care should be exercised in choosing and proportioning the constituent materials, especially with a view to reducing the water requirement, as excessive water weakens the ferrocement.

B.4 Reinforcement

The reinforcing mesh (with mesh openings of 6 to 25 mm) may be of different kinds, the main requirement being flexibility.

Galvanizing, like welding, reduces the tensile strength, and the zinc coating may react with the alkaline environment to produce hydrogen bubbles on the mesh. This can be prevented by adding chromium trioxide to the mortar.

The volume of reinforcement is between 4 and 8% in both directions, ie between 300 and 600 kg/m^3; the corresponding specific surface of reinforcement ranges between 2 and 4 cm^2/cm^3; the corresponding specific surface of reinforcement ranges between 2 and 4 cm^2/cm^3 in both directions.

Hexagonal wire mesh, commonly called chicken wire mesh, is the cheapest and easiest to use. It is very flexible and can be used in very thin sections, but is not structurally as efficient as meshes with square openings, because the wires are not oriented in the principal (maximum stress) directions.

Square welded wire mesh is much stiffer than chicken wire mesh and provides increased resistance to cracking. However inadequate welding produces weak spots.

Square woven wire mesh has similar characteristics as welded mesh, but is a little more flexible and easy to work with than welded mesh. Most designers recommend square woven mesh of 1 mm (19 gauge) or 1.6 mm (16 gauge) diameter wires spaced 13 mm (0.5 in) apart.

Expanded metal lath, which is formed by slitting thin gauge sheets and expanding them in the direction perpendicular to the slits, has about the same strength as welded mesh, but is stiffer and hence provides better impact resistance and better crack control. It cannot be used to make components with sharp curves.

Skeletal steel, which generally supports the wire mesh and determines the shape of the ferrocement structure, can be smooth or deformed wires of diameters as small as possible (generally not more than 5 mm) in order to maintain a homogenous reinforcement structure (without differential stresses). Alternatively, skeletal frameworks with timber or bamboo have been used, but with limited success.

Fibres, in the form of short steel wires or other fibrous materials, can be added to the mortar mix to control cracking and increase the impact resistance.

B.5 FERROCEMENT PRODUCTS

B.5.1 Trusses and Rafters

Clay tiled roofing is commonly adopted for houses using timber trusses or rafters as supporting structures and timber reapers as secondary members. Ferrocement trusses or rafters are better in performance and durability than those of timber. They are also free from fire hazards and termite attack. The increasing cost of timber has made the use of ferrocement for trusses and rafters a very attractive alternative. Ferrocement trusses and rafters are cheaper than those made of timber by about 30%.

Ferrocement trusses and rafters can be fabricated by using cement mortar (1 : 3) and steel skeleton. The steel skeleton consists of a straight rod at top and bottom of the section with 3 mm diameter mild steel wire spiraled as stirrups. This is clothed by 26 gauge chicken wire mesh. For a 3 m long ferrocement rafter, the recommended cross section is 40 mm × 120 mm.

B.5.2 Ferrocement Rafters

Ferrocement rafters are used for supporting tiled roofing. These rafters spaced at approximately 750-1000 mm and in span ranges of 3.6 - 6m will have a conventional timber scantling of size 25 mm x 50 mm. They are made to the profile of I - Section and assembled together by using steel bolts and nuts. Spanning between these rafters, the normal reapers can be replaced by ferrocement trough units which can directly receive the country tiles. Serving as the water proofing element, the trough units also provide good ceiling finish. Such rafters can be mass-produced. The calculated steel as reinforcement is provided in the body of the rafter in addition to providing normal weld mesh and chicken mesh distribution steel. It is even possible to improve the performance of such sections by marginally stressing the welded mesh element on a long line stressing bed. With the availability of concrete cutting saws, it should even be possible to make these rafters on long line stressing bed, like rolled steel joists and make them available in the market as ready to use elements. The prestressing will be effective only for preventing cracks due to handling stresses at early age to avoid excessive deflection.

B.5.3 Ferrocement Trusses

If the requirements are for small housing schemes with lean-to-roofs, monolithic ferrocement trusses can be made in any desired profile. These trusses and their sections and reinforcement can be designed to match the loading requirements.

B.5.4 Casting Procedure

The welded mesh required for the truss is cut and fabricated to the desired profile. Around this skeleton two layers of chicken mesh are woven and placed in an appropriate mould. Into the mould, placed horizontally on a casting platform, fine cement mortar is placed and compacted. Unlike for structures in contact with water, like boats, and water tanks, higher cement ratio is not required for the mortar used in such flexural members. 1:3 or leaner mix can be employed. If vibration facilities are available such mortar could gain strength upto 30.0 N/mm² at 28 days.

B.6 FERROCEMENT ROOFS

Ferrocement roofs can be made in situ or with precast components, the former being useful for free forms, the latter being appropriate for modular and repetitive constructions.

Depending on the design, ferrocement roofs can be made to span large areas without supporting structures, thus saving costs and providing unobstructed covered areas. If the ferrocement surface is properly executed (complete cover of wire mesh, dense and smooth finish, cracks sealed) no surface protection is needed, thus saving further costs. However, it is advantageous to apply a reflective coat on the outer surface to reduce solar heat absorption.

B.6.1 Framed Ferrocement Roof

Once the walls are erected, no reinforced concrete ring beam is required, as the roof is designed to clamp the walls together.

Around the top, outer edge of the walls, a timber frame (6x6 cm) is fixed, as well as two tripod frames above the floor area. The surface described by these frames are hyperbolic paraboloids (hypars), which are made up of straight lines. This simplifies the fixing of the wire mesh.

The mesh (2 or 3 layers) is stretched over the frame and nailed or stapled onto it. The frame is only needed to hold the mesh during construction, as the structure will be self-supporting once plastered.

Reinforcing bars are fixed around the wall and along the folds of the roof.

The roof is plastered by a team on top forcing the mortar through the mesh, while another team below recovers the falling mortar to plaster the inside.

This curved roof system, permits the wind to blow around smoothly, making it very suitable for hurricane prone areas.

B.6.2 Attic Type Ferrocement Core Units

This system consists of a three dimensional ferrocement core unit of 3 cm thickness. The weight of a core unit is about 900 kg and these core units are supported on 4 precast concrete columns and are fixed with in-situ concrete joints. This simple core unit can be used also as an emergency shelter. The core is designed to be permanent and more hazard resistant.

B.7 DESIGN OF SMALL CAPACITY FERROCEMENT WATER TANK

B.7.1 Design Procedure

The design of ferrocement water tank involves determining the volume of the tank, layers of welded mesh and chicken mesh, and its thickness, stresses and deflection.

The following steps are involved in the design of a tank.

(i) For a given capacity, the dimensions of the tank are determined. Connections for dead storage, outlet, inlet and over-flow are very important. Locations of these are to be finalised as follows:

Free board height - 75 mm (top of tank)

Inlet connection - 50 mm (top of tank)

Outlet connection - 75 mm (from bottom slab)

The bottom slab should have a minimum gradient of 1 in 100 to drain the dirt through the scour pipe. All these fixtures should be well plugged with suitable threads in advance to facilitate free connections of valves, plugs and extensions.

(ii) The properties of locally-available reinforcement are evaluated. Their strengths, and Young's moduli are determined. Permissible stresses and Young's moduli are reduced to suit the ferrocement properties.

(iii) The Young's modulus of the composite section is determined to evaluate the deflection of the bottom slab[5].

(iv) Provide 1 percent steel and distribute it in mesh form and rod type of reinforcement (i.e. small diameter bars, weld mesh and chicken mesh) equally in vertical walls.

(v) The boundary conditions of the vertical walls, bottom slabs are established.

(vi) Compute the maximum stresses at the bottom of the wall and bottom slab, and check the corresponding permissible values at first crack. In the case of compartmental tanks, vertical stiffeners are reduced to the equivalent thickness of the bottom slab to calculate the deflection.

(vii) The deflection of the bottom slab is checked. This should not be greater than 1/400 of the span, as the slab is subjected to prolonged reported loading.

(viii) The cover of reinforcement for water tanks shall be at least 5 mm.

(ix) The volume fraction of reinforcement in the loaded direction shall be between 0.5 and 1.0 percent.

(x) The specific surface of reinforcement shall not be less than 0.5 cm²/cm³.

(xi) The tensile stresses in steel reinforcement shall not exceed 320N/mm², when smaller diameter meshes are used. This can be reduced by a factor of 0.7 for water tanks.

(xii) There shall not be any visible crack in the water tank.

(xiii) The minimum wall thickness of water tank should be 25mm for tanks upto 1,000-lit capacity.

(xiv) Clear cover of reinforcement for ferrocement water tanks is 5 mm.

(xv) For small capacity ferrocement water tanks, welded mesh of 10 g × 10 g - 100 mm × 100 mm and chicken mesh 18 g - 12 mm hexagonal opening are ideal as the reinforcement matrix.

Circular ferrocement water tanks are subjected to predominantly hoop stresses. The first crack strength of ferrocement in bending is Influenced by the quality of the mortar. However, bending cannot be avoided in water tanks. This is taken care of by increasing the thickness of the ferrocement wall and bottom slab of the tank.

Ferrocement can resist reversal of bending moment as the steel is placed on both the faces. As the water tank is subjected to repeated loading, a limiting value of first crack of only 20 micron is suggested, Table B.2.

Table B.2
Limiting values of crack width

Stage	Material state	Width of crack of ferrocement (in microns)	State of the wall
1(a)	Linearly elastic	20	Complete water tightness
1(b)	Quasi-elastic	50	Non-corrosive
2	Plastic	100	Non-corrosive
3	Plastic over	100	Corrosive

B.7.2 Deflection of Tanks

One of the important criteria for the design of a ferrocement water tank is its deflection. In the case of reinforced concrete water tanks, the section has to be uncorked one, which is achieved by restricting the stress in steel to 100N/mm².

B.7.3 Capacity Tank

Design of ferrocement water tank of 100-liter capacity:

Dimensions: 1.0 m x 1.0 m x 1.0 m
Material:
Weld mesh - 100 x 100 mm - 10g x 10g
Chicken mesh - 18g hexagonal, 12 mm opening
Sand passing through 2.36 mm sieve
Water-cement ratio - 0.4
Cement mortar - 1:2
Properties of material
Steel
Modulus of elasticity of reinforcing System, E_s = 2 x 10⁵ N/mm²
Allowable tensile stress of reinforcing system, : σ_s = 320 N/mm²
Diameter of weld mesh, 10 gauge = 3.14 mm
Mortar
Modulus of elasticity of cement mortar, E_m = 0.25 x 10⁵ N/mm²
Tesile stress of cement mortar, : σ_m = 3.0 N/mm²
Ferrocement
Effective modulus of ferrocement composite, E_t = 0.5 x 10⁵ N/mm²
Allowable tensile stress of ferrocement composite,: σ_t = 4.0 N/mm²

B.7.4 Dimension of the Tank

Provide 7.5 cm free board. Water outlet connection is provided at 7.5 cm above the bottom slab. Scour is provided in bottom slab with a slope 1/100. The tank is cast with bottom slab of 3cm. The top slab is 2.5cm thick. The bottom slab is assumed to be simply supported, while the wall is free at the top and fixed on all the three sides. Thickness of the wall varies from 2.5 cm at the top to 3 cm at the bottom.

Properties of Steel Skeleton

The steel skeleton consists of one layer of weld mesh 100 x 100 - 10g x 10g at the center and two layers of chicken mesh of 18g hexagonal 12 mm opening and tied together to form the skeleton with fixtures.

Thickness of wall = 3 cm

Thickness of steel skeleton

(0.125 x 4 + 0.314 x 2) = 1.128 cm

$$\text{Thickness ratio} = \frac{1.128}{3} = 0.376$$

B.8 PREFABRICATED FERROCEMENT HOUSES

B.8.1 Introduction

In India there is a huge shortage of residential building due to past backlog, high population growth rate, lack of resources and deterioration of existing houses. Thus, there is an urgent and compelling need for a construction breakthrough in housing through the introduction of appropriate techniques such as prefabrication. Prefabrication connotes that a structure is assembled in total or in part from factory or site-made components.

Industrialization of houses are achieved through the development of the building system. Building system is defined as the application of modern management techniques to coordinate design, manufacturing, site operations and overall financial and managerial administration into a disciplined method of building.

Ferrocement, a labor intensive highly versatile construction material, is suitable for prefabrication. The basic prefabricated ferrocement elements are light so that they can be erected manually or with light equipments. The reduction of self-weight also results in savings in supporting structures. In addition, ferrocement technology can be both adaptable for mass production in factory and on-site fabrication for labour intensive schemes.

B.8.2 Prefabricated Ferrocement Houses

Castone - Mixed Prefabricated System

This building system consists of 25 mm thick mass-produced wall panels coupled with planks and hollow block slabs. The panels are bolted together and the slab is fixed on the panels by means of similar nuts and bolts. Any type of roofing material could be used. The stability of the system is achieved by a box type configuration with one side open.

The castone panel is a prefabricated concrete element of 2.5 m x 0.46 m reiforced with chicken wire mesh and one layer of jute fabric. It is cast using a special fabricating technique with unique features. The panel weights about 100 kg making it possible to be assembled manually. A four-man team can erect about 20 m² in a single day. An advantage of the system is its flexibility and relocatability. The house could be dismantled and and erected else where.

B.8.3 Ferrocement Modular Thin Shell Houses

The structure developed consists of two barrel type shell which cut each other at right angles to form a modular double curved room element. The plan dimensions are 7.20 m x 7.20 m corresponding to an internal floor space of 6 m x 6 m per unit. This modular thin shell house was developed by Architect Florentino and the first house built in Manila in 1977 was his residence.

Basic concept of two barrel type shells that cut each other at right angles and thus form a modular double curves room element.

The reinforcement system consist essentially of two layers of welded hexagonal wire mesh supported by steel bars. The reinforcement is prefabricated either on site or in the shop. The wire mesh and the insulating material are cut according to a predetermined pattern conforming to the configuration of the module divided into eight identical sections. Two layers of reinforcements are laid out with styropor, the insulating material, sandwiched between them.

The identical sections of prefabricated reinforcements are brought to the site and erected on a standardized system of wood scaffolding. The sections are then properly secured and joined together to form the entire assembly of a module. The mortar with cement - sand ratio of 1:2 is applied either by machine or by hand on both faces of the reinforcement - insulation assembly.

The erection cost Compared with the conventional concrete vault construction and concrete lintel roofing, there is a saving of 35% and 138% respectively.

Owners attitude towards the new type of housing is positive, with the relatively low price of the house and low amount of maintenance as major reasons. Complaints are about poor acoustical and thermal insulation, and difficulty to enlarge the house.

B.8.4 Hyparoof Modular Assembly

The structure is the product of economy in applying modular coordination in the design and shop fabrication of standard components using low-cost ferrocement. Its basic modular assembly is a symmetrical monopod consisting of a thin ferrocement shell roof supported by a precast reinforced concrete column erected on an in situ concrete footing. A number of monopods when joined together in modular coordination and complemented with an enclosure of precast ferrocement wall panels between columns forms the roof-wall shell of the building. The adapted f)or module, bounded by columns, is 3.80 m x 3.80 m. Various arrangemei ts of floor plans for specific functions are possible such as waiting sheds, clinic, shopping marts, hospitals and variations of residential dwelling. This system is patented in the Philippines and the exclusive right to build was assigned to the PBM Construction Services, Inc.

A unique feature of this structure is the roofing system. The roof module consists of four quadrants of concave-convex surface whose configuration is that of a hyperbolic parboiled. The sloping roof surface converges towards the center where it is supported by a column. Surface water discharges at the hollow of the column acting as a downspout flowing into a ground drainage system. Along the 3.80 m x 3.80 m perimeter of the shell roof are reinforced edge beams framing horizontally the four quadrants bounded by equally reinforced sloping ribs. The reinforcing terminals are extended and joined with other modules when installed as a roofing system.

Another ferrocement component is the 0.825 m wide by 2.05 m long precast wall panel, 25 mm thick. Precast horizontally either with or without window or door opening, all reinforcement are extended for joining with other structural members.

A roof module is cast with seven bags of cement using a cement-sand ratio of 1:2 and from 0.4 to 0.6 parts of water. The module reinforcement assembly are shop fabricated. Continuity of modules is achieved by welding the main steel and hand joining the wire mesh of adjoining modules. With a single moving roof-module form, five roof modules can be cast in six working days by an eight-man crew.

B.8.5 Individual House Building System - Ferrocement Metodo

In Brazil, the Metodo Engenharia Ltd. has developed a prefabrication system based on thin panels of ferrocement.

The panels, factory prefabricated, are constructed using a reinforcing mesh covered with two layers of hexagonal wire mesh on which concrete is poured to a thickness of 30 mm. An insulating layer of 20 mm exfoliated vermiculite is cast on the concrete. The panels are bolted together by means of joining steel plates. Erection requires a light mobile crane. An individual house can be erected by a working team within three or four hours. Usually, the roof of the house is of corrugated asbestos cement on a wooden frame and the ceiling is made of insulating wood wool panels. Electrical ducts and plumbing are cast in the elements.

B.8.6 Multiform Ferrocement House Building System

The core structure of this system is the triangulated latticed beam composed of galvanized steel wires electrically welded to apex rods at crossing points. The depth of the girder varies from 70 mm to 165 mm and the diameter of the apex rod is either 8 mm or 10 mm.

A portal frame is formed using this wire girder as column and beam, bonded and welded to give continuity. The portals are so constructed that the apex rod faces outwards. These portals are set in a horizontal girder member located true and welded to stubs set in pile foundations. Horizontal girders are provided between frames to provide stiffness to the entire structure.

A special welded-mesh, galvanized lath which may incorporate a bitumen craft vapour barrier, are fixed by means of hog rings and staples to the internal face of the frame to form the internal component of the wall. Services are then run through the wall cavity and brought out through the lath. A flash coat of cement plaster is then sprayed on the face of the lath which will ultimately carry the internal plaster. The completed mortar thickness is 20 mm.

A similar lath is then fastened to the outward face of the lath structure. The external skin, a minimum thickness of 40 mm, is then sprayed in one or more passes, and with or without the use of a flash coat. External finishes and colours can be achieved in leveling coats using normal plastering techniques and suitable additives.

The same technique is used for the roof except that the thicker coat is applied to the inside instead of outside.

The core structure is planned on a horizontal module of 1,200 mm and a vertical module of 300 mm which is sufficient for the incorporation of doors and windows. For single story structures, it has been calculated that requirements will be within a free span of 12 modules (14.4 m) and a vertical height of 13 modules (3.9 m).

B.8.7 Precast Ferrocement Housing System

The system was developed to meet the demand for earthquake and storm resistant low-cost housing. The walls of the houses are assembled from panels, 3 m wide and 2.4 m high, while the roof elements are from 6 m x 3 m panels. The roof panels, weighing 680 kg, were the heaviest element in the system. The ferrocement panels, 13 mm thick, are made from one layer of Waston mesh. Windows and door frames are cast in the factory.

Precasting is done using concrete modules and hand operated gantries for stripping and stacking.

Advantages of this housing system are that: it is competitive in cost with traditional houses, it is relatively indestructible and highly resistant to storm damage, the walls are easy to clean and the atmosphere inside the house is equivalent to the outside shade temperature of the area.

B.8.8 Stucanet Building System

The structural system utilizes, Poutrafil, a built-up column and Stucanet panel. The latter consists of a Stucanet wire mesh with a hard paperboard "woven" within the mesh. The paperboard has regular voids at the intersection of the wires and serves as ground for the plaster applied on the mesh. The applied plaster penetrates through the paperboard voids and surrounds the wire producing a solid wall the reinforcement.

The entire supporting frame of the house consists of Poutrafil spaced at 450 mm maximum intervals. The Poutrafil is fastened to U-steel plate cast in the floor slab. The frame is then covered first on the exterior side with the Stucanet panels. The exterior side receives two layers of mortars of 30 mm each projected by machine or applied by hand. These two layers are projected on both sides of the exterior panel. The interior Stucanet panel is then put in place and is plastered on the inner side. The space between the two panels is filled with insulating material prior to fixing the internal panel. The Stucanet are fixed by means of a stapling instrument either hand operated or power stapler.

The mortar consists of 450 kg of Portland cement for one cubic meter of sand. The sand should be free of salts. To obtain specially plastic mortars up to 5% plastification, additives are used. Cement-lime mortars can also be used with a mixed of 1 part cement + 3/4 parts lime + 4.5 parts sand. For a square meter of wall used 2 m² of Stucanet wire mesh and 4 m of Poutrafil. The corresponding labor input is one man hour each for the erection of Poutrafil and Stucanet, and for spraying the mortar.

B.8.9 W-Panel Building System

The W-Panel Building System utilizes the structural capacity of the W-panel. The panels consist of a three-dimension No. 14 gauge wire frame utilizing a truss concept for strength and stiffness. Each surface of the wire space frame has a 50.8 mm square welded mesh pattern of longitudinal and transverse wires. Individually welded spreader wires, designed to maintain the surface wires of the space frames at the required distance apart, extend between the two panel faces and have a module of approximately 50.8mm.

The wire frame, after fabrication, is provided with a polyurethane foam core. The foam is held back approximately 12.7 mm from each face of the wire frame to permit the wire to be embedded in an application of a 22 mm thick Portland cement plaster finish in the field. Portland cement mortar is applied by hand or gun after installation of plumbing, electrical utilities and other utilities.

Code-recognized and fire rated, the standard panel is 1.20 m wide by 2.5 m long and weighs approximately 12 kg making it extremely easy to handle in delivery and at construction site. as a structural component, panels are also supplied in lengths from 1.8 m to 4.3 m. The panel is not sensitive to close tolerances and the cutting or redimensioning of the panel for varying design does not destroy its structural capability.

Housing and commercial construction using the W-panel building system contain all the advantages of:

1. Simplicity of erection.
2. Long range structural life of reinforced masonry constructions.
3. Reduced cost of heating and cooling due to insulation value of polyurethane foam core.
4. Fire resistant structure using fire-rated W-panel.
5. Low maintenance qualifications of cement plaster.
6. Elimination of termite, fungus and dry rot.
7. Excellent resistance to hurricanes, earthquakes, tornadoes and floods.
   (a) Roof to wall and wall intersection.
   (b) Roof to beam and typical lintel.

B.8.10 Advantages

The materials required to produce ferrocement are readily available in rural/urban areas.

It can take almost any shape and is adaptable to almost any traditional design.

Where timber is scarce and expensive, ferrocement is a useful substitute.

As a roofing material, ferrocement is a climatically and environmentally more appropriate and cheaper alternative to galvanized iron and asbestos cement sheeting.

The manufacture of ferrocement components requires no special equipment, is labour intensive and easily learnt by unskilled workers.

Compared with reinforced concrete, ferrocement is cheaper, requires no formwok, is lighter, and has a ten times greater specific surface of reinforcement, achieving much higher crack resistance.

Ferrocement is not attacked by biological agents, such as insects, vermin and fungus.

Flexural strength and modulus of elasticity are higher than conventional reinforced concrete and timber

Possesses excellent water resistant properties

Highly durable and further increased when impregnated with a polymer

Precast elements can be trucked to site and can be easily erected

Easily prepared

Ferrocement products are cheaper than the products fabricated with other materials

Economically feasible.

B.8.11 Problems

Ferrocement is still a relatively new material, therefore its long-term performance is not sufficient known.

Although the manual work in producing ferrocement components requires no special skills, the structural design, calculation of required reinforcements and determination of the type and correct proportions of constituent materials requires considerable know-how and experience.

Galvanized meshes can cause gas formation on the wires and thus reduce bond strength.

The excessive use of ferrocement for buildings can create unhealthy living conditions, as the high percentage of reinforcement has deleterious electromagnetic effects.

B.8.12 Remedies

Research on the condition of older ferrocement structures.

Development of simple construction guidelines and rules of thumb which can be applied without special technical knowledge.

Galvanized mesh can be immersed in water for 24 hours and then dried for 12 hours, in order to allow the salts used during galvanizing to come to the surface. the residue can then be brushed off.

Problems with galvanized mesh can be reduced by adding chromium trioxide to the mixing water.

Complete enclosure of dwelling units with ferrocement components (ie for floor, walls and roof) should be avoided.

The technology of using organic fibres such as sisal coconut fibre and date-palm fibre to reinforce concrete for producing thin sheets larger than tiles was developed in 1977-78, because the galvanised corrugated iron sheets had a major draw back, that they used to get corroded and began to leak after

3-4 years in coastal and polluted climates.

The production process in most wide spread use for the manufacture of corrugated sheets from FRC was developed by the IT building materials workshop in Cradley Health in Britain.

Other FRC based products suitable for self builders, developed by the cradley Health workshop, include arch sheets as permanent shuttering for floors to save the inordinate consumption of timber used for this work in the role of framework, FRC shells have been used successfully in the installation of otherwise conventionally constructed brick arches . In these applications the FRC occupies a structural role only in the few hours following the installation of the maisonary super structure. After that the masonry itself acting in compression takes over the role leaving the FRC form work components just to provide a tidy finish.

For the purpose of calculating the cost of FRC sheet composites, table given below provides a general guide lines to the quantities of materials used.

Materials consumed per m^2 of flat screed (thickness)

Material	6mm	8mm	10mm	12mm
Cement (Kg)	7.5	10	12.5	15
sand (Kg)	7.5	10	12.5	15
fibre (g)	225	300	375	450

With a fully manual process a three man team can prepare a square meter of material in about 15 to 25 minutes. A vibrating table can cut the time down by as much as half.

Fibre reinforced sheets/pantiles can be made with fibres from steel, carbon, glass, plastic and natural fibres. For description of natural fibres, their properties, constituents materials and mechanics, of fibre composite.

CORRUGATED FIBRE CONCRETE ROOFING SHEETS

Special properties - Local, low-cost method

Economical aspects - One of the cheapest durable roofing materials

Stability - Good, if properly manufactured and installed

Skills required - To be developed

Equipment required -Simple, locally made, transportable moulds

Resistance to earthquake - Uncertain - Not known

Resistance to hurricane - Good, if well installed and secured

Resistance to rain - Good

Resistance to insects - Good

Climatic suitability - All climates

Stage of experience - Fairly nature technology, but research still needed

B.10 SHORT DESCRIPTION

Corrugated FC sheets

- Were the first FC roofing elements to be developed, as the aim was to substitute gci and ac sheets;
- require fairly simple, locally made equipment and a very well coordinated working team of at least two workers;
- consume about the same amount of cement as asbestos cements sheets (15 kg per m^2), on account of their greater thickness and production method by manual tamping, but require no electricity;
- are difficult to handle when fresh and to cure in water tanks, because of their large size;
- are difficult to transport and install without breakage, and do not tolerate inaccurately constructed supporting structures;
- withstand strong wind forces because they are heavy and have few overlaps.

In most cases FC pantiles are easier to produce and install than FC sheets and therefore represent the more appropriate solution.

B.10.1 Production of Corrugated FC sheets

Materials and equipment

- Cement: ordinary Portland cement (9.8 kg per 10 mm thick corrugated sheet of 100 x 78 cm) corresponding to cement : sand ratio of 1 : 1; a pozzolana (eg rice husk ash) can be added to improve fibre durability and reduce cement content, but causes slow setting, which necessitates a larger number of moulds and larger workspace.
- Sand: (10 kg per sheet) preferably with angular particles and good grain size distribution between 0.06 and 2 mm, free from silt and clay.
- Fibre:(0.2 kg per sheet) mainly natural, such as sisal, jute, coir, or banana fibre, but also synthetic fibres, eg polypropelene or glass fibre, can be used. Long fibres can be used, but require a different (more difficult) manufacturing process and result in weaker products. Short fibres, chopped to lengths of 12 to 25 mm, are easy to process, provide cohesiveness to the wet mortar, permitting reshaping without cracking, and also help to prevent cracking due to drying shrinkage.

- Water : Preferably drinkable water, just enough to make the mortar mix workable.
- Admixtures : Such as waterproofers may be used, if the sand is not well graded, and colorants, if the gray cement color is not desired.
- Screeding board : a flat horizontal board with outer frame, to define the FC sheet size and clamp down the polythene interface sheet.
- Corrugated setting moulds: gci or ac sheets, enough for two days production. All sheets should be obtained from a single batch made from a single master mould, as sheets from different batches or different producers are likely to have dissimilar corrugations. Accuracy in the corrugations is vital for proper installation and trouble free performance.
- Other equipment : standard workshop tools.

B.10.2 Moulding and Curing

- The correctly proportioned and well-mixed mortar is troweled evenly onto the polythene sheet, which is fixed on the Screeding board; the mortar is tamped, leveled to a uniform thickness of 10mm and smoothed off with the trowel.
- The frame is removed, the edges of the mortar layer trimmed and the screeding board titled, such that the polythene sheet with the wet fibre concrete is allowed to gradually slide onto the corrugated mould held below.
- The fresh FC sheet and mould is placed on a stack for primary curing for 24 hours, after which they are hard enough to be demoulded and placed upright for further curing (by regular watering), or completely immersed in water tanks for about 2 weeks.
- Demoulding should not be done later than 48 hours after moulding, as the sheets tend to shrink on drying, and will crack if resisted by the setting mould.

B.10.3 Production of FC Ridge Tiles

- Materials and equipment : same as for sheets, but different shape of frame, and screeding board made with hinges, so that it can be bent and used as the setting mould, held in a template.
- Moulding and curing : same as for sheets.

B.10.4 Installation of FC Roofing with Corrugated sheets

The corrugated FC sheets are laid on timber roof structures in much the same way as gci and ac sheets. However, FC sheets are less flexible and can be damaged if the loads are not evenly distributed. Therefore, care must be taken in constructing the substructure, to ensure that the top edges of all members are properly aligned. If nails or bolts are used, holes (of slightly larger diameter) should be drilled beforehand. Alternatively, nibs with wire loops can be cast-in during moulding, avoiding the need for drilling. Mitred corners are essential for a weather-tight fit.

B.11 SHORT DESCRIPTION :

FC pantiles

- were developed to overcome most of the problems encountered in producing and installing corrugated FC sheets ;
- are made most efficiently on a small vibrating table (hand powered or run by a car battery), which can be operated by a single trained worker;
- can be made thinner (6mm) than FC sheets (10mm), and their cement: sand ratio (between 1 : 3 and 1 : 6) is less than for FC sheets (1 : 1), so that the cement used for making tiles is less than 5 kg per m² of roofing.
- are easy to handle when fresh and to cure in water tanks;
- do not tend to break as easily as sheets during transport and installation, and minor inaccuracies in the supporting structure have no negative effects;
- are easily torn off by strong wind forces, if they are not well fixed to the substructure.

B.11.1 Production of FC Pantiles

Materials and equipment

- Cement : same as for FC sheets, but less than 0.4 kg per 6mm thick pantiles of 50 x 25 cm, corresponding to cement : sand ratio of 1 : 3; with superior cement, up to 1 : 6 is possible.
- Sand : same as for sheets, but 1.2 kg per Pantiles.
- Fibre : same as for sheets, but 0.02 kg per pantile.
- Water and admixtures : same as for sheets.
- Screeding machine : comprising a vibrating screeding surface and interchangeable, hinged frame (for products of different shapes and thicknesses), whereby the vibrating mechanism is either powered by a 12 volt car battery or hand-powered. (A variety of models, depending on different user requirements and desired output rates are available from the intermediate Technology Workshops, United Kingdom).

- Setting moulds : these are part of the pantile production kit, and are generally made of impact-resistant pvc, with rib markings (for accurate positioning of the tile edge) and supporting frame for stacking.
- Other equipment : same as for sheets.

B.11.2 Moulding and Curing

- The wetmix is trowelled onto the polythene interface sheet on the screeding machine and, under vibration, smoothed with a trowel to the same level as the surrounding steel frame. At a predetermined spot at the top end of the pantile, a match box-size nib is formed and a wire loop pushed into it (required for fixing to the roof).
- The steel frame is lifted off the screeding source and the polythene sheet slowly pulled over the pvc setting mould, ensuring correct positioning of the tile edge to achieve uniform curvature.
- The mould with the fresh tile is then placed on as tack of moulds for initial setting and curing (24 hours), after which the tiles can be demoulded and cured for 2 weeks in water tanks.

B.11.3 Production of FC Ridge Tiles

- Materials and equipment : same as for pantiles, but with a different steel frame and setting moulds.
- Moulding and curing : same as for pantiles, but with nibs and wire loops fitted after the tile is placed on the setting mould.

B.11.4 Installation of FC Roofing with Pantiles

The FC pantiles are laid on timber laths (spaced at 40 cm centres) in the same way as clay roof materials. Sight inaccuracies do not cause major problems. As in the case of corrugated FC sheets, the use of nails requires predrilling, but is not normally required if strong winds do not occur. The wire loops, nailed or tied onto the timber laths, generally hold well.

Section C

LIGHT WEIGHT CONSTRUCTION TECHNOLOGY

C.1 Introduction

Light weight concrete can be produced by various methods. All depend on either on the presence of air voids in aggregates or in the matrix, or by omitting fine aggregates. We can classify light weight concrete as

(a) Light weight aggregate concrete
(b) Aerated concrete
(c) No fines concrete

Light weight aggregate concrete is produced using light aggregates such as pumice, expanded slag, clinker etc. which are described in detail.

Aerated concrete is produced by using foaming agents, such as aluminum powder that produce low unit weight through generation of gas while the concrete is still plastic.

No fines concrete made with gravel aggregates is not strictly, light weight concrete though its weight is about 70% of normal concrete.

Light weight concrete is now an established building material. It is used not only on account of its light weight but also because of high thermal insulation compared with normal concrete. Lighter the concrete larger is the insulation properties in general. Earliest applications of light weight concrete was in building industry and for producing insulating screeds. The principal structural use of light weight concrete is in construction of under bed of floos and roof slabs, where substantial economy can be achieved by decreasing the dead load. Indirect saving in steel and decreased size in foundation are advantages.

The disadvantages are greater cost, need for care in placing, greater porosity and greater drying shrinkage.

Structural 'lightweight' aggregate concrete can be produced with a strength in excess of 310 N/mm² and even higher strengths have been attained in certain cases, although at the expense of increased densities. The aggregates used include sintered pulverized fuel ash, expanded shales, clays and slates, foamed slag, and pumice or scoria. As most of these aggregates absorb considerable quantities of water (up to 80 per cent by volume), the effect on workability within a few minutes of mixing is such that a wet mix can become too dry. It is therefore necessary to wet, but not saturate, the aggregate before mixing. A good portion of the mixing water is also best added before introducing the cement. Rich mixes containing 350 kg/m³ cement or more are usually required to give a satisfactory strength.

The cover to reinforcement when using lightweight concrete should be 10 mm more than that used for normal dense concrete. The increased cover is necessary because, besides being more permeable, lightweight concretes carbonate more quickly than dense concrete and the protection afforded to the steel by the alkaline lime is lost.

C.2 Properties of Lightweight Aggregate Concrete

A density of 1850 kg/m³ may be considered as the upper limit for a true lightweight concrete although this value is sometimes exceeded. Lower densities than those stated will be obtained in cases where the concrete is only partially compacted.

In general the lower densities can only be achieved at the expense of lower strengths. The range of strengths given in each case are typical of what may be achieved in practice, but it should be noted that rather rich mixes are required for the higher strength values in the case of the lightweight fines to achieve these higher strengths. The tensile and shear strengths of lightweight aggregate concrete are less than that for natural aggregate concrete of the same compressive strength. The reduction in the case of tensile strength may be as much as 30%.

The modulus of elasticity of lightweight concrete is about 0.5 to 0.75 times the value of natural aggregate concrete of the same compressive strength varying from 7 to 21 kN/mm². Values of elastic deformation, shrinkage and creep are, therefore, greater for lightweight concrete. Extra reinforcement is also necessary.

C.3 Clinker and Breeze

Clinker and breeze aggregates have been in use for many years in the production of blocks and slabs for internal partitions and other interior walls. These aggregates are cheap and plentiful, and provide a very useful product after they have been crushed and graded.

In general, clinker is regarded as a well-burnt fused or sintered mass containing little combustible material; whereas breeze is a more lightly sintered and less well-burnt residue and therefore contains more combustible matter.

Increasing quantities of combustible matter causes corresponding increases in moisture movement. The combustible content is determined approximately by igniting a small sample of the clinker at specified temperatures and finding the percentage loss in weight. the specified limits for combustible content are varied according to the position in which the concrete is to be used. They are given in BS 1165 : 1966 as follows:

Class A1: Not more than 10 per cent General purposes, plain concrete.

Class A2: Not more than 20 per cent Interior work not exposed to damp conditions, cast in situ.

Class B3: Not more than 25 per cent Precast clinker blocks.

Clinker and breeze aggregate is quite unsuitable for reinforced concrete due to its porosity and absorptive properties, which maintain a more moist condition than the surroundings. The sulphur content is also a factor in accelerating the corrosion of embedded steel.

The main sources of clinker aggregates are the older power stations where the boilers are fired by solid fuel in chain-grate stokers. However, many of the more modern power stations now use pulverized fuel for firing their furnaces. The use of pulverized fuel produces two forms of burnt product, namely pulverized fuel ash (pfa) and furnace bottom ash. The latter residue may occur partly in the form of a clinker, similar to that from the older power stations, but it is very variable in quality and is accompanied by a great deal of useless dust. The better grades can however be-used as lightweight aggregate for blocks.

C.4 Pumice Concrete (Natural Aggregates)

Pumice is most widely used natural lightweight aggregate in common use. Provided it is free from fine volcanic dust and materials not of volcanic origin, such as clay, pumice produces a satisfactory lightweight concrete with a density of between 720 kg/m^3 and 1440 kg/m^3.

Pumice provides better thermal insulation than other types of lightweight concrete.

Scoria is a vesicular glassy volcanic rock. Scoria resembles industrial cinders and is usually red to black in color. Very satisfactory lightweight concrete, weighing from 1440 - 1760 kg/m^3 can be made from scoria.

When obsidian is heated to the temperature of fusion, gases are released which expand the material. The interiors of the expanded particles are vesicular and the surfaces are smooth and quite impervious. Expanded obsidian has been produced experimentally. The raw material was crushed and screened to size and coated with a fine material of higher melting point to prevent agglomeration.

The rock from which perlite lightweight aggregate is manufactured has a structure resembling tiny pearls compacted and bound together. When perlite is heated quickly, it expands with disruptive force and breaks into small expanded particles. Usually, expanded perlite is produced only in sand sizes. Concrete made with expanded perlite has density ranging from 800 - 1280 kg/m^3. It is a very good insulating material.

C.5 Foamed slag

Foamed slag is made by rapidly quenching blastfurance slag produced in the manufacture of pig iron. Its texture and strength are dependent on the chemical composition and the method of treatment, but in general the structure is similar to natural pumice.

Foamed slag aggregates are supplied in two sizes, coarse 15-3 mm and fine, 3 mm down. They should comply with BS 877:Part 2:1973. Foamed slag may be regarded as satisfactory if it fulfills the following requirements:

1. Freedom from contamination by heavy impurities, including air-cooled slag.
2. Freedom from volatile impurities such as coke or coal.
3. Freedom from an excess of available sulphate.

The properties of foamed slag concrete are given in Table 1. The thermal conductivity when dry can be almost as low as pumice concrete.

In common with other lightweight concretes, foamed slag has a higher moisture movement than ordinary concrete. The value is however, as good as the best of other lightweight concretes.

C.6 Expanded Minerals

Naturally occurring clays, shales, and slates may be used to produce lightweight porous material of a cellular texture by suitable treatment and heating up to temperatures of about 1000°C-1200°C. Material with similar characteristics may also be obtained from pulverized fuel ash or 'fly ash'. After crushing and screening to the desired size, these processed materials form good lightweight aggregates. The properties of the lightweight concrete, they are used to make are given in Table C.1.

Table C.1.

Typical Properties of Different Types of Light weight concrete

Type of lightweight concrete	Air-dry density kg/m³	Compressive strength N/mm²	Drying shrinkage per cent	Thermal conductivity W/m°C	Working properties	Nail and Screw holding
Sintered palverized fuel ash (Lytag)	1360-1760[1]	14.2-42.0[1]	0.04-0.07	0.32-0.91	Easily worked	Satisfactory
Expanded shale or clay (Aglite & Leca)	1360-1340[1]	14.0-42.0[1]	0.04-0.07	0.24-0.91	Easily worked	Satisfactory
Foamed slat	1680-2080[1]	10.5-42.0[1]	0.03-0.07	0.24-0.93	Easily worked	Satisfactory
Pumice	720-1440	2.0-14.0	0.04-0.08	0.231-0.60	Easily worked	Satisfactory
Clinker	1040-1520	2.0-7.0	0.04-0.08	0.35-0.67	Easily worked	Satisfactory
Aerated cement mortar	400-960	1.4-4.9	0.05-0.18	0.10-0.22	Easily worked	Satisfactory
No-fines concrete (a) gravel aggregate 1:8 aggregate/ cement ratio (by volume)	1600-1840	3.5-11.0	0.02-0.03	0.65-0.80	Difficult	
(b) lightweight aggregate 1 : 6 aggregate/cement ratio (by volume)		2.4-3.1	Depends on aggregate used	Depends on aggregate used	Easily worked	
Dense concrete containing gravel or crushed stone ?	22-2480	14.0-70.0	0.03-0.05	1.40-1.80	Hard	Good when plugged

1. These higher values of density and strength are obtained by replacing the lightweight fines by a natural sand

Expanded clay aggregate is produced on the Continent and at one plant in Britain where it is produced under the trade name of Leca (lightweight Expanded Clay Aggregate). It is made by a patented process using a special grade of blue clay which bloats readily when heated. The resulting material consists of light, hard, rounded pellets with a dense skin and a honeycomb interior. Angular material may also be present if oversized material has been crushed and added. The material is supplied in three nominal sizes: coarse, 20-10 mm; medium, 10-3 mm; and fine, 3 mm down. Expanded shale is also produced on a relatively small scale, the manufacture being based on an American process of sintering colliery shale and clay. This material, which is known as Aglite in Britain, has similar properties to Leca except that it is crushed and is, therefore, angular. It is also slightly more porous.

A somewhat similar lightweight aggregate known as Lytag is produced by pelletizing fly ash and burning it on a sinter strand at temperatures up to 1400°C. A bed of pellets, on a moving steel grid, passes under a furnace and the pellets become sintered into a clinker-like mass. After cooling, it is screened and graded into three sizes: coarse, 13-8 mm; medium, 8-5 mm and fine, 5 mm down.

Since these lightweight aggregates are porous, they absorb considerable quantities of water. The amount of water absorbed by Aglite in 24 hours can exceed 30% by volume and the ultimate water absorption is of the order of 50% by volume. The corresponding figures for Leca and Lytag are generally about two-thirds of these values, indicating that a considerable amount of water can be present within these aggregates. This can have a considerable influence on the thermal insulation, shrinkage, and creep of the concrete which is made with them.

Although the use of these processed aggregates produces a saving in weight of the concrete as compared with the use of natural aggregates, the cost of the lightweight concrete (of equivalent strength) is 30-40% more. Their use does not, therefore, normally result in a saving in the cost of the structure.

Exfoliated vermiculite is very light in weight, but is so soft that the action of mixing with cement and water is sufficient to cause collapse of particles. Vermiculite concrete is, therefore, likely to be of uncertain density and to have little compressive strength.

C.7 Aerated Cement Mortars

Aerated cement mortars (often referred to simply as aerated concrete) are made by introducing air, or specially foamed gas, into a cement slurry so that, after setting, a hardened mortar with a cellular structure is formed. The slurry usually consists of a mixture of cement and siliceous material such as sand and/or pulverized fuel ash.

There are two main methods of forming the cellular structure, namely (1) the addition of powdered aluminum or zinc, which combines with the lime in the cement to generate hydrogen gas and (2) using a foaming agent. In the first of these processes, the aluminum or zinc powder is added to the cement slurry during mixing, the quantity of powdered metal being about 0.1 to 0.2 per cent of the weight of the cement. Within a few minutes, hydrogen gas begins to evolve causing the slurry to rise, the action continuing for an hour or so. The slurry then sets to form a material consisting of multitude of closed bubble holes surrounded by hardened cement mortar. The density of the material depends on the quantity of powder metal used and the temperature at the time of manufacture. For most purposes densities within the range 550-950 kg/m^3 are satisfactory.

While aerated cement mortar made in this way can be relatively impermeaable to moisture owing to the closed type of pores, it has the disadvantage of a high drying shrinkage and moisture movement. Precast units, which can be matured well before use, thus materially reducing effective shrinkage movements when the units are built into a structure, do not suffer from cracking to the same extent as aerated cement mortar cast in situ. Curing in an autoclave, using steam under pressure, considerably reduces drying shrinkage and moisture movements. This procedure is generally adopted for the production of blocks and precast units.

In the proprietary systems which incorporate the autoclave treatment, the cement is mixed with sand or blast-furnace slag which has been ground in a ball mill to the same fineness as the cement. Pulverized fuel ash or lime may also be added.

Aerated cement mortars with a similar texture can also be produced by the use of foaming agents such as resin soap. The foaming agent is either mixed with the cement, sand and water and the air entrainment is achieved by whisking in a high speed mixer, or is turned to foam by the use of compressed air using a special foam producing apparatus, and the foam then added to a cement: sand mortar in an ordinary concrete mixer. The latter method provides a more uniform density, provided the foaming time is carefully controlled.

Foamed cement mortar can be made with a density as low as 320 kg/m^3, but then has no appreciable strength and is only of use for insulation purposes in dry situations.

Some foaming agents are sufficiently stable to permit the use of cement: sand mixes for the production of foamed concrete at densities above about 800 kg/m^3. The average crushing strength varies with the density as indicated in Fig. 1. As with other lightweight concretes the relatively high drying shrinkage presents problems in the use of this material in situ, and ample provision of 'expansion joints' is essential. Precast blocks must be cured for several weeks before use unless steam curing is adopted. Even then it is advisable to build in a cement: lime : sand mortar not richer in cement than 1:1:6 if visible cracking is to be avoided. The inclusion of steel reinforcement has extended the use of autoclaved aerated cement mortars for structural units such as wall units and lintels. Because of the danger of corrosion of the reinforcement, however, such units should be protected from the weather and the reinforcement should be given a protective coating of cement latex compound or similar material.

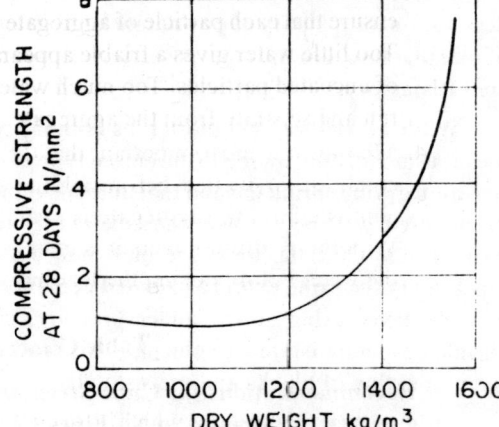

Fig. C.1 Compressive strength-density relationship for aerated cement mortar.

C.8 Aerated mortar Concrete

Coarse natural or lightweight aggregate can also be used in conjunction with aerated cement mortars. Natural aggregates increase the density, and to some extent reduce the moisture movements. Lightweight aggregates do not cause the same increase in density, but the moisture movements are more nearly those of aerated mortars.

C.9 No-Fines Concrete

No-fines concrete is composed of cement and coarse aggregate only, the fine aggregate being omitted in order to leave uniformly distributed voids through-out the mass. The aggregate may be gravel or crushed stone, blastfurance slag, crushed brick, or one of the light aggregates already mentioned. The quality of the aggregate should accord with the requirements already given, and the relevant British Standard.

The following comments are based on experience in the production of no-fines concrete on a large scale for houses and multistorey flats.

1. **Grading of aggregate:** The aggregate should be as nearly single

sized as practicable. It is usual to use an aggregate of which not more than 5 per cent is retained on a 20 mm mesh sieve, and not more than 10 per cent passes a 10 mm mesh sieve.

2. **Recommended proportions:** Normal dense aggregate, 50 kg cement to 0.28 m³ aggregate; light aggregate, 50 kg cement to 0.21 m³ aggregate.

3. The water/cement ratio should be the minimum necessary to ensure that each particle of aggregate is coated with cement grout. Too little water gives a friable appearance with a large proportion of uncoated particles. Too much water causes the cement grout to run and separate from the aggregate.

4. **Mixing:** It is most important that the aggregate is used in a damp condition or wetted before adding the cement. Mixing should continue until the aggregate is evenly coated with cement grout. With rotary-drum mixers it is generally advisable to add some of the water before raising the loading skip.

Table C.2

Recommendations for Cube Strengths and Permissible Stresses in No-Fines Concrete

Nominal mix	Volume of coarse aggregate per 50 kg of cement m³	Cube strength within 28 days after mixing N/mm²	Maximum permissible stresses N/mm²
Special	0.28 or less	7.0++	
		3.5 +	0.6
1:8	0.28	2.8	0.5

The recommendations cover the use of natural aggregates to BS 882 and air-cooled blastfurance slag to BS 1047, mixed with Portland cement, Portland blastfurance, or other cement included in CP 110.

These requirements may be deemed to be satisfied if two-thirds of the value is obtained at 7 days.

The average cube strength for mix design purposes should be 2-10 N/mm² in excess of the cube strength specified unless more definite information is available.

The attainment of this strength increases the density of the concrete to an extent where the thermal-insulation properties may be impaired.

5. **Placing:** No-fines concrete will not segregate and can therefore be readily placed in deep lifts of up to three storey high in one operation. Pressures on the framework are not high and a comparatively light timber framework serves to support plywood, hardboard, sheet metal, or expanded metal coverings. No-fines concrete should not be rammed but careful rodding should be done wherever obstacles occur, since bridging is more likely to occur than with ordinary concrete.

   Delays in placing the no-fines concrete, particularly during hot weather, must be avoided since the progressive stiffening of the cement paste around the aggregate can cause a serious reduction in the density and strength of the no-fines concrete in the structure.

6. **Joints:** Since the bond between new and existing work is weaker than in ordinary concrete, construction joints should be horizontal and as few as possible. Vertical joints weaken the structure, especially if they occur near an external angle, and except where necessary for the provision of expansion joints they are better omitted altogether. Expansion joints should be used at suitable spacings to take care of shrinkage and thermal movements.

7. **Cube strength:** Experience has shown that with good site control a minimum crushing strength of 2.8 N/mm^2 or more at 28 days is easily practicable with the mix given under (2), above, enabling the building of unreinforced structures five or six storeys in height. Greater strengths can be obtained by using richer mixes and by compacting to a higher density.

Other characteristics: No-fines concrete presents some difficulty in the fixing of various fittings and it is necessary to embed nailing blocks of timber, sawdust-cement, or foamed slag. Cutting away for services should be avoided by forming suitable openings and chases at the time of placing.

No-fines concrete has little resistance to the penetration of water, but there is also very little capillary action. Thus, there is no tendency for the water to be drawn into the wall; the provision of a rendering, with care in arranging and fixing flashings at various openings, is sufficient to waterproof the structure satisfactorily.

The rendering should preferably have a rough surface texture, dry dashings being particularly suitable. A 1:1:6 or 2:1:9, cement: lime : sand rendering is normally considered suitable, except perhaps in cold climates where a somewhat richer mix may be used to provide increased resistance to disintegration by frost and continually wet conditions.

The thickness of no-fines concrete is more often determined by the requirements of thermal and acoustic insulation than stability. For instance, in order to obtain insulation similar to that of a 275 mm closed-cavity brick wall, plastered internally, it is necessary to use 250-300 mm of no-fines concrete rendered externally and plastered internally.

A general indication of other important properties of no-fines concrete is given in Table 1.

Construction Control of Lightweight Concrete: Commercially available lightweight aggregate is usually supplied in the three principal sizes depending upon its application. These are fine, medium, and coarse and range in size to 20 mm maximum. Production of uniform concrete with lightweight aggregate involves all the procedures and precautions as for ordinary concrete. However, the problem is more difficult where lightweight aggregate are used because of greater variations in absorption, specific gravity, moisture content, and amount and grading of undersize. If unit weight and slump tests are made frequently and the cement and water content of the mix are adjusted as necessary to compensate for variations in the aggregate properties and condition, reasonably uniform results can be obtained.

Concretes made with many lightweight aggregates are difficult to place and finish because of the porosity and angularity of the aggregates. In some of these mixes the cement mortar may separate from the aggregate and the aggregate float toward the surface. when this occurs, the condition can generally be improved by adjusting aggregate grading. This can be done by crushing the larger particles, adding natural sand, or adding filler materials. The placeability can also be improved by adding an air-entraining mix; as sand content is increased, the optimum amount of fines is reached when the concrete no longer appears harsh at the selected air content. From 4- to 6-percent air is best for adequate workability, and the slump should not exceed 150 mm.

To ensure material of uniform moisture content at the mixer, lightweight aggregate should be saturated 24 hours before use. This wetting will also reduce segregation during stockpiling and transportation. Dry lightweight aggregate should not be fed into the mixer; although this will produce a concrete which can be readily placed. immediately after being discharged, continuing absorption by the aggregate will cause the concrete to segregate and stiffen before placement is completed.

It is generally necessary to mix lightweight concrete for longer periods than conventional concrete to assure proper mixing. Workability of lightweight concrete with the same slump as conventional concrete may vary more widely because of differences in type, porosity, and specific gravity of the materials. For the same reason, the amount of air-entraining agent required to produce a certain amount of air may also vary widely. Continuous water curing, by covering with damp sand or use of a soil-soaker hose, is particularly advantageous where concrete is made with lightweight aggregate.

C.10 Nailing Concrete

C.10.1 Definition, Use, and Types: Concrete into which nails can be readily driven and which will hold the nails firmly is called nailing concrete. Such concrete is used for constructing cants to which roofing material and flashing can be nailed. Among the aggregates that produce good nailing concrete are sawdust, expanded slag, natural pumice, perlite, and volcanic scoria.

C.10.2 Sawdust Concrete: Good nailing concrete can be made by mixing equal parts, by volume, of portland cement, sand, and pine sawdust with sufficient water to give a slump of 125 to 50 mm. Nailing is easier if the sand passes a No. 16 or No. 8 screen. Rigid adherence to the stated mix proportions is not essential. If the concrete is too hard, the amount of sawdust may be increased as much as 100 percent while keeping the quantities of cement and sand the same. Concrete proportioned on this basis is very workable and bonds well with the base concrete. After sawdust concrete is 3 days old, nails can easily be driven into it and have excellent holding power.

The concrete should be mixed thoroughly, preferably in a mixing machine unless the quantity is very small. It should be moist cured for 2 days and then allowed to dry for a day or more before any nailing is done.

C.10.3 Types and Grading of Sawdust: Sawdust should be clean, free from chips and lumps that will not pass a 6 mm screen and not so fine that all will pass a No. 16 screen. Concrete made with coarse sawdust requires about 24 hours to harden, whereas that made with fine sawdust requires about 48 hours. An increase in the fineness of sawdust (greater surface area of the wood particles) may result in extraction of a larger percentage of organic acids and consequently result in retarded set and reduced strength.

The following tabulation gives results of tests of different types of sawdust. The tests show that some types are entirely unsuitable for use.

C.11 Material   Notes

Material	Notes
Sugar pine	Set hard at 1 day. Good nailability.
Pine	Set hard at 2 days. Good nailability.
Pine and fir mixture	Set hard at 3 days. Good nailability.
Hickory, oak, or birch	No set at 3 days, some at 14 days. Never satisfactory.
Oregon fir at 28 days.	The sawdust is very fine. Partial set
Cedar	No set at any time.

Analyses of pine and cedar sawdust for tannin (tannic acid) showed the cedar to contain 2 - percent tannin and the pine to contain none. Appreciable amounts of bare in the sawdust retarded setting and weakened the concrete.

In view of the variable behaviour of different kinds of sawdust, it is advisable to try a sample of the material before procuring the quantity required.

C.12 Porous Concrete

Definition and Use: Porous concrete is a special type that is commonly used either where free drainage is required or where lighter weight and low conductivity are to be provided without the use of lightweight aggregates. (Sometimes the use of lightweight aggregates is not practicable or desirable.) Porous concrete is ordinarily produced by gap grading or single-size aggregate grading. In special draintile, a No. 4 to 9 mm or 9 to 12 mm aggregate is frequently used alone; a low water cement ratio and the minimum amount of cement are required to merely cover and cement the aggregate particles together into a mass much resembling that obtained in a popcorn ball. Occasionally, inserts of porous concrete may be installed as weep holes or drains in hydraulic structures such as canal linings to prevent back pressure or uplift from breaking the lining upon dewatering. Such concrete may require type V cement, especially for drainage structures or special draintile where soluble sulfate conditions exist.

Occasionally, porous concrete is placed upon rock foundations under split sewer pipes to drain ground water. Specifications call for 7-day strengths, as determined by 150 by 300 mm cylinders, of not less than 45 KN/Sq m and porosity such that water will pass through a slab 300 mm thick at the rate of not less than 100 gallons per minute per square meter of slab with a constant 100 mm depth of water on the slab.

C.13 Design of Lightweight Concrete Mix

Properties of lightweig... aggregates which influence or affect the concrete properties are bulk density, grading, particle shape and water absorption characteristics of aggregates.

The density of the concrete produced depends to a great extent on the bulk density of the aggregates. The unit weight of the aggregates vary depending upon the type of aggregate and the production plant. The coarse fraction, up to 20 mm size may have dry loose unit weights from 500 to 900 kg/m³ and the finer fractions from 700 to 1100 kg/m³. The grading characteristics of the finer fractions are known to affect the properties of lightweight concrete whether used in block type mixes or in fully compacted concrete. The grading limits of lightweight fine aggregate specified in various standards are compiled in Table C.3.

Many of the lightweight aggregates have particle shapes ranging from rounded to very irregular. It has been found that aggregates which are very irregular, generally produce heavier concretes due to the higher apparent specific gravity.

These considerations have an important bearing upon the methods of proportioning lightweight concrete mixes. The bulk density of the aggregates vary considerably depending upon the moisture content and unless the absorption capacity of the lightweight aggregate has been fully satisfied in the mixer, the aggregate will continue to absorb water from the matrix and there will be a disconcerting loss of workability between the mixer and the point of placing. With lightweight concrete mixes, it is suitable for measuring the workability of lightweight concert since most of the mixes tested at the Building Research station had slumps between zero and 25 mm only and yet they were capable of being easily compacted. Compacting Factor and Vebe tests are being increasingly used for measuring the workability of lightweight concrete mixes. The range of compacting factors for structural lightweight concrete is reported to be smaller than for ordinary concrete and compacting factor values imply differing degrees of workability. The cohesiveness of lightweight concrete mixes can be improved by replacing the finer fractions of the lightweight aggregate by natural sand or by the use of air entraining agents.

C.14 MIX PROPORTIONING BY EMPIRICAL METHOD

The empirical approach to the design of normal density concrete is well established and lightweight concrete is no exception to this. Based on the extensive experimental investigations at the Building Research Station, U.K., has developed empirical graphs relating the important parameters in a lightweight concrete mix and these can be conveniently used to estimate concrete.

The significant parameters influencing the mix proportions are:
- (a) Type of aggregate
- (b) Cement content
- (c) Total water/cement ratio
- (d) Workability
- (e) Strength
- (f) Relative density

Table C.3

Grading limits of lightweight fine aggregate

I.S. and B.S. test sieve	Percentage by weight passing I.S. or B.S. sieve size	B.S. 3797			Requirements of A.S.T.M. Standard C. 330-64T	
	B.8 877	Grading Zone-L1	Grading Zone-L2	Sieve size	Percentage by weight passing Sieve	
4.76 mm	90-100	90-100	90-100	No. 4	85-100
2.38 mm	70-100	55-95	60-100	—	—
1.18 mm	45-90	35-70	40-80	No. 16	40-80
600	20-60	20-60	30-60	—	—
300	10-30	10-30	25-40	No. 50	10-35
150	5-20	5-19	20-35	No. 100	5-25

The compressive strength depends primarily upon the total water/cement ratio and the type of lightweight aggregate used in the mix. The relationship between the compressive strength of water stored cubes and the total water/cement ratio is shown in Fig. C.2 for four different types of most commonly used lightweight aggregates. Lytag generally produces the highest strength for a given water/cement ratio in comparison with other aggregates and Leca producing the least.

The investigations regarding workability showed that for a particular aggregate, the compacting factor is best related to the total water content in the mix. Fig. C.3 and C.4 shows the relationship between the total water/cement ratio and the cement content, the results for a particular aggregate being bounded by curves of constant water content, which also represent lines of approximately equal workability. A compacting factor of 0.8 and above, with a Vee Bee time of less than 12 seconds indicated good workability of the lightweight aggregate concrete mixes. The upper bound curves in the figures

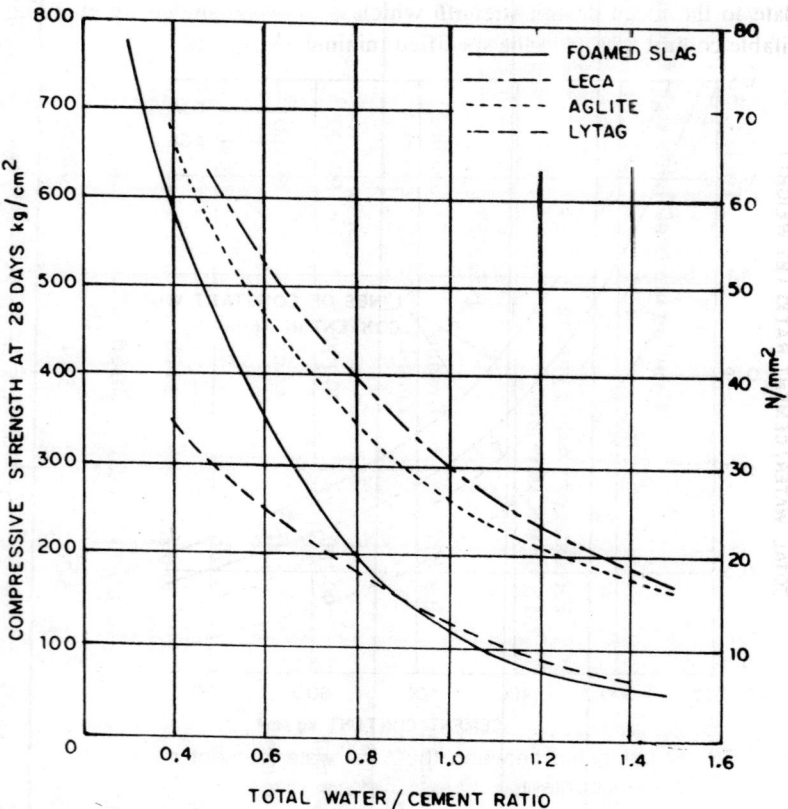

Fig. C.2.   Relationship between the compressive strength of water stored cubes and the total water/cement ratio

correspond to a range of workability from 'Medium to High' with compacting factor values exceeding 0.8.

The relative density depends primarily upon the type of aggregate, the cement and moisture content in the concrete. The relative densities shown in Figs. C.5 and C.6 relate to freshly placed concrete and to concrete stored in air at 17°C and 65 per cent relative humidity for 28 days. Generally the relative density increases more or less linearly with the cement content for fresh concrete, but it becomes still more pronounced as the concrete dries out since mixes with lower cement content lose a larger quantity of water.

A critical examination of the empirical graphs indicates that it is not possible to attain certain combinations of strength and density with all the types of aggregates. Very low relative densities of the order of 1.3 and less can be obtained only with a particular aggregate like Leca, using comparatively lower cement contents. Alternatively high strength concrete suitable for prestressed work can be produced using foamed slag, aglite or lytag with varying densities. The compressive strengths shown in the empirical gaphs relate to the mean design strength which is to be computed by applying suitable control factors to the specified minimum strength.

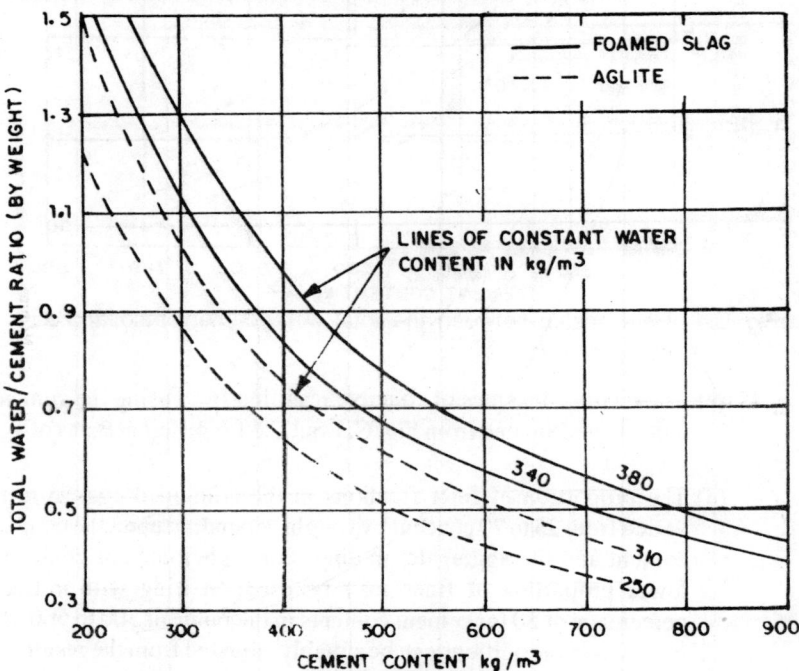

Fig. C.3. Relationship between the total water/cement ratio and the cement content

The mix design procedure is summarized as:

(a) The total water/cement ratio required to achieve the mean design strength is selected from Fig. C.2 depending upon the type of light-weight aggregate available.

(b) Using the water/cement ratio and the required workability the cement content required for different types of aggregates is determined from Figs. C.3 and C.4.

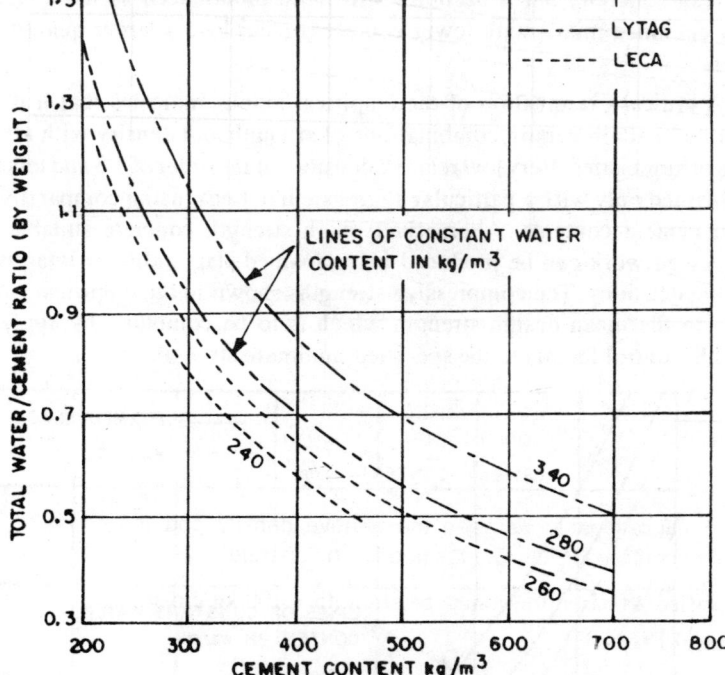

Fig. C.4. Relationship between the total water/cement ratio and the cement content

(c) The relative densities of concrete resulting from using the cement content is estimated from Fig. C.5 and C.6 for different types of aggregates.

(d) The proportion of finer fractions in the combined aggregate is varied from 25 to 70 per cent by weight depending upon the cement content and the aggregate grading. For higher cement contents, lower proportion of fines may be used. Starting with a fines percentage of 50 for cement contents in the range of 300 to 500 kg/m³, the mix proportions can be suitably adjusted from the results of trail mixes as for natural aggregates.

Example 1:

Design a lightweight aggregate concrete mix with the following data:

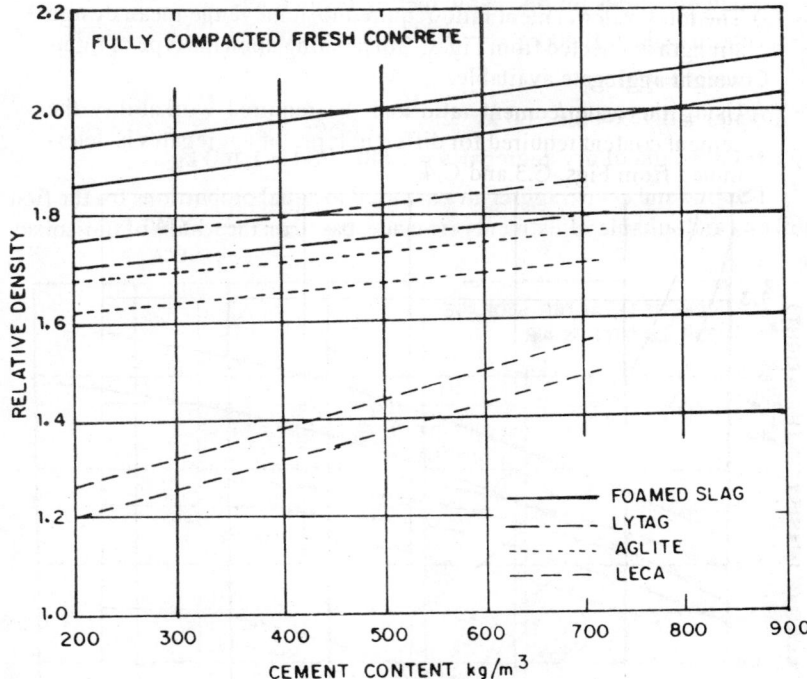

Fig.C. 5. Relationship between the relative density and the cement content for fully compacted fresh concrete

Specified 28 day minimum cube strength = 100 kg/cm²

Control Factor = 0.75

Type of aggregate available - foamed slag

Required workability = Medium to high

Maximum relative density = 2.0

(Air dry concrete)

Design

Mean design strength = 100/0.75 = 133.3 kg/cm²

Total water cement ratio = 0.95

For the desired workability and water/cement ratio of 0.95, the cement content required = 400 kg/m³

Using this cement content, the relative density of air dry concrete obtained is 1.72 to 1.84 using foamed slag.

The corresponding relative density of fully compacted fresh concrete = 1.9 to 1.98

Mean relative density of fresh concrete = 1.94

Batch quantities for one cubic metre of concrete:-

Total weight = 1940 kg

Cement = 400 kg

Water = 0.95x400 = 380 kg

Total weight of dry aggregates = 1940 - 780 = 1.160 kg

The fine and coarse aggregates are used in equal proportions for the first trail mix and suitable adjustments are made, based on the results of trail mixes

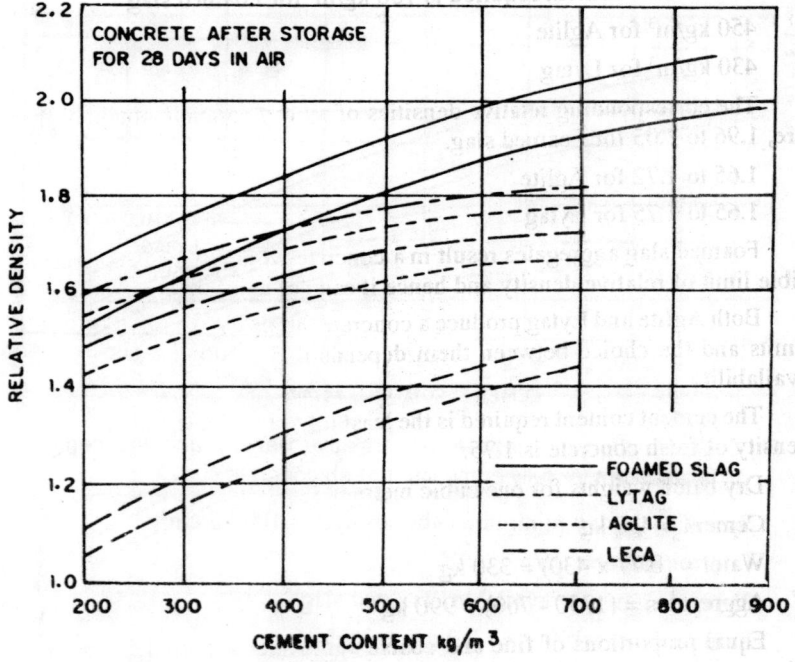

Fig.C.6. Relationship between the relative density and the cement content for concrete stored in air for 28 days

Example 2:

A lightweight concrete mix is required for structural concrete work. A minimum 28 day cube strength of 300 kg/cm^3 is required based on structural considerations. Control factor = 0.75. The relative density of the concrete, not to exceed a value of 1.75.

Workability required is Medium to High.

Available aggregates are Foamed slag, Lytag, Aglite and Leca.

Design the most economical mix and set out the dry batch weights and also the field mix quantities per cubic metre of concrete, if the fine and coarse aggregates contain 5 and 3 per cent of moisture by dry weight respectively.

Design

Mean design strength = 300/0.75 = 400 kg/cm²

Total water/cement ratio = 0.56 for Foamed slag

= 0.69 for Aglite

= 0.77 for Lytag

The required strength can not be achieved by using Leca.

The cement content required is 700 kg/m³ for Foamed slag

450 kg/m³ for Aglite

430 kg/m³ for Lytag

The corresponding relative densities of air dry concrete obtained from are, 1.96 to 2.03 for Foamed slag.

1.65 to 1.72 for Aglite

1.65 to 1.75 for Lytag

Foamed slag aggregates result in a concrete which exceeds the permissible limit of relative density and hence it can not be used.

Both Aglite and Lytag produce a concrete satisfying the relative density limits and the choice between them depends upon their relative cost and availability.

The cement content required is the least if Lytag is used and the relative density of fresh concrete is 1.75.

Dry batch weights for one cubic metre of concrete:

Cement = 430 kg

Water = (0.77 x 430) = 330 kg

Aggregates = (1750 - 760) = 990 kg

Equal proportions of fine and coarse aggregates are used for the trail mixes.

Field mix quantities after adjusting for moisture content in aggregates are as follows:

Cement = 430 kg

Water = 290 kg

Coarse aggregate = 510 kg

Fine aggregate = 520 kg

Section D

NON-CONVENTIONAL STRUCTURES

D.1 INTRODUCTION

Shell Roofs can be defined as a structural curved skin covering a given plan shape and area where the forces in the shell or membrane are compressive and in the restraining edge beams are tensile. The usual materials employed in shell roof construction are insitu reinforced concrete and timber. Concrete shell roofs are constructed over formwork which in itself is very often a shell roof making this format expensive since the principle of use and reuse of formwork can not normally be applied. The main factors of shell roofs are :

1. The entire roof is primarily a structural element.
2. Basic strength of any particular shell is inherent in its geometrical shape and form.
3. Comparatively less material is required for shell roofs than other forms of roof construction.

Fig. D.1. shows some basic nonconventional structural forms.

Barrel Vaults : These are single curvature shells which are essentially a cut cylinder which must be restrained at both ends to overcome the tendency to flatten. A barrel vault acts as a beam whose span is equal to the length of the roof. Long span barrel vaults are those whose span is longer than its width or chord length and conversely short barrel vaults are those whose span is shorter than its width or chord length. In very long span barrel vaults thermal expansion joints will be required at 30 000 centres which will create a series of abutting barrel vault roofs weather sealed together. See Fig. D.5

Stiffening Beams short barrel valuts often have large chords of over 12000 . In such cases stiffening beams should be placed at 3000 to 6000 centres to prevent buckling.

Ribs not connected to support columns will set up extra stresses within the shell roof, therefore extra reinforcement will be required at the stiffening rib or beam positions. See Fig. D.7

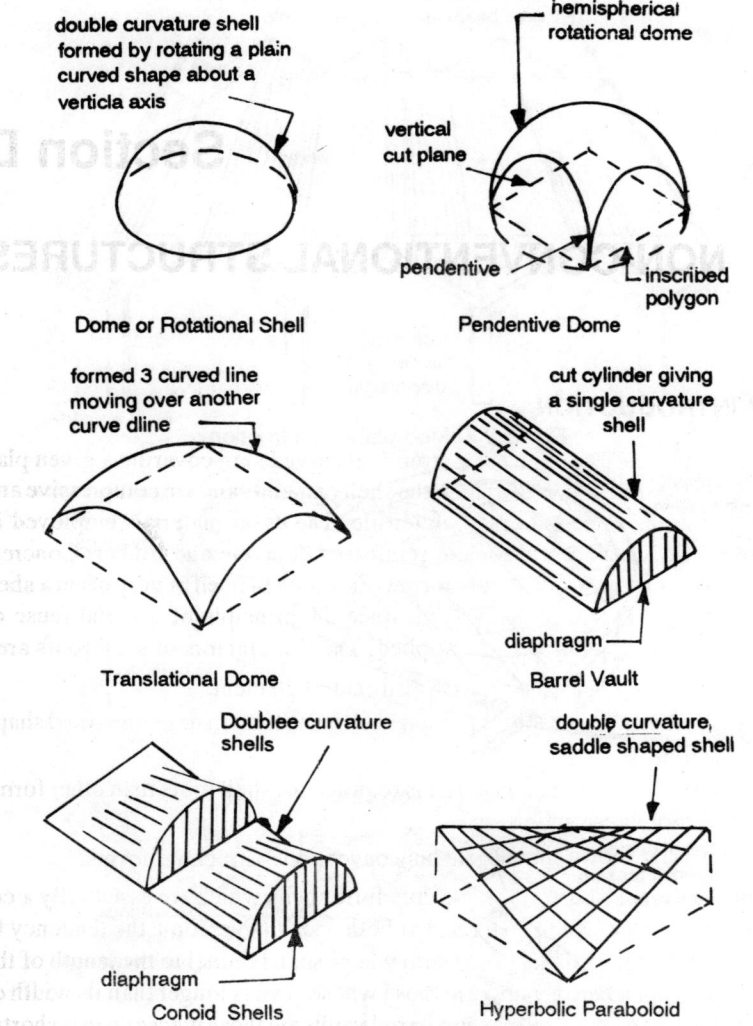

double curvature shell formed by rotating a plain curved shape about a verticla axis

hemispherical rotational dome

vertical cut plane

pendentive

inscribed polygon

Dome or Rotational Shell

Pendentive Dome

formed 3 curved line moving over another curve dline

cut cylinder giving a single curvature shell

diaphragm

Translational Dome

Barrel Vault

Doublee curvature shells

double curvature, saddle shaped shell

diaphragm

Conoid Shells

Hyperbolic Paraboloid

Fig. D.1 Various Shell forms

Other Forms of Barrel Vault by cutting intersecting and placing at different levels the basic barrel vault roof can be formed into a groin or northlight barrel vault roof See Fig. D.8

Conoids : These are double curvature shell roofs which can be considered as an alternative to barrel vaults Spans up to 12 000 with chord lengths up to 24 000 are possible. Typical chord to span ration 2 : 1.

Hyperbolic Paraboloids : The true hyperbolic paraboloid shell roof shape is generated by moving a vertical parabola (the generator) over another vertical parabola (the directrix) set at right angles to the moving parabola. This forms

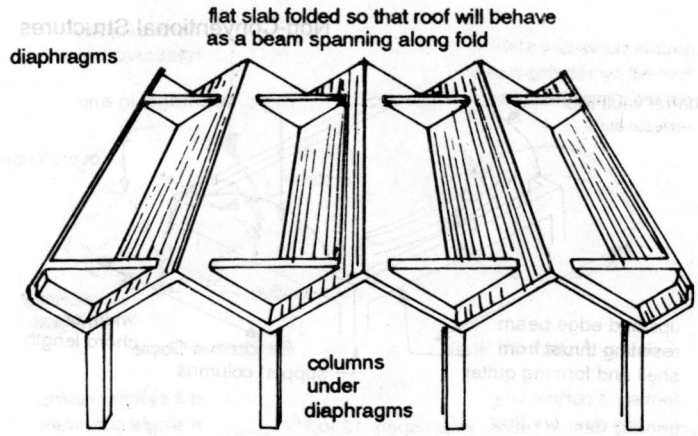

flat slab folded so that roof will behave
as a beam spanning along fold

diaphragms

columns
under
diaphragms

Fig. D.2 Folded plate construction

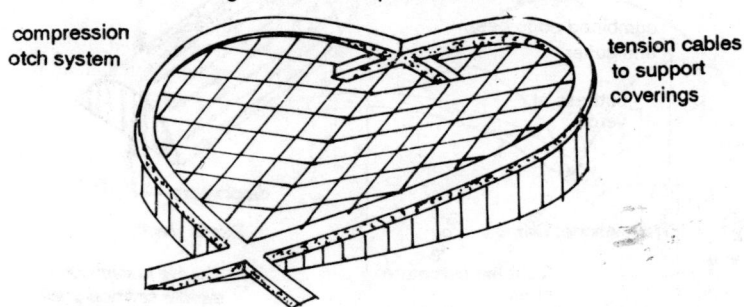

compression
otch system

tension cables
to support
coverings

Fig. D.3 Tension cable structure

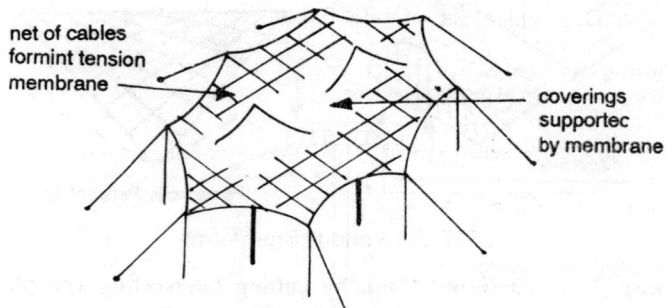

net of cables
formint tension
membrane

coverings
supportec
by membrane

Fig. D.4 Tension membrane structure

a saddle shape where horizontal sections taken through the roof are hyperbolic in format and vertical sections are parabolic. The resultant shape is not very suitable for roofing purposes therefore only part of the saddle shape is used and this is formed by joining the centre points thus to obtain a more

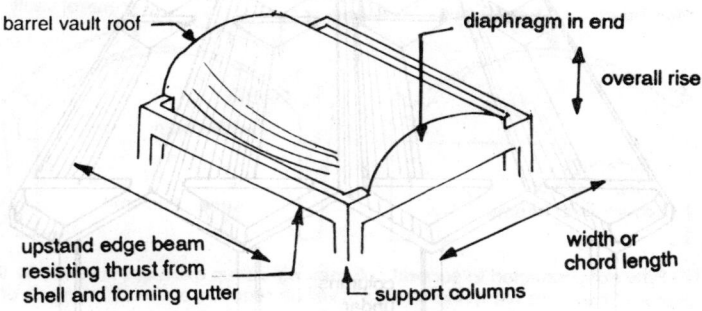

economic design ratios width span 12 to 15
rise span 110 to 115

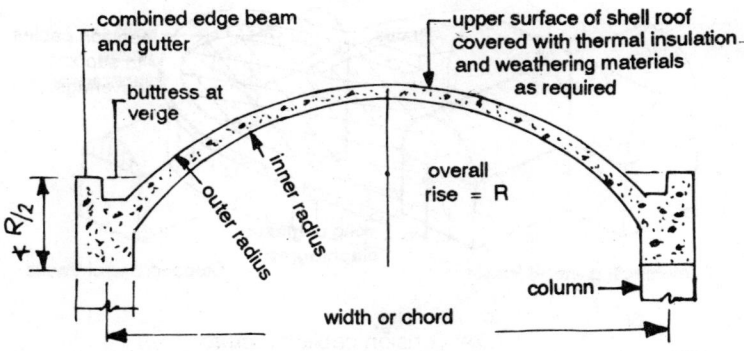

Fig. D.5 Typical Barrel Vault Expansion Joint Details

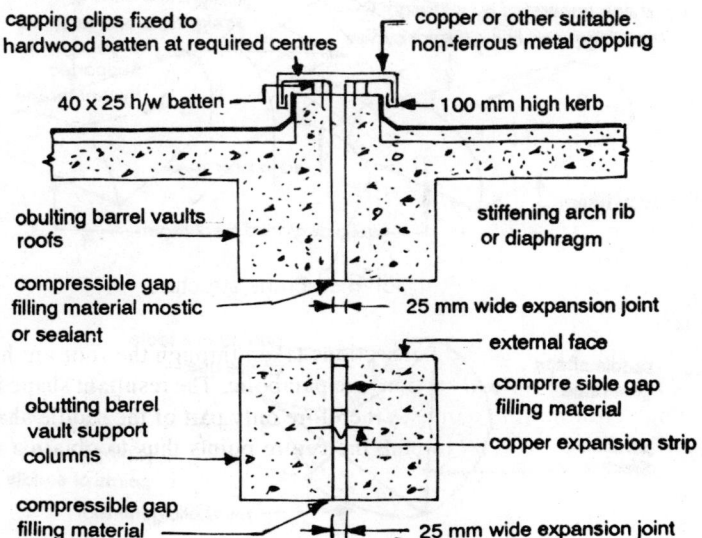

Fig. D.6. Typical Barrel valut expansion joint details.

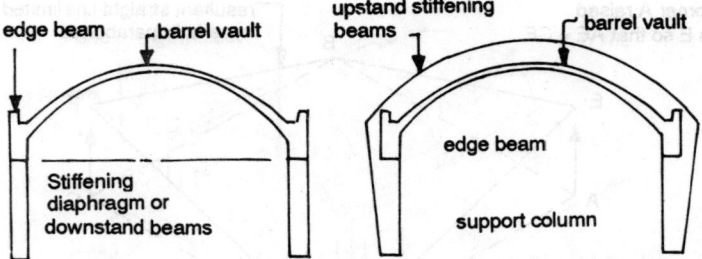

NB Ribs not connected to support columns will set up extra stresses within the shell roof therefore extra reinforcement will be required at the stillening rib or beam positions

Fig. D.7

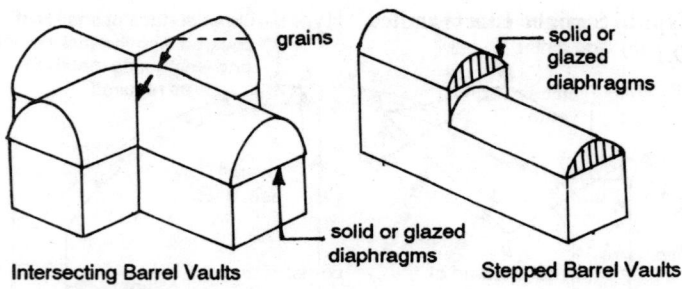

Intersecting Barrel Vaults   **Stepped Barrel Vaults**

Fig. D.8

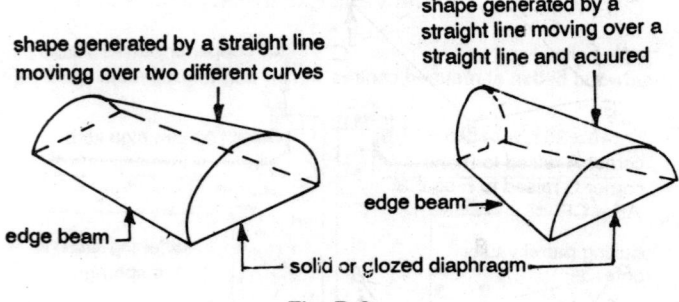

Fig. D.9

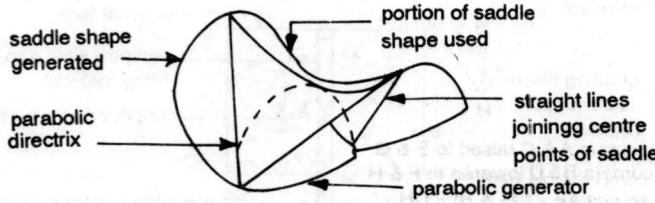

Fig. D.10

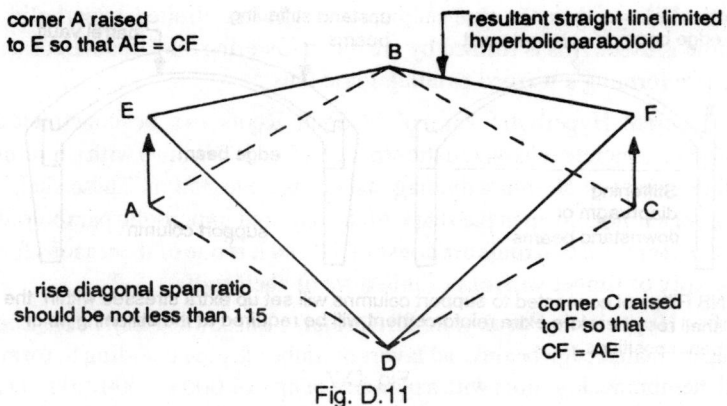

corner A raised
to E so that AE = CF
resultant straight line limited
hyperbolic paraboloid
B
E
F
A
C
rise diagonal span ratio
should be not less than 115
corner C raised
to F so that
CF = AE
D

Fig. D.11

Typical Straight Line Limited Hyperbolic Paraboloid Forms. See Fig. D.12

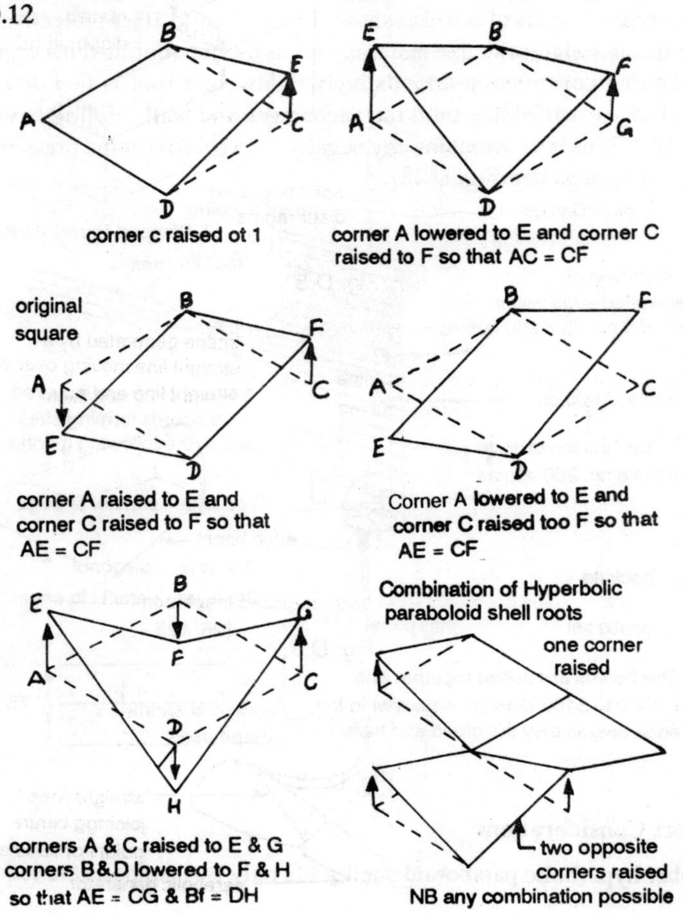

B
E
A
C
D
corner c raised ot 1
E
B
F
A
G
D
corner A lowered to E and corner C
raised to F so that AC = CF
original
square
B
F
A
C
E
D
corner A raised to E and
corner C raised to F so that
AE = CF
B
F
A
C
E
D
Corner A lowered to E and
corner C raised too F so that
AE = CF
E
B
G
F
A
C
D
H
corners A & C raised to E & G
corners B&D lowered to F & H
so that AE = CG & Bf = DH
Combination of Hyperbolic
paraboloid shell roots
one corner
raised
two opposite
corners raised
NB any combination possible

Fig. D.12

practical shape than the true saddle a straight line limited hyperbolic parabotoid is used. This is formed by raising or lowering one or more corners of a square forming a warped parallelogram thus :

Concrete Hyperbolic Paraboloid Shell Roofs can be constructed in reinforced concrete (characteristic strength 25 or 30 N/mm²) with a minimum shell thickness of 50 mm with diagonal spans up to 35.0m. These shells are cast over a timber form in the shape of the required hyperbolic paraboloid. In practice therefore two roofs are constructed and it is one of the reasons for the popularity of timber versions of this form of shell roof.

Timber Hyperbolic Paraboloid Shell Roofs : These are usually constructed using laminated edge beams and layers of timber & glue boarding to form the shell membrane. For roofs with a plan size of up to 6.000 x 6.000 only 2 layers of boards are required and these are laid parallel to the diagonals with both layers running in opposite directions. Roofs with a plan size of over 6.000 x 6.000 require 3 layers of board as shown below. The weather protective cover can be of any suitable flexible material such as builtup roofing felt, copper and lead. During construction the relatively lightweight roof is tied down to a framework of scaffolding until the anchorages and wall infilling have been completed. This is to overcome any negative and positive wind pressures due to the open sides. See Fig. D.13

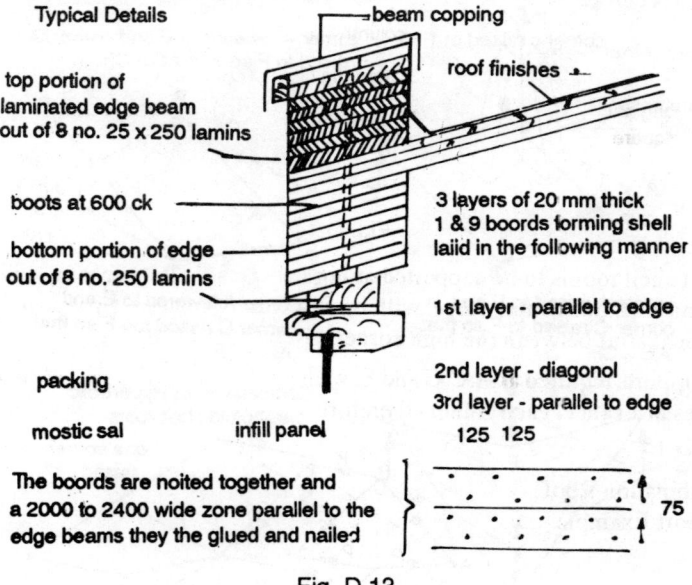

Typical Details

beam copping

top portion of laminated edge beam out of 8 no. 25 x 250 lamins

roof finishes

boots at 600 ck

bottom portion of edge out of 8 no. 250 lamins

3 layers of 20 mm thick 1 & 9 boords forming shell laiid in the following manner

1st layer - parallel to edge

packing

2nd layer - diagonol
3rd layer - parallel to edge

mostic sal   infill panel

125   125

The boords are noited together and a 2000 to 2400 wide zone parallel to the edge beams they the glued and nailed

75

Fig. D.13.

Support Considerations

In timber hyperbolic parabotoid shell roofs only two supports are required :

Edge beams are in compression forces P are transmitted to B and D resulting in a vertical force V and a horizontal force H at both positions therefore support columns are required at B and D.

Vertical force V is transmitted directly down the columns to a suitable foundation. The outward or horizontal force H can be accommodated in one of two ways:

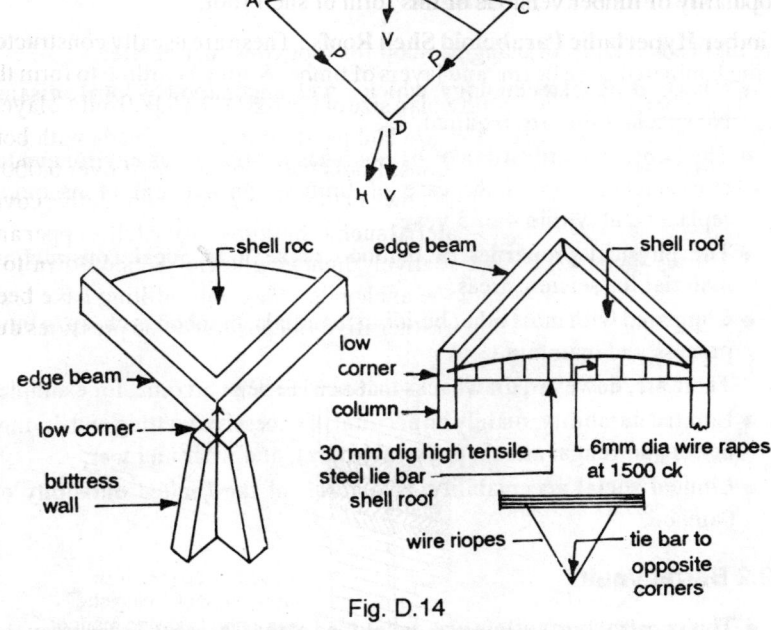

If shell roof is to be supported at high corners the edge beams will be in tension and horizontal force will be inwards. This can be resisted by a diagonal strut between the high corners. Refer Fig. D.14.

Supports required at A:C:G and E. with ties between AC: CE: EG and GA. Forces at J cancel each other therefore no support required at J. See Fig. D.15

Combination Roof Support Example

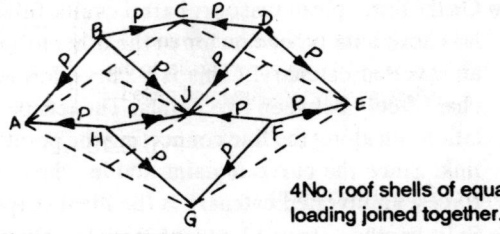

4No. roof shells of equal loading joined together.

Fig. D.15

D.2 BAMBOO ROOF STRUCTURES

Special properties : High strength, flexibility, variety of forms, Economical aspects, Low costs, Stability Good, Skills required Traditional bamboo craftsmanship,Equipment required Tools for cutting, splitting, typing bamboo Resistance to earthquake,Very Good Resistance to hurricane,Good Resistance to rain Depends on protective measures,Resistance to insects,Low Climatic suitability,Warm humid climates.

D.2.1 Bamboo Roof Structures

The main advantages of using bamboo for roof construction are:

- It is a traditional technology, which is well understood by local artisans. No special tools are required.
- The largescale utilization of bamboo has no disastrous environmental consequences (as in the case of timber), on account of its quick replacement within 4 or 5 years.
- The physical properties of bamboo make it an ideal construction material for seismic areas.
- Compared with most other building materials, bamboo is cheap to buy, process and maintain.
  There are, however, drawbacks that need to be overcome, for example:
- Limited durability, mainly on account of excessive wetting and drying, insect and fungal attack, physical impact, and wear and tear;
- Limited social acceptability, as a result of the limited durability of bamboo.

D.2.2 Barrel Vault

- This construction system was developed at the Research Laboratory for Experimental Construction, Kassel College of Technology, Federal Republic of Germany, headed by Prof. Gernot Minke.
- It demonstrates an unusual use of bamboo, in which the construction obtains its stability by compressive forces, acting perpendicularly to the bamboo's axis.
- On the principle of masonry barrel vaults, fullsection bamboo culms are laid horizontally, one on top of the other following a curve, defined by an inverted catenary. (This is a curve formed by hanging a uniform chain freely between two points. The tensile forces induced by gravitation run along the line connecting the points of contact of each chain link. Since the curve remains stable when reversing the direction of forces, an inverted catenary is the ideal shape of a barrel vault.)
- Split bamboo strips of equal length are hung such that their ends are

exactly the same distance apart as the ultimate roof span. The fullsection bamboo culms are laid horizontally forming an inverted vault. Split bamboo strips are then laid on the inside, exactly opposite the outer ones. Holes are drilled through the split and whole bamboo and fixed by bolts or rivets.

● The whole structure is then turned over and fixed on the top of the walls, which preferably should have a timber or concrete ring beam, onto which the roof is connected.

● The roof should be covered with a waterproof membrane for rain protection. This can be covered by a suitable local thatching material, or more appropriately by a 10 cm layer of soil on which grass can grow. For initial reinforcement (to prevent slipping) the soil should be held down by a strong net (as used for fishing). The dense structure of the grass roofs will give the soil cover its ultimate stability.

D.2.3 Small Geodesic Dome

● This construction system was also developed and tested by Prof. Minke and his team.

● The supporting structure of the dome is made up of approx. 1.5 m long pieces of fullsection bamboo culms, connected in a series of triangles, making it rigid. The lengths of the bamboo members are determined by a geometrical design, which requires fairly accurate cutting to achieve a uniform shape. However, the simple connection system allows for adjustments during assembly. For a tighter fit at the connecting points, at which in alternate succession six and five members meet, the bamboo ends are bevelled (slanted).

● In the example described, the span of the dome was 5 m, a size that is easy to prefabricate and transport manually with 5 people.

● Sand filled tin cans served as footings, providing simple adjustment to differing loads.
These were placed in foundations made of old steels drums, which were filled with building rubble and lean concrete.

● A strong waterproof membrane is needed to cover the dome, on which several roofing materials may be used, eg palm leaf or soft stem grass thatch, or wooden shingles on lathing. Such a structure erected at the Kassel College of Technology had a grass roof.

D.2.4 Grid Shell on a Square Base

● This structural concept is the result of research carried out at Achan Technical College, Federal Republic of Germany, with view to develop a lowcost, earthquake resistant roof structure for developing countries, using only local materials and tools. The result was a bamboo grid shell, which is prefabricated on a flat surface and later lifted in the centre to give it its ultimate shape.

- The bamboo cane used has an average diameter of 30 mm and length of approximately 4 m. For the required length 7.2 m, each grid bar required the joining of two canes. Tests showed that the strongest joints were obtained by inserting thinner bamboo pieces in the cavities at the connecting ends and fixing them by means of short dowels.

- With these lengthened bars, a grid is laid out on the ground forming grid sectors of 50 x 50 cm. Each cross point had a dowel connection which was tied with string to prevent slipping, but to allow a scissorlike movement. After lifting the centre of the grid to the required height, 1 m cane pieces are placed approximately diagonally to the rhombic grid sections, in the direction of slope, and firmly tied to the grid, giving it stability.

- The edges of the grid form a square of 6 x 6 m, corresponding to the wall dimensions. A vertical bamboo piece is embedded in each corner of the walls and a kind of bamboo ring tie beam is fixed to them. This in turn holds the grid shell roof in place. The roof is covered by a waterproof membrane and a suitable local thatching material, other than stiffstem grass. A possible alternative to thatch is a ferrocement cover, which would remain in place even if the bamboo grid shell should cease to support it. See fig. 18

D.2.5 Irregularly Shaped Grid Shells

- In order to construct spatially curved loadbearing structures using relatively thin bars, the same principle of inverting catenary lines, as described under "Barrel Vault", is applied. The shape of such grid shells is, therefore, not designed, but determined by using suspended models (eg with chain nets). Several such structures using split bamboo have been developed and erected on a joint project of the Institute of Lightweight Structures, Stuttgart, Federal Republic of Germany, and the School of Architecture, Ahmedabad, India.

- Corresponding to the chain net, the grid is assembled on the ground and tied at each cross point. For irregular base plans, each bar will have a different length, which is measured off the suspended model. Since the split bamboo gets more twisted, the steeper the slope of the grid shell, dowel joints cannot be used, while rope tie joints maintain a harmonious curvature of the structure.

D.2.6 Bamboo Trusses

- In many regions, bamboo is traditionally used for truss constructions, but often use more bamboo than necessary and are not always structurally sound.
- Joint 1: plywood on both sides of the bamboo and held by steel bolts.
- Joint 2: the diagonal member rests against pins inserted through the upper member, whereby the pins support both the purlin and the diagonal member. An intermediate layer (a kind of washer) considerably improves the strength.
- Joint 3: two "horns" at the end of the diagonal fit into two holes in the upper member. (Disadvantage: requires craftsmanship, time and excludes prefabrication).
- Joint 4: bamboo pin passing through three bamboo members, the outer two being parallel.
- The improved bamboo truss, built with joint 2 and a free span of 8 m, was tested in the laboratory by placing it on the floor and simulating vertical roof loads, by a system of hydraulic jacks acting horizontally.

GYPSUMSISAL CONOID

Special properties	Innovative material and design
Economical aspects	Low to medium costs
Stability	Good
Skills required	Special training
Equipment required	Simple wooden framework
Resistance to earthquake	Good
Resistance to hurricane	Good, if protected from rain
Resistance to rain	Low
Resistance to insects	Good
Climatic suitability	Dry climates
Stage of experience	Experimental

While vaults and domes are selfsupporting structural forms when completed, they normally need support and centering while under construction. This usually involves first building an identical vault in wood over which the masonry vault rests, until complete and dried.

In countries where timber is scarce, this type of vaulting is hardly advantageous. A system of building vaults and domes, without this framework, or shuttering, evolved in countries like Egypt and Iran.

The drawings overleaf show the sequence of construction of a small house, which the founder members of Development Workshop and some friends built in New Gourna, Upper Egypt, in 1973. They worked as apprentices alongside two Nubian master masons, skilled in the techniques being used.

The house was built with mud bricks and served, amongst other objectives, as a practical opportunity to master and evaluate the Nubian techniques of building without the use of shuttering, and to obtain a clear guide regarding the relationship of roof span to wall thickness and height for mud brick walls.

The house stands amongst the buildings designed by Hassan Fathy, who revived this building technique in the 1940s.

New Developments

Arches constructed with old car tyres

Simple arches can be constructed over openings by using old car tyres as formwork. This was tried out on a project in India (1986) and found extremely easy to carry out. The sides of the opening, which has the width of the tyre, are erected up to the level at which the arch begins. The tyre is placed on a dry stack of bricks, such that the axis is in line with the top brick layer. The bricks should be laid alternately on each side of the tyre, since excessive load on one side can deform the tyre and distort the shape of the arch. care must be taken that the lower edges of the bricks touch each other without leaving any gaps. Since the tyre is flexible, it can be removed with ease.

Section E

ENERGY SAVING BY USE OF DIFFERENT MATERIALS

E.1 Introduction

Steel, cement and bricks are the most popular building materials. These materials generally produce high quality construction. However, they are high energy intensive, expensive and based on heavy industry. For cement industry large infrastructural facilities are also required. Thus part or full replacement of cement & steel in production of building materials, or methods which could replace conventional bricks would result in substantial saving of energy and thus would result in energy conservation. Generally use of waste material for production of pozzolanic material such as rice husk ash and labour intensive manufacturing methods are likely to result in saving energy and cost of production.

Building industry consumes, sizeable fraction of the material energy budget. Energy is consumed in production as well as in the use of materials in various types of construction. It is estimated that about 18 percent of national energy production is directly utilised by building industry.

Research all over the world has indicated that substantial saving energy / cost can be achived by using materials other than cement in the manufacture of building materials. These materials are

- (a) Fly ash
- (b) Rice husk ash
- (c) Surkhi
- (d) Burnt clay pozzolana
- (e) Blast furnace slag
- (f) Lime
- (g) Other industrial wastes, such as coal washery rejects, sludge, mining wastes etc.

Table E.1 shows energy inputs of materials.

Table E.1

	Material Fuel K.cal/kg	Solid Energy K.cal/kg	Electrical Energy K.cal/kg	Total
1.	Fly ash	0	0	0
2.	Surkhi (Broken Bricks)	0	50	50
3.	Rice Husk Ash	0	50	50
4.	Burnt clay	228	56	284
5.	Pozzolana			
6.	Hydrated Lime	547	50	597
7.	Lime	722	3 7	25
8.	Portland Cement	1440	120	1560

From the above table it is clear that replacement of cement by low energy input material can conserve considerable energy.

E.2 Energy Computations

The energy consumed in manufacturing various building materials in India as computed by Central Building Research Institute, Roorkee, India are given in table E.2 (adopted from reference 3).

Table E.2
Energy Saving in Brick Making

Sl. No.	Process Description	Coal Saving (%)
1.	High Draught Kiln in Brick Firing	15-20
2.	Use of Flyash in Making Bricks	10-30
3.	Use of Coal washery Rejects in Making Bricks	25-75
4.	Use of Rice Husk in Firing Bricks	20-25

Table E.3

Energy Saving Processes in Building Materials

Sl. No.	Process	Use of Industrial Wastes	Approximate (%) saving in Energy
1.	Portland-Poz-zolana	Flyash, Mining Wastes, and Blast Furnace Slag Portland Blast Furnace Slag Cement	15-20
2.	Lime-Pozzo-lana Mixtures	Lime kiln Rejects In-dustrial Ashes	30-40
3.	Lime-Rice-Hu sk-Ash Pozzo-lana	Rice-Husk Ash or rice Husk	30-40
4.	Lime-Sludge, Rice Husk Po-zzolana	- do -	40-50
5.	Sand-Flyash-Bricks	Flyash, Mining Wastes	30-50
6.	Masonary	Mining Mineral Wastes, Cement Lime Sludges	10-20
7.	Cement-Conc-rete Tiles	Mining & Mineral Wastes	10-15

E.3 Other Energy Saving Methods

In Addition to what discussed above a building material which is low energy intensive is discussed below.

E.3.1 Latertic Soil - Lime Blocks (Latoblocks)

The name laterite includes all reddish, residual and non-residual tropically weathered soil, which are derived from decomposed rocks through clays to sesqui-oxide rich crusts. The main constituents of Laterites are oxides of Aluminiuim, Iron and Silicon. Laterites occur in various forms ranging soils, pea gravel to soft rock masses. The present method of using laterite as a construction material suffer from the disadvantage that it has poor weathering qualities and requires heavy protection. They also possess low comperessive strength not more than 2.5 MPa(Mega Pascal. 1 Pa = 1 N/m sq.).

Research at SERC, Madras has shown that a high quality building block can be produced from Laterite soil and lime, by a low energy intensive process. According to the process a mixture primarily containing Lateritic soil and lime is mixed and moulded under pressure. The blocks are then subjected to low temperature of 100°C to produce strong and good quality bricks. These blocks are called Latoblocks.

An important advantage in favour of Latoblocks is that their manufacture is not energy intensive. This is in contrast to conventional clay bricks, which are produced by burning at 1000°C to 1200°C for several days. The energy requirements for the production of Lato bricks is far less than requirements for autoclaved products, which are produced by high pressure steam curing (autoclaving). Thus Lato bricks can be said to be an ideal energy saving building material. Some typical properties are

(a) Wet Compressive strenth ranges from 5 to 15 MPa
(b) Water absorption is about 12%
(c) The blocks can be finished well with smooth faces and sharp corners.
(d) No warping of the blocks is possible as in the case of burnt clay bricks.
(e) The blocks have good dimensional stability.
(f) Inspite of smooth surface the blocks have very good plasterability.
(g) The blocks pass the drop lest i.e when dropped from a height of 2 meters on hard surfaces.

For more details the reader should refer to reference 4.

In addition to the energy conservation aspects discussed above, there are other ways and means by which energy can be conserved. These are

1. Use of natural fibres in the form of twines as reinforcement in place of steel in reinforced concrete.
2. Use of natural fibres in production of roofing material.
3. Alternative fuels for firing bricks.

In a country where production of steel is far behind the demand, any saving of steel how ever small is welcome and desirable. In this context the role of natural fibres, has importance.

E.3.2 Fibre-Cement Roofing Sheets

Of all the difficulties facing self-help builders, the cost of buying the roof cladding is probably the worst. The principal material in present-day use, the galvanised corrugated iron sheet, is also,for most countries,an imported product. Even with this cost penalty there is no guarantee of durability.

A relatively new intermediate technology has been developed for producing sheets,thinner but much larger than tiles. This technology emerged from the various efforts by researchers to use organic fibres such as sisal, coconut coir, and data palm fibre to reinforce concrete-FRC. Corrugated roof sheets of natural fibre reinforced cement first began to be manufactured experimentally in 'village-scale' production plants in 1977-78.

The production process consists of a simple tilting table on which a wet screed of proportioned cement, sand and natural fibre is prepared inside a steel frame, resting on a polythene sheet. When level and tamped to remove air bubles, the screed is transferred from the laying-up board by a tilting action which enables it to slide on to the corrugated mould. It is left to set hard for 24 hours before removal to begin curing.

The process is deceptively simple. Normal workshop techniques of process and quality control have to be applied. Otherwise may lead into serious trouble with porous, cracking and badly fitting sheets. Another mistake is to ignore the importance of erecting firm, regular roof frames to carry the new products which, being concrete, are brittle with little capacity to flex. The potential longevity of the roof sheets has been established. The earliest test samples, after 5 years exposure to alternate periods of frost and snowed, heavy rain, storm-force winds, and hot summer spells, show no sign of deterioration. Minor discoloration in industrially polluted atmosphere was noted as well as rust staining around fixings, but otherwise no defects had emerged in the early years.

Other FRC-based products suitable for self builders, include arch shells as permanent shuttering for floors to save the inordinate consumption of timber used for this purpose, as described earlier. Also in the role of formwork, FRC shells have been used successfully in the installation of otherwise conventionally constructed brick arches. In these applications the FRC occupies a structural role only in the few hours following the installation of the masonry superstructure. After that the masonry itself, acting in compression, takes over the role leaving the FRC formwork components just to provide a tidy finish.

Table E.4

Materials consumed per square meter of flat screed (thickness)

Material	6 mm	8 mm	10 mm	12 mm
Cement (Kg)	7.5	10	12.5	15
Sand (Kg)	7.5	10	12.5	15
Fibre (g)	225	300	375	450

For the purpose of calculating the cost of FRC sheet composites, The thickness chosen for the most of the corrugated roof sheets currently being produced is about 8 mm. Arches used as formwork are usually made to a thickness of 10-12 mm. With a fully manual process a three-man team can prepare a square metre of material in about 15 minutes. A vibrating table can cut the time down by as much as half. The indicates that no steel is used hence considerable cost lowering and energy conservation is the benifit.

E.3.3. Use of rice husk as fuel for firing building bricks

An estimated 9.8 million tonnes of coal is required annually for burning 5000 billion building bricks. The problem of supply of such a large amount of coal is difficult and its use is undersiable in terms of energy conservation.

In our country, it is estimated that about 26.5 million tonnes of rice husk is obtained as agricultural waste every year. This waste material can be utilised as renewable source of fuel and thus help in conserving the precious fossil fuel coal.

Rise husk is a hard fibrous material, with a calorific value of about 2800 K cak/Kg again 7000 K.cal/Kg for the best quality coal. It has about 20% silica as mineral content.

Experiments conducted in various research institution has indicated potential for using rice husk as an alternative fuel for burning building bricks. Which can be done in three ways.

Fuel Consumption

- For firing in bull's trench klin normally 18 tonnes of coal is required per 1 lakh of bricks. With 25 tonnes of rice husk restituted for coal only 7 tonnes of coal is required. Alternatively 7 tonnes of coal, 12 tonnes of fine wood and 50 tonnes of rice husk can be used. The last option completely eliminates the need for coal. The quality of bricks is not altered by using rice husk.

These figures indicate that rice husk be used in conjunction with coal can result in substantial saving of coal. Upto 30 - 40% saving can be achieved in a commercial Bull's Trench Kiln. Also as an alternative use of rice husk and fire wood can completely eliminate the use of coal.

It can be said that the technique descussed above can be used in rice growing areas and substantial saving of coal can be effected.

Section F

CONSTRAINTS

F.1 ENVIRONMENTAL CONSTRAINTS

F.1.1 Orientation Of Building

Orientation is defined as a method of setting or fixing the direction of plan of the building in such a way that it derives maximum benefit from the elements of nature such as sun, wind and rain. Therefore once the site is chosen or accepted as available for the construction of building, the Architect's first aim should be proper orientation prior to planning and design of building. Proper orientation means to utilize the natural gifts in achieving functional comfort inside the building through the planned aspects of the building units. The knowledge of orientation is the first pre-requisite for good planning, no matter whether it depends upon the circumstances or it is to be decided by choice.

Alternatively, good orientation means proper placement of plan units of the building in relation to sun, wind, rain, topography and outlook and at the same time providing convenient access both to the street and backyard.

Orientation in case of non-square buildings is indicated by the direction of the normal to the long axis. For example, if the length of the building is East - West, its orientation is North -South. It should be remembered that poor orientation of the building results in discomfort conditions inside the building. Though the comfort conditions in such buildings can be created through the mechanical devices or other means but the operational cost of such devices is very high.

F.1.2 Factors Affecting Orientation

The various factors affecting orientation are as follows:

1. Solar heat gain, for which it is essential to know sun's path throughout the year and its relative position with respect to the locality.
2. Prevalent breeze or wind,i.e. the direction of prevailing wind in sum-

mer when it is required and in winter when it is to be avoided.
3. Rainfall i.e. the direction and intensity of rain.
4. Site conditions i.e. location of site either rural, urban or sub-urban, neighbourhood and surroundings.

In view of the above data, particularly solar and climatic data, the orientation is made based on the need of the summer or winter comfort. For places, where summer causes greater thermal discomfort, the building as a whole should be oriented to intercept minimum solar radiation in summer and vice-versa.

F.1.3 Orientation Criteria Under Indian Conditions

Solar heat and humidity are the two controlling factors in the design of a building, particularly of residential type. Indian climate (i.e, combined effect of natural agencies such as sun,wind and rain) for design purpose is generally classified either hot-arid or hot-humid. Accordinly, India can be divided into zones from climate point of view, viz

1. Hot Arid zones (or Dry Arid zones)
2. Hot-humid zones (or Wet zones)
1. **Hot-Arid zones:** Such zones having hot dry climate are found mostly in the interior of the country, away from the coastal belt. Hot dry climate is characterized by the high summer day-time temperature, low relative humidity and wide range of temperatures between day and night, and between summer and winter such as for example prevails generally in the Northern India and Central India.
Therefore, comfort requirements call for the removal of the hot air through walls, roofs, windows etc for orientation in such Hot Arid Zones.
2. **Hot-Humid zones:** Such zones having wet climate are found generally along the coastal belts of India. Hot-Humid climate is characterized by low summer day-time temperature. High relative humidity due to violent monsoon conditions, prevail generally in West coast (Bombay), East coast (Bengal) etc.

Therefore, comfort requirements call for the free movement of air through the doors, windows and other openings and at the same time protection from the violent monsoon during four months for orientation in such Hot-Humid zones.

Based on available data in India, on prevailing monsoon winds the following orientations have been suggested by C.B.R.I, Roorkee keeping in view the indoor comfort conditions for a dwelling.

(A) For Hot - Arid zones

1. **Northern India (like Punjab):** Orientation should be done along the direction East and West facing North.

2. **Central India:** Orientation should be done along the directions E-SE and W-NW facing N-NE.

3. **Delhi proper:** The best position of a building from orientation point of view is considered when the longer side makes an angle of 22.5 degrees on the East - West line towards East-South.

(B) For Hot-Humid Zones (or wet zones):

1. **West coast Regions (like Bombay):** Orientation should be along the direction S-E and N-W, facing S-W.

2. **East coast Regions (like Madras):** Orientation should be along the direction S-E and N-W, facing N-W.

3. **Bengal:** The best orientation is considered to be along East and West facing South.

At hill stations like Kashmir, Shimla,etc. the winter season causes more discomfort, and therefore deserves greater consideration in orientation of buildings. The sole criterion for optimum orientation, therefore, is to obtain maximum solar energy on the building in winter. The orientation of the buildings in hill stations should be such that the living rooms are open on the South and West sides of the sun.

F.1.4 Suggestions For Optimum Orientation

The investigations carried out at C.B.R.I, Roorkee and Railway Testing and Research Centre, Lucknow, have arrived at following conclusions, regarding orientation of buildings:

1. The best orientation from solar point of view requires that the building as a whole should receive the maximum solar radiation in winter and minimum in summer. For practical evaluation, it is necessary to know the duration of the sunshine and hourly solar intensity on the various external surfaces on representative days of the seasons. Fig. F.1 gives the computed solar energy per hour on unit surface area, normal to the rays under standard atmospheric condition for different altitudes of the sun. The total heat intake is calculated for all possible orientations of the buildings for the extreme days of summer and winter and final orientation is decided.

   In order to calculate the solar energy on any surface other than normal to the rays, one should know the altitude of sun at that time and use Fig. F.1 to get the corresponding value of direct solar radiation (I_n). The solar radiation incident on any surface (I_s) can then be computed with the help of the following relationship (see Fig. F.2).

$I_s = I_n (\sin \beta \sin \phi + \cos \beta \cos \alpha \cos \phi)$ where α is the wall solar azimuth angle, β is the solar altitude and ϕ is the angle of cut of the surface from the vertical. ($\phi = 0$ for vertical surfaces and $\phi = 90$ for horizontal surfaces).

2. In hot climates, living rooms on the South and West sides should be protected by verandahs, baths, stores. Verhandas may be eliminated on the south by the use of chhajjas or sun breakers and then varandahs may be provided on the East and West facings. Verandahs should not be provided on the north facing as for as possible. In long buildings, such as hospitals, schools, one of the long sides should face North and South, and West protected by verandahs. Drawing offices and darkrooms should be located on the north side.

3. The exposure of the house to the sun is also reduced by shady trees or bushes on the sunny side and by keeping the shorter walls East and West, so that the minimum wall area is exposed to the sun rays.

4. In hot and humid areas as already discussed, the orientation is governed by the direction of the breeze.

5. Generally, all the rooms which are occupied in the day time should be placed on North and East side while the other rooms are placed in the direction of prevailing wind and at the same time protected by verandahs from the heat. Eastern or North-Eastern corner with cross ventilation is regarded best for kitchen. Latrines (or water closets) should be so located that the wind passing through them should blow in a direction away from the house.

6. The judicious location of rooms, specific shape of the building with its external surfaces and proper ventilation are also imperative with the choice of proper orientation. From the ventilation point of view, the height of a house should not be more than twice the width of the street. This is called 63½ rule.

Accoording to this rule, the height is fixed by two imaginary lines, namely, the horizontal line and the diagonal line. The horizontal line is drawn at right angles to the road, through the centre of the front line. The location of this horizontal line is taken at the higher point along the line.

The diagonal line is drawn in the direction of the building at 63½ degrees from where the horizontal line meets the rear boundary. No part of the building is allowed to project beyond the diagonal line as a rule except that for minor parts such as chimneys, turrets, etc.

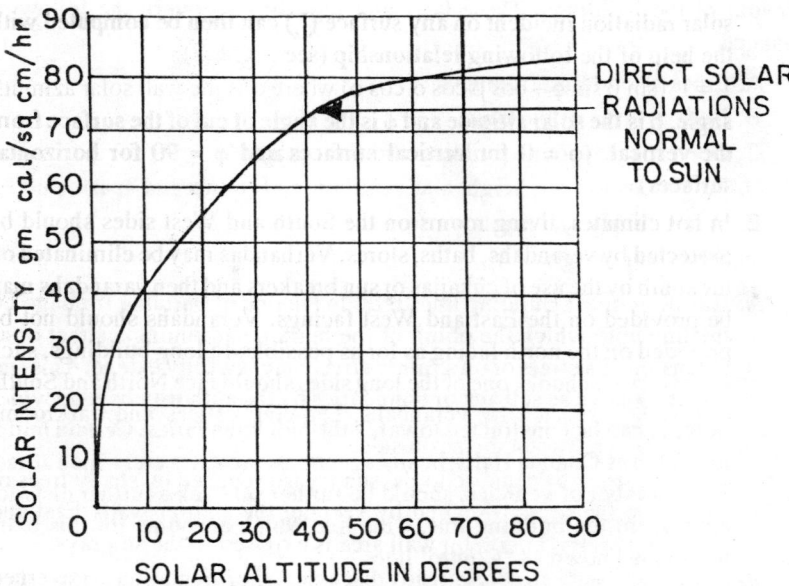

Fig. F.1. Solar altitude in degrees

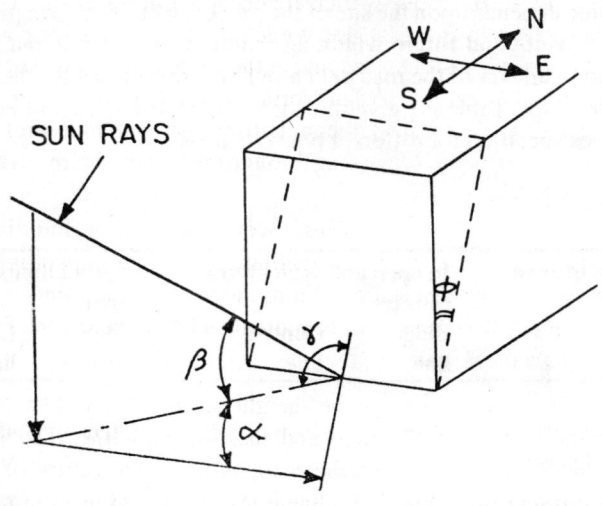

Fig. F.2

4.2 Legal Constraints

There are certain rules and regulations laid down by the municipalities or town planning authorities or urban improvement boards in their jurisdiction. These have to be considered by an architect while planning and designing the layout of the buildings. These laws and regulations govern the following aspects.

A. Lines of building frontages.
B. Built-up area of building.
C. Open spaces around building and their heights.
D. Provisions to size, height and ventilation of rooms and apartments.
E. Water supply and sanitary provisions.
F. Structural design.

A. **Building lines:** Building lines refer to the line of building frontage i.e. the line upto which the plinth of the building adjoining a street or an extension of a street or on a future street may lawfully extend. This line is often known as setback or boundaries by the authority beyond which nothing can be constructed towards the plot boundaries. Certain buildings such as Cinema Halls, business centres, factories etc. which attract large number of vehicles, should be further set back a further distance apart from the building line. This line which accounts for this extra margin is known as "Control line."

Sometimes a line is fixed known as the "General Building Line" and no building or its portion should project beyond this front line. The fixation of building line depends upon the site of the proposed building, keeping in view the present width and future widening requirements. A minimum distance either from boundary of the road or centre line is prescribed for the line of the building frontage. Table F.1 gives the idea of such distances for building and control lines specified for different types of areas.

Table F.1

Types of road	In open and agricultural country		Actual limits in urban areas.	
	Bldg. line	Control line	Bldg. line	Control line
1. National and state highway	30 m	56 m	30 m	45 m
2. Major district roads	24 m	45 m	15 m	24 m
3. Other district roads	15 m	24 m	9 m	25 m
4. Village roads	12 m	18 m	9 m	15 m

National building code specifies a minimum frontage of 6 m on any street. Advantages of fixing such building lines are:

1. It facilitates future widening of streets.
2. It keeps away the noise and dust of the street.
3. It prevents the formation of blind corners at the intersection of the streets and provides open spaces.

B. **Built up area of buildings:** The built up or covered area equals to the plot area minus the area due for open spaces. Generally, Floor area ratio (F.A.R.) for different occupancies and types of construction are laid down by the authority based on various factors such as occupancy class, type of construction, width of street fronting the building and the traffic loads; locality and density, parking facilities and local fire-fighting facilities.

$$\text{F.A.R.} = \frac{\text{Covered area of all floors}}{\text{Plot area}} \times 100$$

The floor area ratio values are specified in National Building Code for different occupancies and types of constructions. The covered area is governed by F.A.R. or F.S.I. (Floor Space Index). The following limitations for built up areas have been recommended:

1. In a business area, the covered area shall not exceed 75 per cent of the area of the site, provided sufficient space for parking etc. is available at the site.
2. In an industrial area, the built-up area or covered area shall not exceed 60 per cent of the site area.
3. In residential area, maximum permissible built up area is 60 per cent of plot area for plot of area less than 200 sq m; 50 per cent of the plot area for plot of area 200 to 500 sq m; 40 per cent for plot having area 500 to 1000 sq. m. and 33.33 per cent for plot of area greater than 1000 sq. m.

C. **Open space around buildings and their heights:** The open spaces should be left inside and around a building, particularly of residential type, to meet the lighting and ventilation requirements of the rooms abutting such open spaces. In case of building abutting streets in the front, rear or sides, the open space provided shall serve the purpose of future widening of such streets. All such open spaces whether interior or exterior shall be kept free from any erection there on and shall be open to sky and no cornice, roof or weather shade more than 0.75 m wide shall overhang for project over such open spaces. Open spaces are classified as Exterior open spaces and Interior open spaces.

Exterior Open Spaces

1. **Front open space:** Every building fronting a street shall have a front

yard forming an integral part of the site, of a minimum width of 3 m and in case of two or more sides fronting a street, an average width of 3 m but in no case less than 1.8 m.

2. **Rear open space:** Every building shall have a rear yard, forming an integral part of the site, of an average width of 3 m and at no places measuring less than 1.8 m except in case of back to back site the width of the rear yard shall be 3 m through out. Subject to the condition of free ventilation, the open space left up to half the width of the plot shall also be taken into account from calculating the average width of the rear yard.

3. **Side open space:** Every semi-detached and detached building shall have a permanent open air space, forming integral part of the site, of not less than 3 m in width on the sides.

For height or buildings above 10 m and upto 25 m, in addition to the minimum open spaces required for height of 10 m, there shall be an increase in the minimum open space at a rate of 1 m per every 3 m or fraction there of, for heights above 10 m.

For height of building above 25 m and upto 30 m there shall be a minimum open space of 10 m.

For heights of buildings above 30 m, in addition to the minimum open space required for the height upto 30 m, there shall be an increase in minimum open space at the rate of 1 m for every 5 m or fraction there of for heights above 30 m subject to a maximum of 16 m.

Interior Open Spaces

The whole of the one side of one or more rooms intended for human habitation and not abutting on either the front, rear or side open spaces shall abut on an interior open space whose minimum width shall be 3 m.

D. **Provisions to size, height and ventilation of rooms and apartments:**

(I) **Size of rooms:** From the view of health and ventilation, certain absolute minimum areas for individual rooms and apartments have been laid down by National Building Code as follows:

(a) **Habitable rooms:** The area of habitable room shall not be less than 9.5 sq. m., where there is only one room with a minimum width of 2.4 m. If there are two rooms, one of these shall not be less than 9.5 sq m and other shall be not less than 7.5 sq m with a minimum width of 2.4 m.

(b) **Kitchen:** The area of kitchen shall be not less than 5.5 sq m with minimum width of 1.8 m. If there is a separate store, the floor area of kitchen may be reduced to 4.5 sq m. A kitchen which is intended for use as a dining room also, shall have a floor area of not less than 9.5 sq m with a minimum width of 2.4 m.

   (c) **Bath rooms and water closets:** The size of a bathroom shall not be less than 1.5 x 1.2 m or 1.8 sq m; if it is combined bath and water closet, its floor area shall not be less than 2.8 sq m with a minimum width of 1.2 m. The minimum floor area of water closet shall be 1.1 sq m.

(II) **Height of buildings and rooms:** The height of the building is decided by two factors, either by the width of the street on which it fronts or the minimum width of rear spaces. In National Building Code, the height and number of storeys are related to Floor Area Ratio and provisions of open spaces. The maximum height of building is generally decided on the basis of width of the street. But general criteria is:

$$\text{Height} = 1.5 \times \text{width of street} + \text{front open space.}$$

   (a) **Habitable rooms:** The height of all rooms for human habitation shall not be less than 2.75 m measured from the surface of the floor to the lowest point of the ceiling. In case of air conditioned room, the height shall not be less than 2.4 m upto false ceiling.

   (b) **Kitchen:** The height of a kitchen measured from the surface of the floor to the lowest point in ceiling shall not be less than 2.75 m except for the portion of accommodate floor trap of the upper floor.

   (c) **Bath room and water-closet:** The height of a bathroom or water closet measured from surface of the floor to the lowest point in the ceiling shall not be less than 2.2 m.

(III) **Lighting and Ventilation of Rooms:** Room should have, for the admission of light and air, one or more apertures, such as windows and fan lights, opening directly to external air or into an open verandah or gallery. The area of such window openings exclusive of doors and inclusive of frames is specified as follows:

   (a) 1/10th of the floor area for dry not hot climate.

   (b) 1/6th of the floor area for wet hot climate.

   (c) The aggregate area of doors and window shall not be less than 1/7th of the room.

In addition to the above means of ventilation, every such room shall have ventilation of atleast 0.3 sq. m. in area near the top of each of two of the walls of such room and these ventilators preferably placed opposite to each other, for through ventilation. Generally, the aggregate area of such ventilators is provided at the rate of 0.1 sq. m. for every 10 cubic meters of space of such rooms.

E. **Water Supply and Sanitary Provisions:**

(a) **Residential building:** Dwellings with individual conveniences should have at least one bathroom with a tap, one water closet and one sink or nahani. But when only one water closet is provided in a dwelling, the bathroom and water closet should be separately accommodated.

In case of dwelling without individual conveniences, there should be one water tap with draining arrangement in each tenement, one water closet and one bath room for every two tenements and water taps in common bathrooms and water - closets.

(b) **Buildings other than residential:** For public buildings meant for offices, schools, colleges etc, the total number of persons for which sanitary services should be provided is determined on the basis of one person for each 5 sq. m. of floor area of each room. For every 25 persons or part thereof, there should be a water closet. For every 100 persons, there should be a urinal. The sanitary unit for either sex should be separate.

For buildings meant for cinemas, theatres and public assembly halls, one water closet for every 200 males or females should be provided. Urinals may be provided at the rate of one for 100 persons.

For factory and workshop buildings, there should be an absolute minimum one water closet for either sex and one water closet for every 40 persons and one urinal for 100 persons.

All these sanitary fittings should be located in an accessible position and provided with sign indicating their purposes and sex for which they are meant for. Adequate lighting and ventilation of these sanitary units is highly essential.

F. **Structural Design:** Regulations and Bye-laws also dictate the design stresses, safe loads and bearing capacities which should be considered in the structural design of the building. Design requirements of following component of building should be taken into account.

1. **Depth of foundation:** Depth of foundation is determined by the engineering formula, but leaving aside this, the minimum depth of foundation should be taken as below:

   (a) For single storeyed building = 0.75 to 1 meter below finished ground level.

   (b) For double storeyed building = 1.0 to 1.3 m below finished ground level.

2. **Width of foundation:** The thickness of wall in spread foundation (or stepped footing) is extended by off-sets on each side equal to half brick width that is 5 cms. The thickness of concrete in foundation should nearly be equal to 5/6th of the thickness of wall in the superstructure.

3. **Plinth:** This is the portion of building between the surface of the surrounding ground and ground floor level. The Plinth level of the building is kept higher (generally 30 cms or more) than the surrounding ground level such that adequate drainage of the site is assured.

4. **Damp-proof course (D.P.C.):** D.P.C. is provided to prevent moisture rising up the wall. The horizontal D.P.C. usually 4.0 cm thick is provided about 10 to 20 cm above final ground level.

F.3 Material Constraints

Construction materials control is fully one half of what construction inspection is all about. Many inspectors understand quality construction but fail to understand fully what their authority and responsibility requires them to do. The implication often too firmly implanted in some inspector's minds is that they are on the job to assure that the project will be constructed with only the "best" quality materials.

In short, the inspector is not on the job to enforce what he believes to be proper construction, but rather to obtain the type and quality of construction that has been called for in the plan and specifications.

Most specifications, in their general conditions, provide that unless otherwise specified, all materials incorporated in the work are to be of the best available grade in the local trade area. Materials called for on the drawing that are not called for in the specifications, but that are known to be required for a complete project are similarly required to be of comparable quality. These phrases are usually found in the specifications prepared by some one who does not really know how to specify proper quality in a product. Generally, an inspector can plan on some arguments over either of these terms, because that which one person considers high quality another may consider substandard. Often the specifications provide that all materials furnished must be free of defects or imperfections, and must normally be all new materials.

F.3.1 Materials and Standard Specifications

One of the most common occurrence during construction is the constant search by the contractor to obtain material that cost less than those actually specified, and offer them as substitutes. The ever present desire to use cheaper materials that frequently have not had the test of time to show that they will perform as well as a specified product frequently leads to claim against the engineer for negligence if it is determined later that the material did not perform as required. Great care must be used in approval of new, substitute materials as well as in the application of some established ones. The architect and the engineer has the duty to see that the material furnished in compliance with his drawing and specifications are actually suitable for the particular uses intended. Reliance on producer's sales literature is hazardous at best. Architect and engineers are liable for failure to have a new material tested. Thus, it is easy to understand his occasional reluctance to try new products. It may even be desirable to require that the manufacturer of such new materials furnish guarantees that extend beyond the usual time. The refusal of a producer to provide such guarantees may be sufficient reason for rejecting the material.

By definitions, a set of standard specifications is a preprinted set of specifications, usually comprising both a set of general conditions and complete Technical specifications for all types of construction materials that the originating agency expects normally to cover in their kind of work. When adopted by a public agency or by a design from working on a project for public agency, the total content of the standard specifications becomes a part of the contract documents.

F.3.2 Requests for Substitutions of Materials

By far one of the most frequent requests received on the job will be request by the contractor to use substitute materials for those actually specified by the architect or engineer. The conditions controlling the use of such alternate choices of materials differ some what between public and private construction contracts, and must be considered separately.

Whenever a substitute is offered, the contractor is obligated to vie adequate notice of an intention to offer a substitute not wait until it is already too late to get delivery in time for the material actually called for. Then, after submitted, sufficient time must be allowed for the architect or engineer to review the technical data submitted by the contractor in support of his claim that the material is the equal of the specified one.

On a private project, the design firm may specify a single proprietary item for every item on the project if it chooses to do so, and there is no obligation to anyone except the owner to consider substitutes unless the architect or engineer wishes to do so. Further more, if in the judgement of the architect or engineer no substitutes may be considered, then only the specific brand or model of the specified material may be used in the work. All materials delivered to the site for the use on the work must be rejected if they fail to confirm to the specific terms of the plans and specifications.

On public works projects, certain limitations exist that limit the power of architect or engineer to specify a single name brand of a material if the equivalent materials are on the market. All specifications for proprietary materials are required to name at least two brand names and the words "or equal" of materials is called for by brand name. In certain cases, the law controlling the specifying of brand names allows an exception to the rule:

1. if the material specified is required to be compatible with existing facilities.
2. if the product specified is unique and no other brand is made. In some cases, there are also provisions for a public agency to specify a single proprietary material as part of a research of experimentation program in which the single product specified is the article being researched.

Basically, the architect or engineer is considered to be final judge of the quality of a material. If in their determination the material offered is not equal, they have the power to reject it summarily. Furthermore, the contractor may be required to carry the burden of the cost that may be necessary to prove equivalency where a laboratory analysis or similar certification is required. A material may be judged as not being equal on the basis of physical or chemical properties, performance, selection of materials, or even due to dimensional incompatibility with the design of the finished structure where the use of the alternative product may require the redesign of portions of the previously designed structure to accommodate the substitute product.

In any case material offered as substitute to a specified article must be submitted to the architect or engineer or to the owner for consideration and approval before such a substitute material may be used in the work. The inspector must reject any material that fails to satisfy one of the following two requirements:

1. It is the specific product called for in the plans and specifications.
2. It is a substitute product that the architect or engineer or the owner has approved in writing to the contractor.

In the absence of either of these conditions, the substitute product must be rejected by the inspector and required to be removed from the construction site.

F.3.3 Time to Consider Substitutions

The construction industry needs generally to be made aware that substitutions proposed by bidders and contractors unduly disrupt the normal bidding and construction processes. Too often, valuable time and efforts to key personnel are wasted in the consideration of such requests that are originated by the proposer solely for his own financial benefit. Two factors should be kept in mind before a contractor proposes a substitution. First there may be several prefectly valid but undisclosed reasons why the selection or specification was established as it was in the first place. Second, and in any case, time and effort will be required for the architector's engineer's or owner's investigation of the proposed substitute material, for which the personnel must be paid, and during which the work and other necessary activities may be delayed.

The claim is often made that all attempts to limit the consideration of substitutions result in a stiffing competition and loss of economy to the owner. This argument might be valid if the efforts of everyone concerned with the problem were without cost, and if economics were the only interest of each owner.

There are several distinct periods during the life of a project when requests for substitution can be expected to be submitted.

1. Design phase.
2. Bidding phase.
3. Time between bid opening and award of contract.
4. Construction phase.

The only time that consideration should be given to evaluation and acceptance of a substitute "or equal" material is either during the design phase or the construction phase just listed. No consideration should be made of any submittal or request for consideration of substitute material during any of the other listed times.

Consideration of an "or equal" material during the bidding phase is not only unfair to the architect/engineer because it allows too little time for a fair appraisal but actually places the specifier in a high risk situation for the benefit of a third party and it is also quite unfair to the vendors who were willing to spend their own time and money during the design phase to provide data and consultation with the designer for the proper application of the specified material.

On public work projects, there should be no consideration of any substitution between the bid-opening time and the date of the actual award of the contract. It is, however both possible and practical in public work construction to limit consideration of proposed substitutes to a 35 to 45 day period just following the signing of the contract, if desired. This would effectively eliminate one of the "squeeze" plays used by some contractors to effect a substitution of a material for their sole personal gain. By this method, a contractor may submit a proposed substitute material at the last minute, claiming that the originally specified material is not available or has a long lead purchase time which would result in delays to project completion. Nevertheless, if it had wanted to do so, the contractor could generally have ordered the specified material months before and met the schedule. Furthermore, the contractor should be informed that it will be subject to payment of liquidated damages for any over run of the project schedule resulting from inability to obtain the material on time due to its failure to place the order early. In practice, it has often been found that this position materially changes the predicted long lead purchase time and often the material arrives right on schedule.

F.3.4 Inspection and Rejection of Materials

It is responsibility of the inspector to inspect promptly all materials delivered to the site prior to their being used in the work. The practice of withholding inspection until the job is done, then announcing to the contractor that the work fails to conform the specifications, is totally unacceptable conduct for a resident project representative. Certain types of intermittent inspections as performed by government agencies, such as building departments, permit this type of inspection, but it is only because the responsibility for on-site quality control is that of the contractor and the owner's representative. The building departments responsibility is limited to assurance that all requirements of the code and the approved plans and specifications have been followed. But in certain cases it may be desirable to perform inspection of materials or fabricated materials prior delivery at site.

As described before, the inspector not only has a right to reject faulty materials, but is obligated to do so. Upon the rejection of nonconforming materials, they should be clearly and indelibly marked in such manner that the article cannot be used in the work without the mark being clearly visible to the inspector. Such marks can be made with an indelible felt-tip pen, paint, or impression type markers. The inspector should require that all rejected material be immediately removed from the construction area and placed in a separate pile to be transported off the site the same day. The inspector should assure himself that the rejection marks cannot be easily erased and the conconforming articles returned to the site as "new materials".

F.3.5 Storage and Holding of Materials

There is no firm formula for the determination of which facilities will be provided to the contractor as a working or staging area for his storage of construction materials. In no case should any contractor assume that he has the right to block public thorough fares or to use public property of any kind, including parking lots, for its construction purposes, unless specific written authority has been granted.

The resident engineer or inspector must be concerned not only with the quality of the material as delivered but should require that all such materials be properly handled during delivery, unloading, transporting, storage and installation so that undue stresses will not result in latent defects. When in doubt, the resident engineer or inspector should have the design firm or the owner to contact the manufacturer of the affected material, who is generally just as interested in its proper handling in the field as the owner. This is because the manufacturer is often the victim of unjust claims for defective materials when, infact, the problem may have been due to improper handling during construction.

Section G

MAJOR CONSTRUCTION PROBLEMS

Following are the major construction problems which cause failure:

1. Inadequate over all supervision and inspection
2. Poor mixing and placing practices
3. Cold and hot weather
4. Erection problem
5. Formwork failure.

Fortunately most failures of concrete structures occur either during the construction stage, or before the structures are fully in use. Ironically many a collapse comes as workers are concreting the roof slab or placing concrete near the top of the multistory structure, when construction loads have reached a maximum on a structure whose strength is only partially developed.

It is the purpose of this chapter to discuss those failures whose primary cause is found in errors or deficiencies of the construction process, whether they occur while construction is actually in progress or not. Naturally, some failures occur during the later life of a structure because of the way in which construction was handled, and some failures which occur during construction can be traced to flaws in design.

"Failure" is used in the broad sense of noncompliance with the designers intent or specifications, and include defects of position, alignment, dimensions, wearing surfaces and the like, as well as partial or total collapse.

Since many cases involve several contributing causes, cases have been assigned to the various categories according to the judgment as to what was the most significant factor leading to the failure. A few early cases presented first under a "multiple causes" heading demonstrate the compounding of problems that often occurred in the early days of the concrete construction. In analyzing these failures, it is of course helpful to consider the state of development of concrete technology at that time, to realize that material strengths were often substantially lower than those available today, and that quality control, proportioning, and curing practices were far different from those now current.

G.1 Field inspection and Supervision

Competent and strict, almost unfriendly, supervision seems to be one key to the problem of how to prevent failures. In many of the cases, there are instances of errors in performance which could have been prevented by proper supervision of the work.

For example, a three tier concrete home was designed where concreting was specified to follow inspection and approval of reinforcement in place. At the arranged date for the inspection, the slab bars were all in place but the beam bars had not yet arrived at the job. On the next day without appointment. The inspection trip was repeated, only to find the concrete already completed and a truck load of beam bars just arrived. It took considerable argument to have a strip of beam bottom form removed and the still soft concrete chopped out to see the beam bars. They were all 3/8 inch and the contractor had merely substituted two 3/8 inch bars for each 3/4 inch or larger bar required. Only after a threat of calling the police was followed by an actual trip toward the station, did the contractor agree to immediately demolish the floor and rebuild it.

G.1.1 How can Competent Supervision be Obtained

The foregoing cases are but a few of more dramatic examples of what may happen to concrete construction when inspection is inadequate, or in some cases completely lacking. In various investigations of structural distress, the smaller sloppy and shoddy work in masonry, steel and timber as well as in concrete have been observed. It can be prevented with competent supervision.

The complete control of concrete work by the designing engineer is necessary, including the preparation of all reinforcement details and all supervision in the field. Admitting the difficulty of an engineer obtaining a sufficient fee for all of the services suggested, a procedure was employed successfully by the structural engineer for over 50 years. There the structural consultant not only receives his design fee from the architect or owner, but also furnishes the concrete and reinforcement shop drawing and detailed supervision of the work as a part of the contractor's construction cost. Possible conflict of interest as well as the restrictions in many professional codes of ethics will probably not permit ready acceptable of such a formula.

Reliance by many owners on supervision control by the municipal building inspectors usually, on the argument that such service has been paid for by the building permit fee, is of questionable value. In some jurisdictions the field supervision is competent and thorough, but in many areas, it is nonexistent, even to the extent of disregarding the fact that an additional story had been built with no change in the approved plans.

Design engineers responded favourably to a plea voiced by 'Engineering News Record' for improved and increased supervision as a means of avoiding collapses. Field inspection is the designer's responsibility. Experience has shown that shirking this responsibility may lead to a tarnished reputation, whether deserved or not. There should not be a design contract if it does not also include the field supervision. Recent decisions of the courts may force acceptance of the principle that there can be no separation of design and supervision. A system, which has been operated successfully in Europe for a number of years, makes use of a technical control bureau which checks all designs before any work is done, then checks construction to see that it proceeds in accordance with design and specifications. Linking such a control bureau for the job would provide financial incentive for eliminating any built-in design weakness as well as halting shoddy or incorrect field practices.

G.2 Mixing and Placing Problems

Good concrete cannot result from the placing of a poor mix nor can a good mix badly placed result in the best quality concrete structure. Concrete technology has greatly advanced from the days when such headlines as "worthless concrete responsible for the failure "appeared in 1916. With accumulated knowledge such as that summarized in ACI standards for proportioning concrete, and for measuring, mixing and placing as well as in ASTM standards for material testing, it is now possible to purchase concrete according to specified strength, with entrained air for extra durability if desired and with additives to retard or accelerate set and enhance workability of the mix. Mechanical vibration and improved techniques in forming have reduced many of the placing problems and yet, as the following discussion indicates, there are many hurdles remaining before perfect concrete can be obtained.

G.2.1 Control of the Concrete Mix

When an industry produces a product whose quality is whole standard batches are thrown away. With concrete the problems are a little different; once the mix is in the forms, it is difficult if not impossible to throw it away.

The widespread use of compression test cylinders as a measure of concrete strength is a great advance over earlier lack of control but still leaves room for improvement. Even though early strength results can be used to predict 28-day strength, the concrete has already hardened in the form before any test results are available. In an attempt to get control checks of concrete before using it, a supplementary system was instituted. Cylinders were made in the field from concrete as delivered before it was deposited in the hopper, and the cylinder were weighed immediately. Low weight cylinders caused question as to the suitability of the concrete, which was specified to be made with traprock aggregate. Several cases of the wrong aggregate as well as over wet-mixes were caught this way. Of course this system only supplemented and did not replace the standard compression tests.

G.2.2 Placing and Finishing

No matter how good the concrete mix, it may be unserviceable or totally defective if not properly placed. Placing problems leading to deterioration or ultimate collapse include those of incorrectly located reinforcement, excessive concentration of reinforcing bars, insufficient consolidation of concrete, badly placed construction or expansion joints, incorrect location or elevation, etc. Occasionally steel location and concentration make it literally impossible to place concrete. Example of this type of invitation to distress or failure is the case of shallow concrete dome with enough layers of reinforcement to occupy more than the thickness called for if the steel were to be coated with mortar. The over-all steel mat was so dense that, even rain would not go through. The resulting shape or dome was not satisfactory although the structure was stable. No amount of argument convinced either side that a better job could be done.

G.3 Hot and Cold Weather

Optimum temperatures for curing concrete are well known today, and precautions for protecting concrete against extremes of high temperature or frost have been well defined in the ACI standards for hot and cold weather concreting and documented by the research of a number of organisations.' Concrete can be safely installed in freezing weather if precautions are taken and frost protection equipment is prepared and available before the work is started. Likewise, quality concrete can be produced in very warm weather if proper protective measures are planned. Because these measures may require extra labour and add to the apparent cost of the works, they are sometimes disregarded unless supervision is strict. A contractor may also be the victim of sudden, unseasonable temperature changes, and be inadequately prepared to protect concrete. Early failures due to freezing were common before builders realized the possibility of demage. Even today, in spite of all of the knowledge about temperature control and protective measures, there are still some cases of distress and collapse caused by temperature extremes.

G.3.1 Cold Weather - Frozen Concrete

Ignorance and carelessness in winter concreting show up quickly with the arrival of warm weather: frozen concrete may have some strength and a good appearance until it thaws out. If its frozen condition is known, very careful provisions for supporting and curing it at higher temperature until it attains strength may save the structure but alternate freezing and thawing will cause disintegration.

A number of serious failures in the early part of the century were caused largely by ignorance or disregard of proper methods and protection in cold weather. The collapse of Reed's Bath House in Atlantic City in 1906 was attributed to frost penetration of the freshly placed concrete. The one story structure had brick walls a concrete slab and girder roof supported on interior columns as well as the bearing wall. Frost penetration of varying depths in the girder was found, and columns were reduced to rubble - the photographed remains looked like a purposely demolished structure. Concrete was of such poor quality that it could be broken off and crushed in the hand. Poor quality brick work and improper placing (roof beams apparently placed to the under side of the slab, with the slab cast separately) made a bad situation even worse.

The dangers to concrete work in freezing weather are well documented. Today the necessary frost protection measures are known but this knowledge is of no use if the needed equipment is not assembled and made available before frost strikes. After concrete is frozen, no frost protection methods will cure the trouble - the concrete must not be permitted to thaw partly and refreeze. Any concrete program in the winter where frost is a possibility must be preceded by a complete assembly of all wind breaks, heating devices and distribution necessary to protect against the coldest day for the largest expected area of work. Only such insurance will protect against weather damage to the work.

G.3.2 Hot Weather

When concreting is undertaken in hot weather without any special precautions, any or all of the following effects may be experienced:

1. Setting is accelerated, shortening the time available for proper placement and finishing.
2. Strength is reduced because of effects on the hydration process.
3. The tendency toward cracking, either before or after hardening, is increased.
4. Adequate curing becomes a more critical requirement.
5. Control of air content in air - entrained concrete is made more difficult.

While these difficulties do not often lead to serious collapse or other evident failures, they do contribute to general lowering of the quality of the concrete produced. If the strength goes far below the level specified, this alone may be a "failure" sufficient to cause rejection of the finished work.

G.4 Handling and Erection Difficulties

Moving or lifting of a prefabricated structural unit - whether a precast pile, the suspended span of a cantilever bridge, a precast, prestressed or plain reinforced beam or a lift slab section - can only be safely accomplished if erection stresses do not go beyond the yield points, and if the lifting equipment provides sufficient continuous and uniform supports. Failures in this class frequently result from insufficiency of small details or nonuniform action of the lifting hardware. Another common cause is the absence of adequate bracing during erection and assembly. Small horizontal forces will upset the neutral equilibrium existing during the upward lifting of a large mass if it is laterally unrestrained.

Horizontal loading during erection - wind on one side of an almost completed job caused the failure of a 5 million gal tank at Littletion, Colorado. The 100 feet diameter water tank was made of vertical precast staves, 6 x 28 ft. set with edges tight. After more than half the height had been wrapped with die-stressed wires and completely grouted, six staves fell inward, releasing all wire tension and all the wires had to be removed. While the discussion of responsibility was going on, a similar tank with all staves in position also collapsed from lateral wind pressure before any wire had been installed.

Handling problems may be especially critical for prestressed members as an incident at Wellington, New Zealand, in 1955. During transfer of 105 ft. prestressed beams from casting bed to stockpile, one of the trucks ran into a depression; a beam tipped over and exploded from the release of the prestress as it deflected laterally. The shock set off a beam standing vertically, and that also broke into fragments.

For hoisting precast members, spreaders or lifting beams which will insure vertical application of loads at the correct points should always be used. The designer should provide additional reinforcement for any reversed stresses that will occur during erection. Lack of adequate bracing to resist lateral forces is a continuing source of accidents and collapses during construction. The designing engineer or architect can contribute to avoid this type of failures by considering methods of erection and carrying out the work at every stage of design, instead of leaving them wholly to the contractor.

G.5 Formwork Failures

Formwork failures and failures caused by the improper reshoring or premature removal of formwork supports have been all too common, throughout the history of concrete construction. Of course when a building collapses during construction it is easy to blame the formwork, since it and much of the other evidence are covered by the wreckage. Such an explanation sometimes conveniently masks the real cause of trouble, but formwork has truly been at fault in so many cases that it must bear strong scrutiny in any treatment of failures.

Formwork today is seldom too weak to carry the direct vertical load of freshly placed concrete, but all too often it lacks sufficient bracing to withstand the various lateral loads that may be imposed during construction. Lateral forces may be set up by the starting and stopping of rapidly moving buggies or by dumping loads of fresh concrete on the deck. Heavy piles of construction material, nonsymmetrical placement of concrete and wind create unbalanced forces which call for bracing.

Premature removal of forms and shores and careless practices in reshoring have caused numerous failures, or defects such as sagging or cracking in the completed structure. Inadequate size or spacing of reshores may bring the danger of collapse during construction. Defective midsills or other base supports for formwork are also responsible for failures.

So many formwork failures or construction failures associated with formwork have been reported that it is possible here only to summarize the types of failures. The cases have been grouped, according to their probable causes in the following categories:

1. Overloading, either vertical or lateral.
2. Inadequate bracing for lateral forces.
3. Unstable bearing support for formwork.
4. Premature form removal.
5. New or untried methods.

1. **Overloading:** Simple vertical overloading of forms is not too common a cause of form failures today, perhaps because it is reasonably easy to estimate vertical loads and to provide support for them. It is difficult to say how many failure cases attributed to other causes may have involved an actual overloading. Certainly unequal settlement of form supports may cause local overloading of individual shores. The same is true when bent or defective individual shores are used: if they cannot take their full load, the adjoining shores may be overloaded. Eccentricity of load application sharply reduces the buckling load of tubular shores thus resulting in overload.

Inadequate provision for the lateral pressure of freshly placed concrete is more likely to cause bulges and deformations which go unreported, although not unnoticed: than to trigger any newsworthy accident. Although the exact variation of lateral pressure is a matter to be resolved by further research, ACI committee 347 has developed formulas for pressure which may be safely used in designing the forms for vertical concrete members.

The temperature and vertical rate of placement in the forms are factors influencing the lateral pressure development, and if the temperature drops during the concreting, the rate of placement often has to be slowed down to prevent a buildup of lateral pressure which will overload the forms. Fortunately it is possible to observe the deflection before it becomes too large; this is another reason why failures from overpressure are seldom reported.

2. **Inadequate bracing for lateral forces:** A system of formwork to receive wet concrete high above grade or above the previously constructed floor is not the most stable structure. The weight is almost entirely at the top and is supported by an array of posts which are not rigidly connected to the form at the top or to the floor at the bottom. The lack of diagonal bracing is a common threat in such formworks where lateral loads such as winds, cable tensions, starting and stopping of equipment and dumping of concrete are likely to occur.

The foregoing incidents are typical of the many illustrating the need for greater stability in the temporary formwork structure. One step in this direction is to cast columns at least a day ahead of the floor. This provides considerable stiffness against lateral sway. The unit labour cost of placing the concrete may be increased, but the small expenditure is cheap insurance. Stability of formwork can also be improved in the form planning stage, by making adequate provisions for bracing to withstand foreseeable lateral loads.

3. **Unstable bearing support:** Post supported on a lower completed floor can be assumed to have equal and uniform bearing. However, forms for the lowest slab level are often supported on "mudsills" which are not on solid ground but usually on back fill recently placed with great probability of softening form flow of water, either natural run-off or wash water from forms or from truck mixers. Unequal settlement of the sills seriously disrupts the designed equality of post reactions and with good possibility of overloading the posts which do not settle as the load carried by exterior line of posts is released when their sills settle. A common cause of structural defects is the practice of placing mudsills on frozen ground. As construction progresses frost protection heaters

and dripping from concrete placement soften the ground, causing unwanted sags in the supported structure.

4. **Premature form removal:** Premature stripping of forms has caused numerous failures, not only in terms of collapses but in the often unreported sagging of partially cared concrete and in the development of hairline cracks that cause serious maintenance problems in the later life of the structure. During the earlier history of concrete construction, when cold weather work was frequently inadequately protected, there were many failures when the forms were pulled. There was a wide-spread failure to recognize that satisfactory temperature as well as time requirements must be met before removing forms.

5. **Reshoring:** Premature reshoring and inadequate size and spacing of reshores have been responsible for a number of construction failures. The problem is particularly troublesome in multistory flat slab structures which are designed for a relatively light live load. The accumulated load of slabs freshly placed and partially cured together with construction live load may be larger than the load which a floor is designed to carry when it has developed its full strength. Usually only part of that strength is developed when construction of the next story begins.

6. **New or untried methods of forming:** Many of the previously described formwork troubles involved designs, systems or details that had been successfully used many times before the one case that brought failure or accident. In contrast are the accidents which occur with relatively new forming materials or method or work on new type of structures, before there has been opportunity to fully prove these techniques. No doubt these are frequently cases of overloading, arising from a failure to understand structural action of the forms.

EXERCISE

1. What are pozzolanas?
2. What are the sources of pozzolanic materials?
3. What are the advantages of using pozzolanic materials?
4. What are the different types of pozzolanas?
5. What is Pulverisd Fuel ash and what are its sources?
6. What are the advantages of using Fly Ash in concrete?
7. What is Granulated Blast Funace Slag?
8. How is Portland Blast Furnace Cement manufactured?
9. Where will you recommend the use of Portland and Blast Furnace Cement?
10. What property is responsible for giving Rice Husk Ash its pozzolanic action?
11. What combustion properties are to be controlled during manufacture of Rice Husk Ash?
12. What are the applications of Rice Husk Ash?
13. Prepare a comparative table listing the various types of pozzolanas, their sources,salient manufacturing features and applications. Also mention the limitations to their use in your locality and try to include such locally available pozzolanic or nonconventional building materials which find application in your area.
14. Discuss briefly the properties of lightweight aggregates.
15. Describe the procedure for mixing lightweight concrete.
16. Describe the influence of water absorption capacity of lightweight aggregates in design of lightweight concrete mix.
17. Describe production & placing of no fines concrete.
18. What do you understand by nailing concrete.
19. Discuss the factors which influence the design of lightweight concrete.
20. Design a lightweight concrete mix using by Lytag as aggregate with the following data.
Minimum 28 day cube strength 320 kg/cm², control factor = 0.80
Relative density not to exceed 1.8
Workability - Medium to High.

Set out the dry batch weights.

21. Do you think broken brick is a suitable light weight aggregate : discuss.
22. Name the various types of fibres that can be used in the production of FC sheets.
23. What are the different materials used for production of FC sheets/ pantiles?
24. Explain moulding and curing procedures for sheets and pantiles.
25. What are the various types of fibres available in your area? Mention their botanical and English name. Discuss in detail their physical and mechanical properties.
26. How FC pantiles, FC Rings and FC sheets are made?
27. What is rice husk?
28. Compare the calorific value of rice husk coal and wood.
29. Establish the equivalence between the rice husk, coal and fuel wood, taking coal as the basis.
30. Which building material has highest energy input in manufacture and which has the lowest?
31. List the naturally available building materials
32. List the industrial / agricultural wastes which can be used in manufacture of building materials.
33. Explain how loads are resisted in domes, shells and folded plates.
34. What is a hypar shell? Explain why it is a very popular shell in civil Engineering construction?
35. Describe the procedure to erect a baboo geodetic dome.
36. Describe a barrel vault construction.
37. Prepare a comparitve table of various shells forms with respect to shape, strength, structural action and ease of construction.
38. How bamboo construction can be useful in replacing convention construction in rural areas. Describe the advantages and disadvantages with respect to economy, & social adaptation.

References

1. "Appropriate Building Materials" Published by Skat Publications, Switzerland, 1988.

2. "Appropriate Building Materials for Low Cost Housing" Proc. of a symposium held in Nairobi, Kenya, 7 to 14 Nov. 1983, Oxford and IBH Publishing Co., New Delhi.

3. "Three Decades of Building Research in India" by Sarita Prakashan, Delhi, 1984.

4. Proc. International seminar on low cost housing and alternate building materials, Feb. 10-11, 1988, New Delhi, Published by CBRI, Roorkee.

5. "Building Materials and Components" Published for CBRI, Roorkee by TMH, New Delhi.

6. Y. Krishna Raju, N., Design of concrete mixes, CBS Publishers & Distributors, Delhi, 1988.

7. Murdock, L.J. and Brook, K.M., Concrete Materials & Practice, 1979.

8. Akroyd, T.N.W., Concrete, Properties & Manufacture, Pergamon Press, London, 1962.

9. Abdul Kareem, E. and Paul Joseph, G. 'Investigation on Flexural behaviour of Ferrocement and its application to long span roofs', Journal of Ferrocement, Vol. 8, No. 1, Jan. 1978, pp. 1-21.

10. Abdul Kareem, E. and Paul Joseph, G. 'Small Capacity Ferrocement water tanks', Structural Engineering Research Centre, Madras.

11. Desai, P., Viswanatha, C.S. and Kanappan, S. 'Some studies on Ferrocement roofing elements; Journal of Ferrocement, Vol. 12, No. 3, July 1982.

12. Paul Joseph, G. Zacharia George and Sreenath, H.G. 'Ferrocement as substitute for timber', Paper presented at Asia Pacific symposium on Ferrocement, application for rural development, Roorkee, India, April 23-25, 1984.

13. Paul Joseph, G. Sreenath, H.G. and Mani, K. Ferrocement Products, Structural Engineering Research Centre, Madras, Feb. 1991.

14. Paul Joseph, G. 'Design of small capacity Ferrocement water tank'. The Indian Concrete Journal, December 1989, pp. 579-584.

15. Rodes - Austria Co. L. Pama, R.P. Valls, J. Sing, C. 'State-of-the-Art of prefabricated Ferrocement Housing', in proceedings of the International Congress on Housing the impact of Economy and Technology, Nov. 15-18, 1981. Ural, O. and Robert Krap Febaver (Edts.).

16. Sing, S.P. and Sofat, G.C. (Edts.). 'Handbook on Building Economics and Productivity; CBRI, Roorkee, 1988.

17. J. Venkataramana et al. Module 3 DLPU notes of Building Materials & Designs (HT ZG 651), Feb. 1992.

18. Roland Stulz, Kiran Kukherji, Appropriate Building Materials, Published by SKAT, Swiss Centre for Appropriate Technology, 1988.

19. Jose Castro and A.E. Namman. Cement mortar reinforced with natural fibres, ACI Journal, Jan.-Feb. 1981.

20. Lewis, G. Natural vegetable fibre as reinforcement in cement sheets. Magazine of Concrete Research, Vol. 31, No. 107, June 1979.

21. O.J. Uzomaka, Characteristics of Akwara as a reinforcing fibre. Magazine of Concrete Research, Vol. 28, Sep. 1976.

22. Annon, Consultation : Lime - pozzolana papers and proceedings, New Delhi, 8th-9th Dec., 1997, National Building Organisation and UN Regional Housing Centre, ESCAP.

23. A.A. Hammond, "Pozzolona Cements for low cost housing in" Appropriate Building Materials for low cost Housing, Proceedings of a symposium held in Nairobi, Kenya, from 7 to 14 November, 1983.

24. Mohan Rai, Energy Conservation in Production of Building Materials, in Energy and Habitat.

25. B.V. Subrahmanyam et al. "A New Low Energy - Intensive Building Material Based on lateritic soils for low cost housing in developing nations in Proceedings of a Symposium held in Nairobi, Kenya from 7 to 14 November, 1983. Oxford and IBH Publishing Co., New Delhi.

26. Stulz, Roland & Mukerji, Kiran, Appropriate building materials Published by SKAT, Swiss Centre for Appropriate Technology, 1988.

27. Chudley, R., Building Construction Hand Book, Heinemann Newnes, 1989.

28. Pant, J.C., CPM and PERT with linear programming, New Delhi, Jain Brothers, 1986.

29. Peurifov, R.B., Construction Planning Equipment and Methods, Tokyo, McGraw Hills, Inc., 1970.

30. National Building Code of India, Indian Standard Institution, 1970.

31. Shrikanth, L.S., PERT and CPM Principles, New Delhi, East West Press Pvt. Ltd., 1975.

32. Jacob Feld, Lesson from Failures of Concrete Structures, American Concrete Institute, 1965.

33. Rodolfo, J. Aguilar, System Analysis and Design, Prentice Hall, Inc. Englewood, 1973.

34. Andrew, B., Templeman, Civil Engineering Systems, London. The Maceillan Press Ltd., 1982.

35. O'Brien, James Jerome, Scheduling Handbook, New York, McGraw Hill, 1969.

36. Varma, Mahesh, Construction, Planning and Management, New Delhi, Metropolitan Book Co. Ltd., 1977.

Appendix I

A case study: Construction of a highway project between two cities A & B, 50 km apart using network techniques.

The case study is not the study of any actual project and all the activities and durations are the assumed ones. The network is analysed using computers program.

Act.Description	Act.num	St. Node	Final Node
1. Reconnaissance	A	1	2
2. Final Survey	B	2	3
3. Profiles of Routes	F	3	4
4. Soil investigation	C	3	5
5. Road materials investigation	D	3	6
6. Rights of way investigation	E	3	7
7. Dummy	D1	4	5
8. Dummy	D2	5	6
9. Float Tenders	G	6	8
10. Dummy	D3	7	8
11. Acquire labour	H	8	9
12. Planning & allotment of work	I	9	10
13. Dummy	I	4 9	10
14. Acquire Coarse,fine aggregant for section 1.	M1	10	11
15. Acquire Bituminous material for section 1.	B1	10	13
16. Start of work	J	10	14
17. Acq coarse, fine aggregate in section 1.	M2	11	12
18. Dummy	I5	11	14
19. Dummy	d6	12	15
20. Acq coarse, fine aggregate in section 3.	M3	12	22
21. Dummy	D7	13	15
22. Acquire bituminous material in sec 2.	B2	13	16
23. Earthwork, levelling with C.A. in sec. 1	K	14	15
24. Construction of Culvert in section 1.	F	15	17
25. Lay bituminous layer in section 1	A	15	18

(Contd.)

Act.Description	Act.num	St. Node	Final Node
26. Earthwork,levelling with C.A.& F.A.sec 2.	G	15	20
27. construction of road Over Bridge in sec 1.	L	15	23
28. Dummy	D8	16	20
29. Acquire bituminous material in section 3.	B3	16	27
30. Dummy	D9	17	23
31. Completd electrlcal flttlngs in section 1.	B	18	21
32. Lay milestones in section 3.	C	18	23
33. Construct Roaddividers in sec 1.	D	18	19
34. Construct speedbreakers in sec. 1.	E	19	23
35. Lay Bituminoous layer in sec. 2.	H	20	23
36. Dummy	D10	21	23
37. Dummy	D11	22	23
38. Construction of R.C culvert in section 2	M	23	24
39. Construction of milestones in section 2.	J	23	25
40. Complete electrical fitting in section.2	i	23	26
41. E/work,levelling with C.A. & F.A..in sec. 3	n	23	28
42. Construction of roaddividers in sec. 2	k	23	29
43. Construction of a birdge in sec. 2	m	23	31
44. Dummy	d12	24	31
45. Dummy	d13	25	31
46. Dummy	d14	26	31
47. Dummy	d15	27	28
48. Lay bituminoous layer in sect. 3	p	28	31
49. Costruction of speedbreakers in sec. 2	i	29	30
50. Dummy	d16	30	31
51. Construction of roaddivders in sec. 2	s	31	32
52. Lay milestones in sec. 3	r	31	33
53. Consrruction of R.C. culvertin sec. 3	q	31	34
54. Complete electrical fitting in sec. 3	u	31	35
55. Construction of a bridge in sec. 3	n	31	36
56. Construction of speedbreakers in sec.3	t	32	36
57. Dummy	d16	33	36
58. Dummy	d17	34	36
59. Dummy	d18	35	36
60. Electrification of the entire nighway.	p	36	37
61. Check for any incomplete work	o	37	38
62. Commission.	r	38	39

No	SN	FN	DUR	EST	EFT	LST	LFT	TF	FF
1	1	2	2	0	2	0	2	0	0C
2	2	3	3	2	5	2	5	0	0C
3	3	4	2	5	7	5	7	0	0C
4	3	5	2	5	7	5	7	0	0C
5	3	6	1	5	6	6	7	1	1
6	3	7	2	5	7	7	9	2	0
7	4	5	0	7	7	7	7	0	
8	5	6	0	7	7	7	7	0	0C
9	6	8	2	7	9	7	9	0	0C
10	7	8	0	7	7	9	9	2	0C
11	8	9	1	9	10	11	12	2	2
12	8	10	3	9	12	9	12	0	0C
13	9	10	0	10	10	12	12	2	2
14	10	11	1	12	13	12	13	0	0C
15	10	13	1	12	13	18	19	6	0
16	10	14	1	12	13	12	13	0	0C
17	11	12	1	13	14	18	19	5	0c
18	11	14	0	13	13	13	13	0	0c
19	12	15	0	14	14	19	19	5	5
20	12	22	1	14	15	48	49	34	0
21	13	15	0	13	13	19	19	6	6
22	13	16	1	13	14	48	49	35	0
23	14	15	6	13	19	13	19	0	0c
24	15	17	4	19	23	45	49	26	0
25	15	18	3	19	22	43	46	24	0
26	15	20	5	19	24	44	49	25	0
27	16	23	30	19	49	19	49	0	0C
28	16	20	0	14	14	49	49	35	10
29	16	27	1 1	4	15	71	72	57	0
30	17	23	0	23	23	49	49	26	26
31	18	21	1	22	23	48	49	26	0
32	18	23	1	22	23	48	49	26	26
33	18	19	2	22	24	46	48	24	0

(Contd.)

No	SN	FN	DUR	EST	EFT	LST	LFT	TF	FF
34	19	23	1	24	25	48	49	24	24
35	20	23	0	24	24	49	49	25	25
36	21	23	0	23	23	49	49	26	26
37	22	23	0	15	15	49	49	34	34
38	23	24	3	49	52	71	74	22	0
39	23	25	1	49	50	7	3 7	4 2	4 0
40	23	26	1	49	50	73	74	24	0
41	23	28	5	49	54	67	72	18	0
42	23	29	2	49	51	71	73	22	0
43	23	31	25	49	74	49	7	4 0	0C
44	24	31	0	52	52	74	74	22	22
45	25	31	0	50	50	74	74	24	24
46	26	31	0	50	50	74	74	24	24
47	27	28	0	15	15	72	72	57	39
48	28	31	2	54	56	72	74	18	18
49	29	30	1	51	52	73	74	22	0
50	30	31	0	52	52	74	74	22	22
51	31	32	1	74	75	94	95	20	0
52	31	33	1	74	75	95	96	21	0
53	31	34	3	74	77	93	96	19	0
54	31	35	1	74	75	95	96	21	0
55	31	36	2	2 7	4 9	6 7	4 9	6 0	0C
56	32	36	1	75	76	95	96	30	30
57	33	36	0	75	75	96	96	21	21
58	34	36	0	77	77	96	96	19	19
59	35	36	0	75	75	96	96	21	21
60	36	37	2	96	98	96	98	0	0C
61	37	38	1	98	99	98	99	0	0C
62	38	39	1	99	100	99	100	0	0C

* Critical activities are amked with C

Critical path

No.	Times for each event	
	Early start	Late start
1	0	0
2	2	2
3	5	5
4	7	7
5	7	7
6	7	7
7	7	9
8	9	9
9	10	12
10	12	12
11	13	13
12	14	19
13	13	19
14	13	13
15	19	19
16	14	49
17	23	49
18	22	46
19	24	48
20	24	49
21	23	49
22	15	49
23	49	49
24	52	74
25	50	74
26	50	74
27	15	72
28	54	72

(Contd.)

No.	Early start	Late start
29	51	73
30	52	74
31	74	74
32	75	95
33	75	96
34	77	96
35	75	96
36	96	96
37	98	98
38	99	99
39	100	100

The duration of the project is 100.0000 weeks

F	T.O	T.P	T.M
A	1.0	3.0	2.0
3	1.0	5.0	3.0
4	0.5	3.5	2.0
5	1.0	3.0	2.0
6	1.0	1.0	1.0
7	1.0	3.0	2.0
5	0.0	0.0	0.0
6	0.0	0.0	0.0
8	1.0	3.0	2.0
9	0.2	4.0	1.5
10	2.0	5.0	3.0
11	1.0	1.0	1.0
13	1.0	1.0	1.0
14	0.5	2.0	1.0
12	1.0	1.0	1.0
15	4.0	9.0	6.0
16	1.0	1.0	1.0

(Contd.)

F	T.O	T.P	T.M
17	3.0	5.0	4.0
18	2.0	6.0	3.0
20	4.0	7.0	5.0
21	0.5	2.0	1.0
22	1.0	1.0	1.0
23	25.0	35.0	30.0
27	1.0	1.0	1.0
23	1.0	1.0	1.0
23	1.0	3.0	2.0
24	2.5	6.0	3.0
25	0.5	2.0	1.0
26	1.0	2.0	1.0
28	3.5	8.05	.0.5
29	1.0	3,0	2.0
30	0.5	2.0	1.0
31	20.0	35.0	25.0
32	1.5	3.5	2.0
32	0.5	2.0	1.0
33	1.0	1.0	1.0
34	1.0	5.0	3.0
35	1.0	1.0	1.0
36	1.0	2.0	1.0
37	1.5	4.0	2.0
38	1.0	1.0	1.0
39	1.0	1.0	1.0

Result of PERT analysis

No	Start event time	Finish event time	Duration	Optimistic time	Pessi-mistic	Mostlikely time
1	1	2	2.0000	1.0000	3.0000	2.0000
2	2	3	3.0000	1.0000	5.0000	3.0000
3	3	4	2.0000	0.5000	3.5000	2.0000

(Contd.)

No	Start event time	finish event time	duration	optimistic time	pessi- mistic	mostlikely time
4	3	5	2.0000	1.0000	3.0000	2.0000
5	3	6	1.0000	1.0000	1.0000	1.0000
6	3	7	2.0000	1.0000	3.0000	2.0000
7	4	5	0.0000	0.0000	0.0000	0.0000
8	5	6	0.0000	0.0000	0.0000	0.0000
9	6	8	2.0000	1.0000	3.0000	2.0000
10	7	8	0.0000	0.0000	0.0000	0.0000
11	8	9	1.7000	0.2000	4.0000	1.5000
12	8	10	3.1667	2.0000	5.0000	3.0000
13	9	10	0.0000	0.0000	0.0000	0.0000
14	10	11	1.0000	1.0000	1.0000	1.0000
15	10	13	1.0000	1.0000	1.0000	1.0000
16	10	14	1.0833	0.5000	2.0000	1.0000
17	11	12	1.0000	1.0000	1.0000	1.0000
18	11	14	0.0000	0.0000	0.0000	0.0000
19	12	15	0.0000	0.0000	0.0000	0.0000
20	12	22	1.0000	1.0000	1.0000	1.0000
21	13	15	0.0000	0.0000	0.0000	0.0000
23	14	15	6.1667	4.0000	9.0000	6.0000
24	15	17	4.0000	3.0000	5.0000	4.0000
25	15	18	3.3333	2.0000	6.0000	3.0000
26	15	20	5.1667	4.0000	7.0000	5.0000
27	15	23	30.0000	25.0000	35.0000	30.0000
28	16	20	0.0000	0.0000	0.0000	0.0000
29	16	27	1.0000	1.0000	1.0000	1.0000
30	17	23	0.0000	0.0000	0.0000	0.0000
31	18	21	1.0833	0.5000	2.0000	1.0000
32	18	23	1.0000	1.0000	1.0000	1.0000
33	18	19	2.0000	1.0000	3.0000	2.0000
34	19	23	1.0833	0.5000	2.0000	1.0000
35	20	23	2.0000	1.0000	3.0000	2.0000
36	21	23	0.0000	0.0000	0.0000	0.0000

(Contd.)

No	Start event time	finish event time	duration	optimistic time	pessi- mistic	mostlikely time
37	22	23	0.0000	0.0000	0.0000	0.0000
38	23	24	3.4167	2.5000	6.0000	3.0000
39	23	25	1.0833	0.5000	2.0000	1.0000
40	23	26	1.1667	1.0000	2.0000	1.0000
41	23	28	5.2500	3.5000	8.0000	5.0000
42	23	29	2.0000	1.0000	3.0000	2.0000
43	23	31	25.8333	20.0000	35.0000	25.0000
44	24	31	0.0000	0.0000	0.0000	0.0000
45	25	31	0.0000	0.0000	0.0000	0.0000
46	26	31	0.0000	0.0000	0.0000	0.0000
47	27	28	0.0000	0.0000	0.0000	0.0000
48	28	31	2.1667	1.5000	3.5000	2.0000
49	29	30	1.0833	0.5000	2.0000	2.0000
50	30	31	0.0000	0.0000	0.0000	0.0000
51	30	32	1.0833	0.5000	2.0000	1.0000
52	31	33	1.0000	1.0000	1.0000	1.0000
53	31	34	3.0000	1.0000	5.0000	3.0000
54	31	35	1.0000	1.0000	1.0000	1.0000
55	31	36	22.1667	15.000	30.000	22.000
56	32	36	1.1667	1.0000	2.0000	1.0000
57	33	36	0.0000	0.0000	0.0000	0.0000
58	34	36	0.0000	0.0000	0.0000	0.0000
59	35	36	0.0000	0.0000	0.0000	0.0000
60	36	37	2.2500	1.5000	5.0000	2.0000
61	37	38	1.0000	1.0000	1.0000	1.0000
62	38	39	1.0000	1.0000	1.0000	1.0000

Activity start and finish times

No	Early start	Early finish	Late start	Late finsh
1	0.0000	2.0000	0.0000	2.0000
2	2.0000	5.0000	2.0000	5.0000
3	5.0000	7.0000	5.0000	7.0000
4	5.0000	7.0000	5.0000	7.0000
5	5.0000	6.0000	6.0000	7.0000
6	5.0000	7.0000	7.0000	9.0000
7	7.0000	7.0000	7.0000	7.0000
8	7.0000	7.0000	7.0000	7.0000
9	7.0000	9.0000	7.0000	9.0000
10	7.0000	7.0000	9.0000	9.0000
12	9.0000	12.1667	9.0000	12.1667
13	10.7000	10.7000	12.1667	12.1667
14	12.1667	13.1667	12.2500	13.1667
15	12.1667	13.1667	18.4167	19.4167
16	12.1667	13.2500	12.1667	13.2500
17	13.1667	14.1667	18.4167	19.4167
18	13.1667	13.1667	13.2500	13.2500
19	14.1667	14.1667	19.4167	19.4167
20	14.1667	15.1667	48.4167	49.4167
21	13.1668	13.1667	19.4167	19.4167
22	13.1667	14.1667	46.4168	47.4167
23	13.2500	19.4168	13.3600	19.4167
24	19.4168	23.4168	45.4167	49.4168
25	19.4167	22.7500	43.0000	46.3333
26	19.4167	24.5833	42.2500	47.4167
27	19.4167	49.4167	19.4167	49.4167
28	14.1667	14.1667	47.4167	47.4167
29	14.1667	15.1667	72.0833	73.0833
30	23.4167	23.4167	49.4167	49.4167
31	22.7500	23.8333	48.3333	49.4167
32	22.7500	23.7500	48.4167	49.4167

(Contd.)

No	Early start	Early finish	Late start	Late finsh
33	22.7500	24.7500	46.3333	48.3333
34	24.7500	25.8333	48.3333	49.4167
35	24.5833	26.5833	47.4167	49.4167
36	33.8333	23.8333	49.4167	49.4167
37	15.1667	15.1667	49.4167	49.4167
38	49.4167	52.8333	71.8333	75.2500
39	49.4167	50.5000	74.1667	75.2500
40	49.4168	50.5833	74.0833	75.2500
41	49.4167	54.6667	67.8333	73.0833
42	49.4167	51.4167	72.1667	74.1667
43	49.4167	75.2500	49.4167	75.2500
44	52.8333	52.8333	75.2500	75.2500
45	50.5000	50.5000	75.2500	75.2500
46	50.5833	50.5833	75.2500	75.2500
47	15.1667	15.1667	73.0833	73.0833
48	54.6667	56.8333	73.0833	75.2500
49	51.4167	52.5000	74.1667	72.2500
50	52.5000	52.5000	75.2500	75.2500
51	75.2500	76.3333	95.1667	96.2500
52	75.2500	76.2500	76.4167	97.4167
53	75.2500	78.2500	96.4167	97.4167
54	75.2500	76.2500	96.4168	97.4167
55	75.2500	97.4167	75.2500	97.4167
56	76.3333	77.5000	76.2500	97.4167
57	76.2500	76.2500	97.4167	97.4167
58	78.2500	78.2500	97.4167	97.4167
59	96.2500	76.2500	97.4167	97.4167
60	97.4167	99.6667	97.4167	99.6667
61	99.6667	100.6667	99.6667	100.6667
62	100.6667	101.6667	100.6667	101.6667

Floats for the activities

No	Total float	Free float
1	1.000	0.0000C
2	0.000	0.0000C
3	0.0000	0.0000C
4	0.0000	0.0000C
5	1.0000	1.0000
6	2.0000	0.0000
7	0.0000	0.0000C
8	0.0000	0.0000C
9	0.0000	0.0000C
10	2.0000	2.0000
11	1.4667	0.0000
12	0.0000	0.0000
13	1.4667	1.4667
14	0.0833	0.0000
15	6.2500	0.0000
16	0.0000	0.0000C
17	5.2500	0.0000
18	0.0833	0.0833
19	5.2500	5.2500
20	34.2500	0.0000
21	6.2500	6.2500
22	33.2500	0.0000
23	0.0000	0.0000
24	26.000	0.0000
25	23.5833	0.0000
26	22.8333	0.0000
27	0.00000	0.0000
28	33.2500	10.4167
29	57.9167	0.0000C
30	26.0000	26.0000
31	25.5833	0.0000
32	25.6667	25.6667

(Contd.)

No	Total float	Free float
33	23.5833	0.00000
34	23.5833	23.5833
35	22.8333	22.8333
36	25.5833	25.5833
37	34.2500	34.2500
38	22.4167	0.00000
39	24.7500	0.00000
40	24.6667	0.00000
41	18.4167	0.0000
42	22.7500	0.0000
43	0.00000	0.0000C
44	22.4167	22.4167
45	24.7500	24.7500
47	57.9167	39.5000
46	24.6667	24.6667
48	18.4167	18.4167
49	22.7500	0.00000
50	22.7500	22.7500
51	19.9167	0.00000
52	21.1667	0.00000
53	19.1667	0.00000
54	21.1667	0.00000
55	0.00000	0.00000C
56	19.9167	19.9167
57	21.1667	21.1667
58	19.1667	19.1667
59	21.1667	21.1667
60	0.00000	0.0000C
61	0.00000	0.0000C
62	0.00000	0.0000C

The critical path marked with C

| No | Time for each event | | Variance |
	Early srart	Late start	
1	0.0000	0.0000	0.0000
2.	2.0000	2.0000	0.1111
3	5.0000	5.0000	0.4444
4.	7.0000	7.0000	0.2500
5	7.0000	7.0000	0.2500
6.	7.0000	7.0000	0.0000
7	7.0000	9.0000	0.1111
8.	9.0000	9.0000	0.1111
9	10.7000	12.1667	0.4011
10	12.1667	12.1667	0.4011
11	13.1667	13.2500	0.0000
12	14.1667	19.4167	0.0000
13	13.1667	19.4167	0.0000
14	13.2500	13.2500	0.0625
15	19.4167	19.4167	0.6944
16	14.1667	47.4167	0.0000
17	23.4167	49.4167	0.1111
18	22.7500	46.3333	0.4444
19	24.7500	48.3333	0.1111
20	24.5833	47.4167	0.2500
21	23.8333	49.6167	0.0625
22	15.1667	49.4167	0.0000
23	49.4167	49.4167	2.7778
24	52.8333	75.2500	0.3403
25	50.5000	75.2500	0.0625
26	50.5833	75.2500	0.0278
27	15.1667	73.0833	0.0000
28	54.6667	73.0833	0.5625
29	51.4167	74.1667	0.1111

(Contd.)

No	Time for each event		Variance
	Early srart	Late start	
30	52.5000	75.2500	0.0625
31	75.2500	75.2500	6.2500
32	76.3333	96.2500	0.0625
33	76.2500	97.4167	0.0000
34	78.2500	57.4167	0.4444
35	76.2500	97.4167	0.0000
36	97.4167	97.4168	6.2500
37	99.6667	99.6667	0.1736
38	100.6667	100.6667	0.0000
39	101.6667	101.6667	0.0000

The expected curation of the project is 101.6667

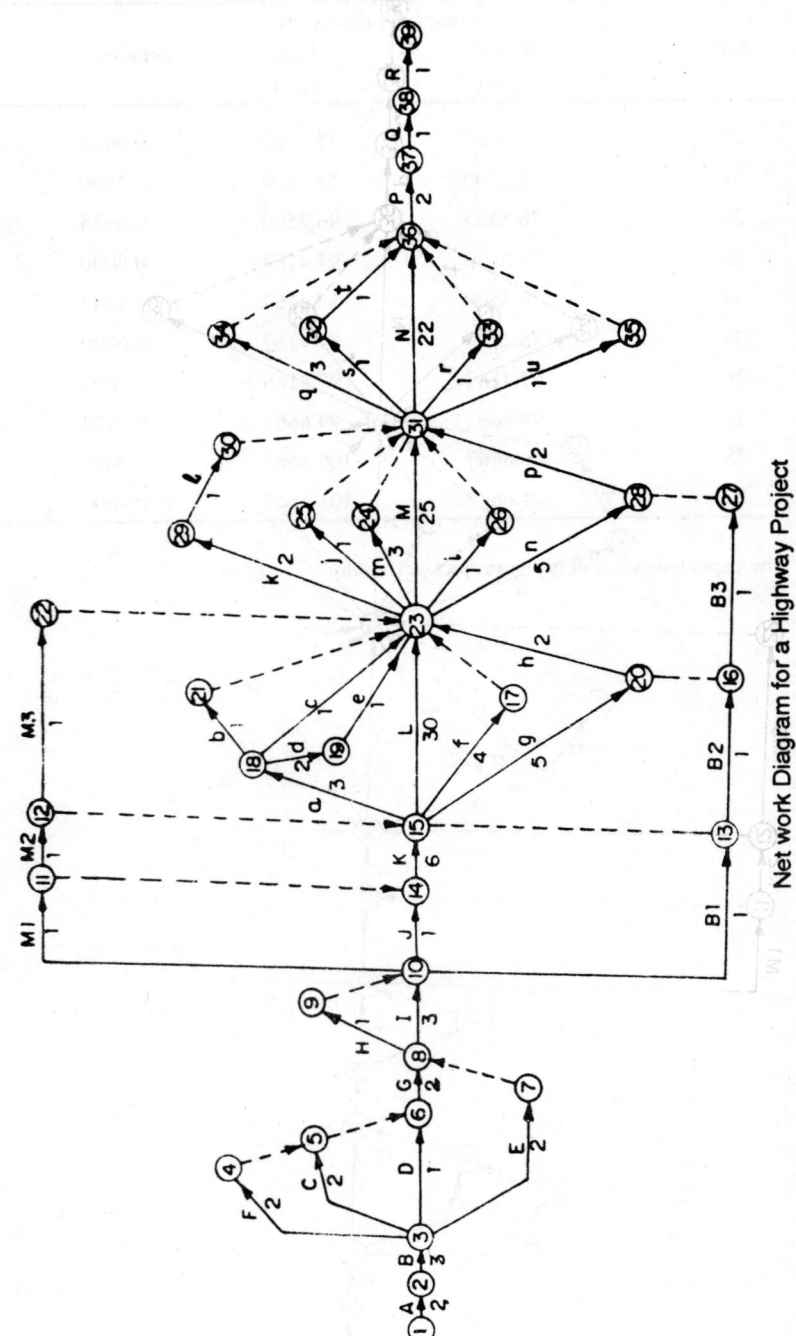

Net work Diagram for a Highway Project

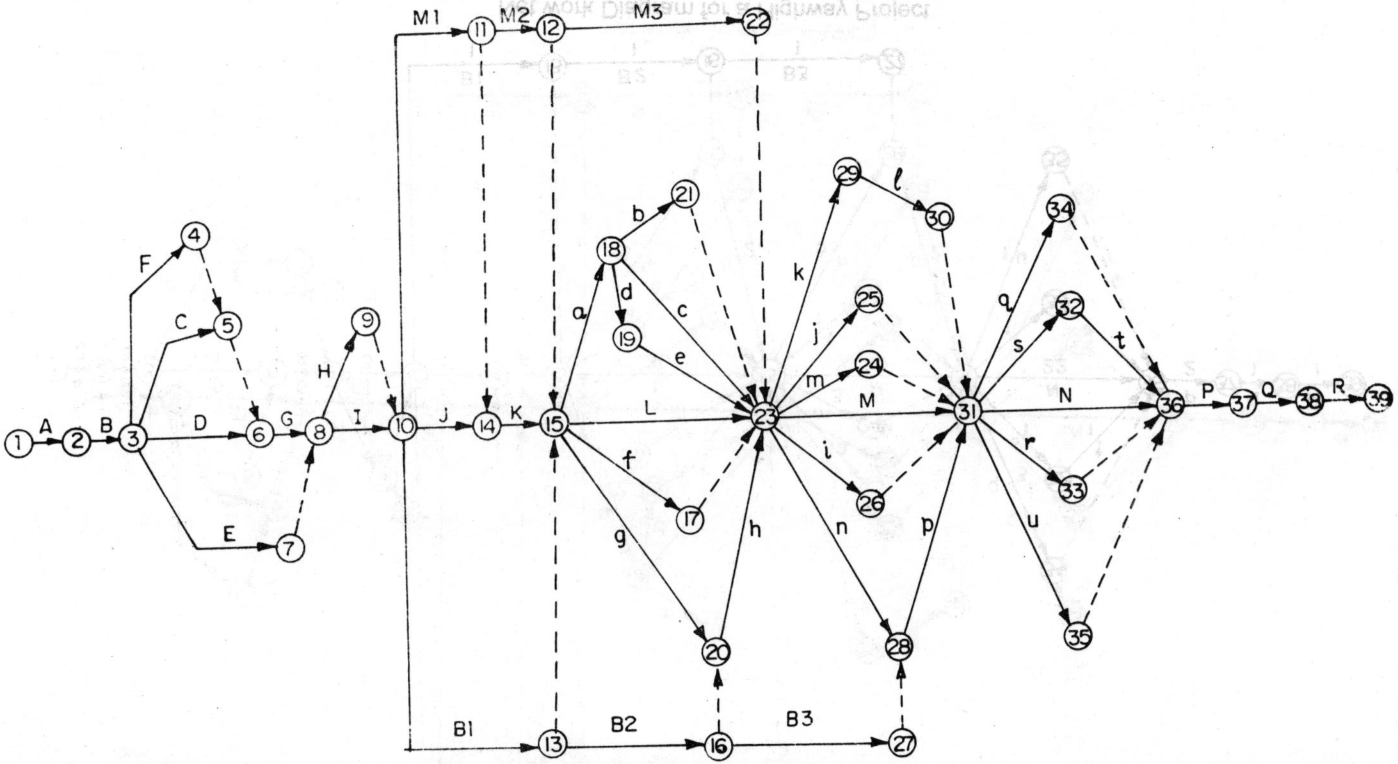

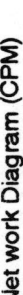

Net work Diagram (CPM)

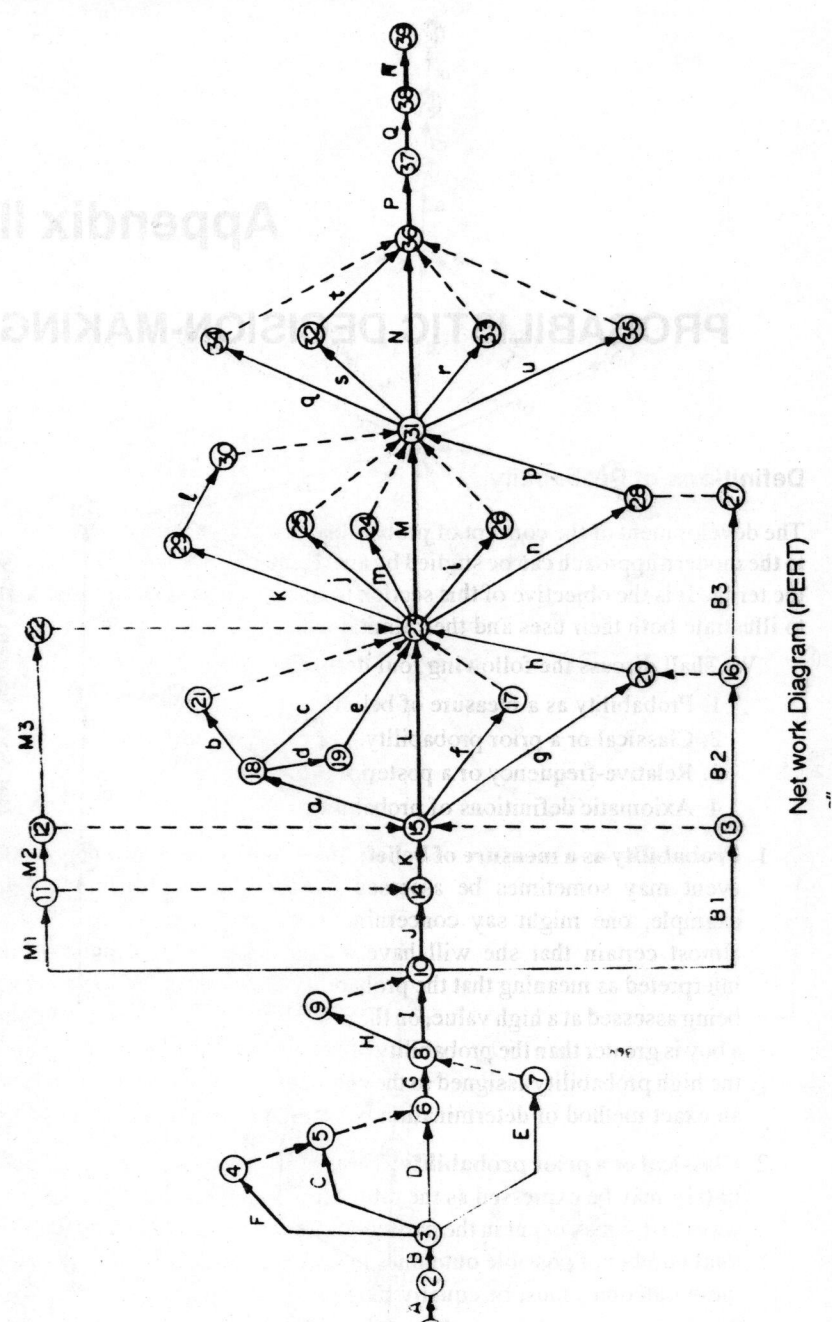

Net work Diagram (PERT)

Appendix II

PROBABILISTIC DECISION-MAKING

Definitions of Probability

The development of the concept of probability from its earliest formulations to the modern approach can be studied by analyzing the various definitions of the terms. It is the objective of this section to introduce these definitions and to illustrate both their uses and their limitations.

We shall discuss the following four definitions of probability:

1. Probability as a measure of belief.
2. Classical or a prior probability.
3. Relative-frequency or a posterior probability.
4. Axiomatic definitions of probability

1. **Probability as a measure of belief:** The value of the probability of an event may sometimes be assessed by one's own judgement. For example, one might say concerning a woman's pregnancy. "I am almost certain that she will have a boy". This statement may be interpreted as meaning that the probability of a woman having a son is being assessed at a high value, on the probability of the woman. Having a boy is greater than the probability of her having a girl. In this example, the high probability assigned to the event involved was not the result of an exact method of determination but was the exercising of a belief.

2. **Classical or a prior probability:** The probability of an event s_i, written $pr(s_i)$, may be expressed as the ratio ns_i/n, where ns_i is the number of ways that s_i may occur in the particular situation considered and n is the total number of possible outcomes in the given situation. In each case, these outcomes must be equally likely.

As the name implies, n, the total number of outcomes of a given situation, and ns_i, the number of outcomes favourable to s_i of the same situation, are found a prior, that is, without actually conducting the experiment. The following example will illustrate the meaning of a prior definition of probability.

Example: In tossing a fair die, these are six possible outcomes of the experiment, each of which is equally likely and three of which result in an odd number. Therefore, the probability of obtaining an odd number is given by

$$\frac{n_i}{n} = \frac{3}{6} = \frac{1}{2}$$

3. **Posterior probability:** The definition of posterior probability is stated as follows:

$$\lim_{n \to 0} \frac{n_i}{n} = P$$

where n is the total number of identical trials in a given problem, n_i is the number of occurrences of the event, and P is the probability of the event.

Random Variables

A random variable describes the possible outcomes of a chance process and it may have discrete or continuous values. The numbers of people watching a football match might be described by a discrete-valued random variable; an anemometer would record wind speed as a continuous random variable. The complete set of all possible values of a random variable is called the sample space. A discrete random variable usually has a finite sample space where as a continuous random variable usually has an infinite sample space. These two types of random variables are considered separately because, their mathematical treatments are very different.

Thus a random variable is a function that assign a number to every outcome of an experiment that is, if, after an experiment, S_1, S_2, S_3, ...S_n are possible outcomes that constitute the sample space S and we wish to assign a number to each of these outcomes by using a rule $X(x_i)$, i = 1, 2, ..., n, then we call this rule a random variable.

Discrete Random Variables

Consider a die with faces numbered 1 to 6. The face-up value can be described by a discrete random variable. With a sample space of six, that is, when the die is rolled one of six values will show. Associated with each of the six possible values is a measure of the likelihood of occurrence of that value. Rolling the die and value constitutes an event. If the die is a fair one then the Probabilities of each possible value emerging face-up after a single event are the same. Further more if the die is rolled many times and the face-up value is recorded it is to be expected that each of the values will be recorded approximately the same number of times. If we choose one particular value, say 2, and let n_2 be the number of times 2 shows up out of a total of N rolls of the die, we can define the probability of event five as

$$P(2) = \lim_{N \to 0} \frac{(N_2)}{(N)} \tag{1}$$

The term $P(5)$ is read as the probability that the event has the value 5. For a fair die it is to be expected that $P(5) = 1/6$ since at each roll any one of the six values is likely to show-up. Another way of interpreting this probability is as a frequency. The event 5 should occur with a frequency of once every six rolls averaged over a large number of rolls. Generalizing the definition (1) to a general event with a value E gives

$$P_E = \lim_{N \to \infty} \frac{(N_E)}{(N)}$$

Example: Consider the experiment of throwing a pair dice once. Let the random variable X be equal to the sum of numbers resulting; and we wish to find its density, $F(X)$; where X may take on the values 2, 3, 4, ..., 12.

Solution:

Here, $F(2) = F(12) = 1/36$,

$F(3) = F(11) = 2/36$, $F(6) = F(4) = F(10) = 3/36$,

$F(5) = F(9) = 4/36$, $6 F(8) = 5/36$

And $F(7) = 6/36$. With algebric manipulation.

$$F(X) = \begin{cases} \dfrac{6 - 17 - X_1}{36} & , x_1 = 2, 3, ..., 12, \\ 0, \text{elsewhere} \end{cases}$$

Using this density function, we can obtain the following probabilities, for example:

(a) $P(5 < x < 8) = \sum_{x=6}^{7} (6 - 17 - X_1/36) = 11/36$,

Probability that the sum is greater than five but less than eight.

(b) $P(x > 9) = \sum_{x=10}^{12} (6 - 17 - X_1/36) = 6/36$

is the probability that the observed sum is greater than nine.

Fig. 1 shows the graph of F (x); the vertical lines indicate the probabilities for the values of x which the random variable assumes.

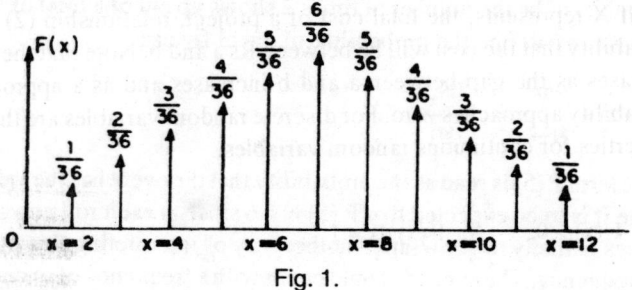

Fig. 1.

It is clear that F (x) > 0 for all x, and one can varify that

$$\sum_{x=2}^{12} \frac{6 - 17 - X_1}{36} = 1$$

Continuous Random Variables

In the preceeding section the random variable was considered to take only discrete values. When the random variable is continuous it may have infinitely many values. For continuous variables we can only give to the probability of achieving a value which lies within a specified range of values in the infinite sample space.

Fig. 2 shows the difference between discrete and continuous probability functions.

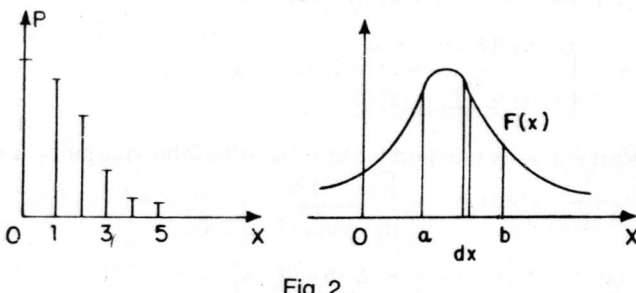

Fig. 2

Let X be a continuous random variable and x the range of values it can have ($-\infty < x < +\infty$). Let F (x) to be the probability density of variable X. F (x) can be thought of as a probability per unit length of x' rather than as a probability. Then the probability that X will have a value between a and b is defined to be

$$P(a < X < b) = \int_a^b F(x)\, dx \qquad\qquad\qquad ....(2)$$

If X represents, the total cost of a project, relationship (2) defines the probability that the cost will be between Rs a and b. Note that the probability increases as the gap-between a and b increases and as a approaches b the probability approaches zero. For discrete random variables are the following properties for continuous random variables.

$$F(x) > 0; \quad \int_{-\infty}^{+\infty} F(x)\, dx = 1 \qquad\qquad ....(3)$$

Example: The number of automobiles that a certain dealer sells per day can be considered a random variable of the continuous type, with the following function for its probability density:

$$F(x) = \begin{cases} B(x+1), 0 < x < 10, \\[6pt] \dfrac{-10(6B-1)}{X_2}, 10 < x < 20 \\[6pt] 0, \text{elsewhere} \end{cases}$$

Find the value of B that makes F (x) a density function and the probability that the dealer will sell between 8 and 14 cars in a given day.

Solution: To determine the value of B, we must integrate F (x over all admissible values of X.

$$F(x)\, dx = \int_0^{10} B(X+1)\, dx + \int_{10}^{20} \frac{-10(6B-1)}{X_2}\, dx = 1,$$

and we obtain B = 1/356 now.

$$P(8 < x < 14) = \int_8^{14} F(x)\, dx$$

$$\int_8^{10} \frac{1}{356}(x+1)\, dx \; + \int_{10}^{14} \frac{690\, dx}{356\, x_2} \quad \text{Solving the equation we get the probability}$$

Distribution Functions for a Random Variable

The two preceding sections have defined probability functions for discrete and continuous random variables. Often it is necessary to determine not the probability of an event or the probability that the variable lies within known limits but the probability that the event has a value less than or equal to one prescribed value. The distribution function defines these probabilities. Figure

3 shows distribution function corresponding to the probability functions of Fig. 2

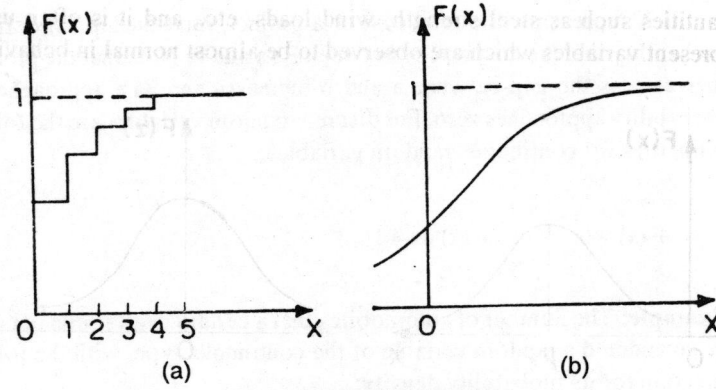

(a)   (b)

Fig. 3. Distribution function.

For a continuous random variable. The distribution function F (x) is defined as the probability that X has a value less than or equal to x, that is

$$F(x) = P(X < x) = \int_{\infty}^{x} F(x)\, dx \qquad (5)$$

Note that equation (4) is obtained from equation (3) of previous article simple by inserting a = −∞ and b = x as the limits. Fig. 3b shows F (x).

For a discrete random variable the distribution function is shown in Fig. 3a and the probability that an event will have a value less than or equal to K is

$$P(E < K) = \sum_{E}^{K} P(t)$$

Thus for the die F (5) is the probability that the number rolled will be less than or equal to 5 is

$$F(5) = P(1) + P(2) + P(3) + P(4) + P(5) = 5/6$$

Gaussian Function

Te normal or Gaussian function is possibly the best known of all probability functions. Its symmetrical bell shape is shown in Fig. 4. The equation of normal curve is

$$F(x) = \frac{1}{\sigma_x \sqrt{2n}} e^{-[(x - \mu_x)^2/2\sigma_x^2]} \qquad \text{For } -\infty < x < \infty \qquad \dots(6)$$

In equation (5) σx and μx are the standard deviation and the mean. The normal function fits the observed behaviour of many practical engineering quantities such as steel strength, wind loads, etc., and it is often used to represent variables which are observed to be almost normal in behaviour.

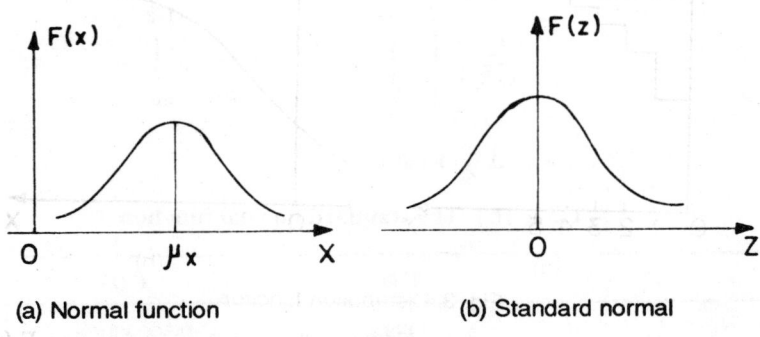

(a) Normal function   (b) Standard normal

Fig. 4

The reason for the usefulness of the normal function is that calculations it can easily be done by using tabulated values. The normal function can be transformed mathematically to the standard normal function which has a mean and a standard deviation = 1 as shown in figure (4 b).

As an example suppose that random variable X is approximated by the normal function, with a mean $μx = 10.0$ and a standard deviation $σx = 4$ and that it is necessary to calculate the probability that X lies between values a = 8.0 and b = 12 from equation (3).

$$P(a < x < b) = \int_{a}^{b} F(x)\, dx = \frac{1}{\sigma_x \sqrt{2\pi}} \int e^{-[(x - \mu_x)^2/2\,\sigma_x^2]}\, dx$$

Substituting $Z = (x - \mu_x)/\sigma_x$ and $= \dfrac{dz}{dx} = \dfrac{1}{x}$ gives

$$P(a < x < b) = \frac{1}{\sqrt{2\pi}} \int_{(a - \mu_x)/\sigma_x}^{(b. - \mu_x)/\sigma_x} e^{-z^2/2}\, dx \qquad \ldots\ldots\ldots(7)$$

The right side of equation (6) is exactly the same as would have obtained by evaluating $P([b - \mu x]/x < z < [x - \sigma x])$ for a standard normal function. For each value of Z a value is listed for $F(z)$. As a result of the summary of z about the zero mean $f(z) = F(-x) = 1 - F(z)$. Then for the numerical values stated above, eqn (6) yields

$$P\,(8 < x < 10) = \frac{1}{\sqrt{2\pi}} \int_{-1}^{-1/2} e^{-z^2/2}\, dz$$

$$= \frac{1}{\sqrt{2\pi}} \left(\int_{-\infty}^{-1/2} e^{-z^2/2}\, dz - \int_{-\infty}^{-1/2} e^{-z^2/2}\, dz \right)$$

$$= F\,(0.5) - F\,(-0.5)$$

Table II.1. The standard normal function

z	F (z)	F (z)
0.0	0.398942	0.500000
0.1	0.396952	0.539828
0.2	0.391043	0.579260
0.3	0.381388	0.617912
0.4	0.368270	0.655422
0.5	0.352065	0.691463
0.6	0.333225	0.725747
0.7	0.312254	0.758036
0.8	0.289692	0.788145
0.9	0.266085	0.815990
1.0	0.241971	0.841345
1.1	0.217852	0.864339
1.2	0.194186	0.884930
1.3	0.171369	0.903196
1.4	0.149727	0.919243
1.5	0.129518	0.933193
1.6	0.110921	0.94520

Thus a simple transformations permit a normal function to be transformed to a standard normal function for which tabulated data makes calculations easy.

Exponential Function

The exponential function is defined by

$$F(x) = \lambda e^{-\lambda x} \text{ for } x > 0$$

where λ is positive constant Figure 5 shows the typical shape of the

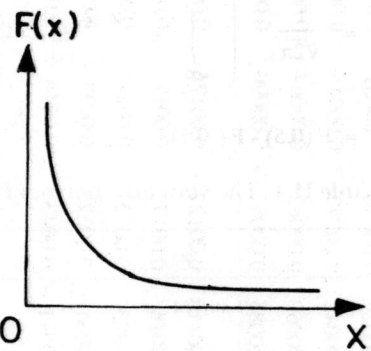

Fig. 5, The exponential function, F (x) = le - lx.

exponental function which, can be used to represent engineering quantities.
In general it represents situations in which probabilities of occurence of small
values of x are high but of high value are low.

Table II.2. Areas under the Normal Curve
(Proportion of total area under the curve from - to designated Z value)

Z	0.00	0.01	0.02	0.03	0.04	0.05	0.06	0.07	0.08	0.09
-3.5	0.00000	0.00022	0.00022	0.00021	0.00020	0.00019*	0.00019	0.00018	0.00017	0.00017
-3.4	0.00000	0.00033	0.00031	0.00030	0.00029	0.00028	0.00027	0.00026	0.00025	0.00024
-3.3	0.00000	0.00017	0.00045	0.00043	0.00042	0.00040	0.00039	0.00038	0.00036	0.00035
-3.2	0.00000	0.00066	0.00064	0.00062	0.00060	0.00058	0.00056	0.00054	0.00052	0.00050
-3.1	0.00000	0.00094	0.00090	0.00087	0.00085	0.00082	0.00079	0.00076	0.00074	0.00071
-3.0	0.00000	0.00131	0.00126	0.00122	0.00118	0.00114	0.00111	0.00107	0.00104	0.00100
-2.9	0.0000	0.0018	0.0017	0.0017	0.0016	0.0016	0.0015	0.0015	0.0014	0.0014
-2.8	0.0000	0.0025	0.0024	0.0023	0.0023	0.0022	0.0021	0.0020	0.0020	0.0019
-2.7	0.0000	0.0034	0.0033	0.0032	0.0031	0.0030	0.0029	0.0028	0.0027	0.0026
-2.6	0.0000	0.0045	0.0044	0.0043	0.0041	0.0040	0.0039	0.0038	0.0037	0.0036
-2.5	0.0000	0.0060	0.0059	0.0057	0.0055	0.0054	0.0052	0.0051	0.0049	0.0048
-2.4	0.0000	0.0080	0.0078	0.0075	0.0073	0.0071	0.0069	0.0068	0.0066	0.0064
-2.3	0.0000	0.0104	0.0102	0.0099	0.0096	0.0094	0.0091	0.0089	0.0087	0.0084
-2.2	0.0000	0.0136	0.0132	0.0129	0.0125	0.0122	0.0119	0.0116	0.0113	0.0110
-2.1	0.0000	0.0174	0.0170	0.0166	0.0162	0.0158	0.0154	0.0150	0.0146	0.0143

(Contd.)

Z	0.00	0.01	0.02	0.03	0.04	0.05	0.06	0.07	0.08	0.09
-2.0	0.0000	0.0222	0.0217	0.0212	0.0207	0.0202	0.0197	0.0192	0.0188	0.0183
-1.9	0.0000	0.0281	0.0274	0.0268	0.0262	0.0256	0.0250	0.0244	0.0239	0.0233
-1.8	0.0000	0.0351	0.0344	0.0336	0.0329	0.0322	0.0314	0.0307	0.0301	0.0294
-1.7	0.0000	0.0436	0.0427	0.0418	0.0409	0.0401	0.0392	0.0384	0.0375	0.0367
-1.6	0.0000	0.0537	0.0526	0.0516	0.0505	0.0495	0.0485	0.0475	0.0465	0.0455
-1.5	0.0000	0.0655	0.0643	0.0630	0.0618	0.0606	0.0594	0.0582	0.0571	0.0559
-1.4	0.0000	0.0793	0.0778	0.0764	0.0749	0.0735	0.0721	0.0703	0.0694	0.0681
-1.3	0.0000	0.0951	0.0934	0.0918	0.0901	0.0885	0.0869	0.0853	0.0838	0.0823
-1.2	0.0000	0.1131	0.1112	0.1093	0.1075	0.1057	0.1038	0.1020	0.1003	0.0985
-1.1	0.0000	0.1335	0.1314	0.1292	0.1271	0.1251	0.1230	0.1210	0.1190	0.1170
-1.0	0.0000	0.1562	0.1539	0.1515	0.1492	0.1469	0.1446	0.1423	0.1401	0.1379
-0.9	0.0000	0.1814	0.1788	0.1762	0.1736	0.1711	0.1685	0.1660	0.1635	0.1611
-0.8	0.0000	0.2000	0.2061	0.2033	0.2005	0.1977	0.1949	0.1922	0.1894	0.1867
-0.7	0.0000	0.2389	0.2358	0.2327	0.2297	0.2266	0.2236	0.2207	0.2177	0.2148
-0.6	0.0000	0.2709	0.2676	0.2643	0.2611	0.2578	0.2546	0.2514	0.2483	0.2451
-0.5	0.0000	0.3050	0.3015	0.2981	0.2946	0.2912	0.2877	0.2843	0.2810	0.2776

(Contd.)

Z	0.09	0.08	0.07	0.06	0.05	0.04	0.03	0.02	0.01	0.00
-0.4	0.3121	0.3156	0.3192	0.3228	0.3264	0.3300	0.3336	0.3372	0.3409	0.0000
-0.3	0.3483	0.3520	0.3557	0.3594	0.3632	0.3669	0.3707	0.3745	0.3783	0.0000
-0.2	0.3859	0.3897	0.3936	0.3974	0.4013	0.4052	0.4090	0.4129	0.4168	0.0000
-0.1	0.4247	0.4286	0.4325	0.4364	0.4404	0.4443	0.4483	0.4522	0.4562	0.0000
-0.0	0.4641	0.4681	0.4721	0.4761	0.4801	0.4840	0.4880	0.4920	0.4960	0.0000

$$Z = \frac{[x_i - E(x_i)]}{\sigma}$$

Areas under the normal curve

(Proportion of total area under the curve from + to designated Z value)

Z	0.00	0.01	0.02	0.03	0.04	0.05	0.06	0.07	0.08	0.09
+0.0	0.5000	0.5040	0.5080	0.5120	0.5160	0.5199	0.5239	0.5279	0.5319	0.5359
+0.1	0.5398	0.5438	0.5478	0.5517	0.5557	0.5596	0.5636	0.5675	0.5714	0.5753
+0.2	0.5793	0.5832	0.5871	0.5910	0.5948	0.5987	0.6026	0.6064	0.6103	0.6141
+0.3	0.6179	0.6217	0.6255	0.6293	0.6331	0.6368	0.6406	0.6443	0.6480	0.6517
+0.4	0.6554	0.6591	0.6628	0.6664	0.6700	0.6736	0.6772	0.6808	0.6844	0.6879
+0.5	0.6915	0.6950	0.6985	0.7019	0.7054	0.7088	0.7123	0.7157	0.7190	0.7224

(Contd.)

Z	0.00	0.01	0.02	0.03	0.04	0.05	0.06	0.07	0.08	0.09
+0.6	0.7257	0.7291	0.7324	0.7357	0.7389	0.7422	0.7454	0.7586	0.7517	0.7549
+0.7	0.7580	0.7611	0.7642	0.7673	0.7704	0.7734	0.7764	0.7794	0.7823	0.7852
+0.8	0.7881	0.7910	0.7939	0.7967	0.7995	0.8023	0.8051	0.8079	0.8106	0.8133
+0.9	0.8159	0.8186	0.8212	0.8238	0.8264	0.8289	0.8315	0.8340	0.8365	0.8389
+1.0	0.8413	0.8438	0.8461	0.8485	0.8508	0.8531	0.8554	0.8577	0.8599	0.8621
+1.1	0.8643	0.8665	0.8686	0.8708	0.8729	0.9749	0.8770	0.8790	0.8810	0.8830
+1.2	0.8849	0.8869	0.8888	0.8907	0.8925	0.8944	0.8962	0.8980	0.8997	0.9015
+1.3	0.9023	0.9049	0.9066	0.9082	0.9099	0.9115	0.9131	0.9147	0.9162	0.9177
+1.4	0.9192	0.9207	0.9222	0.9236	0.9251	0.9265	0.9279	0.9292	0.9306	0.9319
+1.5	0.0332	0.0345	0.9357	0.9370	0.9382	0.9394	0.9406	0.9418	0.9429	0.9441
+1.6	0.9452	0.9463	0.9474	0.9484	0.9495	0.9505	0.9515	0.9525	0.9535	0.9545
+1.7	0.9554	0.9564	0.9573	0.9582	0.9591	0.9599	0.9608	0.9616	0.9625	0.9633
+1.8	0.9641	0.9649	0.9656	0.9664	0.9671	0.9678	0.9686	0.9693	0.9699	0.9706
+1.9	0.9713	0.9719	0.9726	0.9732	0.9738	0.9744	0.9750	0.9756	0.9761	0.9767
+2.0	0.9773	0.9778	0.9783	0.9788	0.9793	0.9798	0.9803	0.9808	0.9812	0.9817
+2.1	0.9821	0.9826	0.9830	0.9834	0.9838	0.9842	0.9846	0.9850	0.9854	0.9857

(Contd.)

Z	0.00	0.01	0.02	0.03	0.04	0.05	0.06	0.07	0.08	0.09
+2.2	0.9861	0.9804	0.9868	0.9871	0.9875	0.9878	0.9881	0.9884	0.9887	0.9890
+2.3	0.9893	0.9896	0.9898	0.9901	0.9904	0.9906	0.9909	0.9911	0.9913	0.9916
+2.4	0.9918	0.9920	0.9922	0.9925	0.9927	0.9929	0.9931	0.9932	0.9934	0.9936
+2.5	0.9938	0.9940	0.9941	0.9943	0.9945	0.9946	0.9948	0.9949	0.9951	0.9952
+2.6	0.9953	0.9955	0.9956	0.9957	0.9959	0.9960	0.9961	0.9962	0.9963	0.9964
+2.7	0.9965	0.9966	0.9967	0.9968	0.9969	0.9970	0.9971	0.9972	0.9973	0.9974
+2.8	0.9974	0.9975	0.9976	0.9977	0.9977	0.9978	0.9979	0.9979	0.9980	0.9981
+2.9	0.9981	0.9982	0.9983	0.9983	0.9984	0.9984	0.9985	0.9985	0.9986	0.9986
+3.0	0.99865	0.99869	0.99874	0.99878	0.99882	0.99886	0.99889	0.99893	0.99896	0.99900
+3.1	0.99903	0.99906	0.99910	0.99913	0.99915	0.99918	0.99921	0.99924	0.99926	0.99929
+3.2	0.99931	0.99934	0.99936	0.99938	0.99940	0.99942	0.99944	0.99946	0.99948	0.99950
+3.3	0.99952	0.99953	0.99955	0.99957	0.99958	0.99960	0.99961	0.99962	0.99964	0.99965
+3.4	0.99966	0.99967	0.99969	0.99970	0.99971	0.99972	0.99973	0.99974	0.99975	0.99976
+3.5	0.99977	0.99978	0.99978	0.99979	0.99980	0.99981	0.99981	0.99982	0.99983	0.99983